T0329535

Dielectric Resonator Antennas

IEEE Press
445 Hoes Lane
Piscataway, NJ 08854

Dielectric Resonator Antennas

Materials, Designs and Applications

Zhijiao Chen
Beijing University of Posts and Telecommunications
Beijing, China

Jing-Ya Deng
Xidian University Shaanxi, China

Haiwen Liu
Xi'an Jiaotong University
Shaanxi, China

Published by John Wiley & Sons, Inc., Hoboken, New Jersey.
Published simultaneously in Canada.

Library of Congress Cataloging-in-Publication Data

Names: Chen, Zhijiao, author. | Liu, Haiwen (Professor), author.
Title: Dielectric resonator antennas : materials, designs and applications / Zhijiao Chen, Jing-Ya Deng, Haiwen Liu.
Description: Hoboken, New Jersey : Wiley, [2024] | Includes bibliographical references and index.
Identifiers: LCCN 2023042910 (print) | LCCN 2023042911 (ebook) | ISBN 9781394169146 (hardback) | ISBN 9781394169153 (adobe pdf) | ISBN 9781394169160 (epub)
Subjects: LCSH: Dielectric resonators. | Microwave antennas.
Classification: LCC TK7872.D53 C485 2024 (print) | LCC TK7872.D53 (ebook) | DDC 621.382/4–dc23/eng/20231107
LC record available at https://lccn.loc.gov/2023042910
LC ebook record available at https://lccn.loc.gov/2023042911

Cover Design: Wiley
Cover Image: © yuanyuan yan/Getty Images

Contents

About the Authors

Prof. Zhijiao Chen received her B.S. degree from Beijing University of Posts and Telecommunications in 2010 and Ph.D. degree from Queen Mary University of London in 2014. She joined the School of Electronic Engineering at Beijing University of Posts and Telecommunications as a lecturer in 2014. Currently, she is an associate professor at the School of Electronic Engineering at Beijing University of Posts and Telecommunications. She was seconded to Ace-Axis Wireless Technology Laboratories Ltd (Essex, UK) in 2012 and joined Northeastern University (Boston, MA) as a visiting student in 2013. From November 2018 to February 2019, she joined the State Key Laboratory of Terahertz and Millimeter-wave in City University of Hong Kong (Hong Kong, China) as a visiting scholar. From September 2019 to December 2019, she joined the National Physical Laboratory (London, UK) as a visiting scholar.

She received the Best Paper Award at *IEEE International Workshop Antenna Technology* (IEEE iWAT2013, Karlsruhe, Germany), the Best Student Paper Award at *IEEE International Symposium on Antennas and Propagation and USNC-URSI National Radio Science Meeting* (IEEE APS/URSI 2013, Orlando, FL), the TICRA Travel Grant at *European Conference on Antennas and Propagation* (EuCAP 2014, Hague, Netherlands), the Third Place at the QMUL Three-Minute Thesis Competition Final 2014, and the Young Scientists Award at *the 2021 International Applied Computational Electromagnetics Society Symposium in China* (ACES-China 2021, Chengdu, China). She has authored/co-authored more than 30 journal articles, one English book, and more than 50 conference papers. She was the 2022 IEEE AP-S Young Professional Ambassador and won the 2023 IEEE AP-S Outstanding Young Professional of the Year. Her research interests include but are not limited to dielectric resonator antennas, millimeter-wave antenna array, semi-smart base station antennas, and antennas for radio astronomy.

Prof. Jing-Ya Deng received his B.E. degree in electronic engineering and Ph.D. degree in electromagnetic field and microwave technology from Xidian University, Xi'an, China, in 2006 and 2011, respectively. He joined Xidian University in 2011 as a lecturer, where he has been a full professor since 2016.

Dr. Deng is a recipient of the National Science Foundation for Outstanding Young Scholars of China. His current research interests include millimeter-wave antennas, devices and circuits, multibeam antennas, phased array antennas, high-frequency/very-high-frequency (HF/VHF) antennas' design and measurement, and digital beam forming.

Prof. Haiwen Liu was born in 1975. He received his B.S. degree in electronic systems and M.S. degree in radio physics from Wuhan University, Wuhan, China, in 1997 and 2000, respectively, and the Ph.D. degree in microwave engineering from Shanghai JiaoTong University, Shanghai, China, in 2004.

From 2004 to 2006, he was a research assistant professor with Waseda University, Kitakyushu, Japan. From 2006 to 2007, he was a research scientist with Kiel University, Kiel, Germany, granted by the Alexander von Humboldt Research Fellowship. From 2007 to 2008, he was a professor with the Institute of Optics and Electronics, Chengdu, China, supported by the 100 Talents Program of Chinese Academy of Sciences. From 2009 to 2017, he was a chair professor with East China Jiaotong University, Nanchang, China. In 2014, he joined Duke University, Durham, NC, USA, as a visiting scholar. In 2015, he joined the University of Tokyo, Tokyo, as a visiting professor, supported by the Japan Society for the Promotion of Science (JSPS) Invitation Fellowship. In 2016, he joined the City University of Hong Kong, Hong Kong, as a visiting professor. Since 2017, he has been a full-time professor with Xi'an Jiaotong University, Xi'an, China. He has authored or co-authored more than 100 papers in international and domestic journals and conferences. His current research interests include electromagnetic modeling of high-temperature superconducting circuits, radio frequency and microwave passive circuits, antennas for wireless terminals, radar system, and radio telescope applications.

Dr. Liu was the recipient of the National Talent Plan, China in 2017. He has served as the editor-in-chief of the *International Journal of RF and Microwave Computer-Aided Engineering* (Wiley), an associate editor of *IEEE ACCESS*, and the guest chief editor of the *International Journal of Antennas and Propagation*. He was the executive chairman of the National Antenna Conference of China in 2015 and the co-chairman of the National Compressive Sensing Workshop of China in 2011 and the Communication Development Workshop of China in 2016.

Preface

Dielectric resonator antennas (DRAs), which benefited from using a dielectric structure for better radiation performance, have already received considerable attention in the past 30 years. Nevertheless, there is currently no comprehensive study on the classification of these dielectric antennas. The concept of DRA might tend to be abused. In this book, DRAs are theoretically studied, and detailed procedures are provided with plenty of design examples. The fabrication process and the potential applications are discussed. Readers can learn the classification, fundamentals, design, and applications of the dielectric antenna from this book.

The existing DRA books (K. M. Luk and K.W. Leung, 2003), (A. Petosa, 2007), (R. S. Yaduvanshi and H. Parthasarathy, 2016), and (R. K. Chaudhary, R. Kumar and R. Chowdhury, 2021) are classic and written by the experts of dielectric antennas. But these books are published before 2007, while the theories and designs of the dielectric antenna have been updated a lot in recent years. The books (R. S. Yaduvanshi and H. Parthasarathy, 2016) and (R. K. Chaudhary, R. Kumar and R. Chowdhury, 2021) focus on rectangular DRA designs and circular-polarized DRA designs, respectively. However, designs proposed in these two books are limited to specific rectangular shape or circular-polarized operation, which is not substantial and needs to be enhanced for real-life applications. Therefore, it is important to publish a new book to conclude the recent developments in dielectric antennas, especially for microwave and millimeter-wave communication applications.

The authors wish to acknowledge the financial support of the National Natural Science Foundation of China under Grant No: 62271067, 62022064, and 62171363; Royal Society through International Exchanges 2019 Cost Share (NSFC) under Grant Ref: IEC\NSFC\19178; Beijing Key Laboratory of Work Safety Intelligent Monitoring (Beijing University of Posts and Telecommunications); International Cooperation Funds of Shanxi Province under Grant No. 2022KWZ-15; the Shaanxi Key Laboratory under Grant No. 2021SYS-04; and the Shaanxi S&T Innovation Team under Grant No. 2023-CX-TD-03. They would also like to acknowledge their national and international collaborators, including Dr. Benito Sanz Izquierdo and Mr. Peter Njogu (both at the University of Kent, UK).

All the works presented in this book are complemented with the help of the authors' graduate students with their hard work on chapter writing and antenna design works under the supervision of the authors. So, the authors would like to thank all the graduate students involved in chapter writing and DRA design, including Hanjing Wang, Jiabin Ma, Ziwei Li, Xuewen Jiang, Jiuyu Zhang, Haixin Jiang, Wei Song, Zeyu Song, Qi Liu, Ming Zhu, Pufan Li, Xiaohan Yin, Hao Xu, and Hongliang Tian. This book cannot be completed without their helps.

November 2023

Zhijiao Chen
Beijing, China
Jing-Ya Deng
Xi'an, China
Haiwen Liu
Xi'an, China

1

Introduction

CHAPTER MENU

1.1 Motivation

In 2019, 5G mobile networks appeared, building the base for operators to provide 5G mobile services to industries, enterprises, and consumers. The 5G wireless communication enables various commercial applications such as autopilot, telemedicine, and Industrial Internet of Things (IIoT). In the Beyond 5G (B5G), communication fabric expands to all intelligent services, allowing people, robots, smart devices, and any intelligent agents to interconnect and collaborate with each other (Figure 1.1).

For example, the IEEE 802.11ad and IEEE 802.11ay standards operating on 60 GHz (covering 57–71 GHz) are the most expected wireless local area network (WLAN) technologies for the ultra-high-speed communications. They provide data throughput rates of up to 6 Gbps. However, after years of research and development efforts, mmWave techniques are still on the brink of delivering high-quality 5G communications.

The low-cost and high-reliability mass production of the mmWave devices will be of great importance to complete the 5G story. But the challenges are higher than they have ever been with previous cellular technologies. Especially, there are still many remaining design issues to tackle for 60 GHz WLAN communications. This includes a low-complexity but efficient mmWave antenna for WLANs that aim to provide low-cost high quality of services (QoS). Waveguide-based antenna array

Dielectric Resonator Antennas: Materials, Designs and Applications, First Edition.
Zhijiao Chen, Jing-Ya Deng, and Haiwen Liu.

Published 2024 by John Wiley & Sons, Inc.

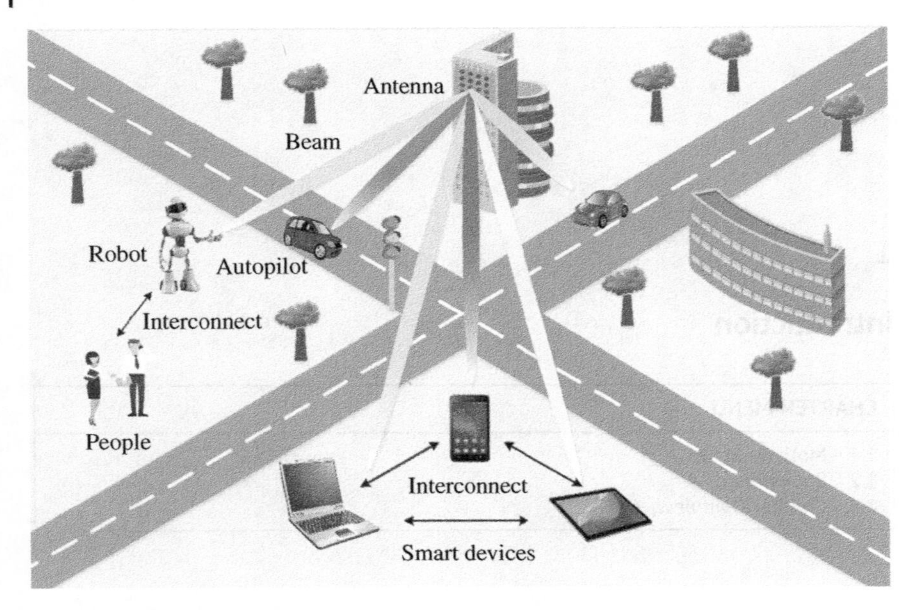

Figure 1.1 The wireless communications for 5G and B5G.

and reflector array antenna have shown extraordinary abilities in tracking the targets and wireless sensing. However, most of these existing designs are based on the high-cost bulky metal structure, which are infeasible for commercialization.

The high-gain radiating beam suffers from coverage shortfalls. To achieve the full coverage for the 5G users, mmWave systems are equipped with multi-beam or phase array antennas. This raised a number of computation and implementation challenges on the antenna array to maintain the anticipated performance gain and coverage of mmWave systems. As a result, the quality of experience (QoE) of the 5G devices highly relies on their antenna performances such as the gain for compensating the path loss and the coverage for seamless connection. Most importantly, the high cost on the hardware and software implementations will lead to poor usability ratings and low adoption by service providers and consumers.

Driven by the demand for miniaturized and low-cost solutions for the mmWave communications, the dielectric-based fabrication techniques have been investigated by many research groups. This includes investigations on novel dielectric materials and fabrication processes to bring the fundamental improvements in mmWave antennas. Functional dielectrics such as low-temperature co-fired ceramic (LTCC) and complementary metal oxide semiconductor (CMOS) materials have expanded the antenna applications to many priority research areas like on-chip implementations. In the commercial marketplace, low-loss printed circuit boards (PCBs) with a wide range of dielectric constant are sold. For example, Rogers corporation provides a series of PCB with *DK* value of 2–10 and *Df* value of 0.0009–0.0045 at 8–40 GHz. Their prices are lower than LTCC and CMOS, but vary from country to country due to different tax rates. The dielectric-based fabrication offers a competitive solution for low-cost and high-reliability mass production of the mmWave antenna. Especially, mmWave antennas based on the substrate integrated waveguide (SIW) technique show advantages of easy fabrication, low cost, and reduced size over the metal waveguide, while overcoming the radiating loss of the microstrip line on the mmWave band.

Promising candidates for the high frequency applications includes the lens antenna and dielectric resonator antenna (DRA). The lens antenna utilized the dielectric lens to control the field distribution at the aperture, providing a high-gain antenna system with a simple architecture. The high-gain radiation is dominated by the focal distance of several wavelengths, thus resulting in a bulky structure of the lens antenna. In comparison, DRA has a low-profile structure but has limited gain enhancement, thus requiring array feeding network for high-gain radiation. Nevertheless, the employed dielectric loading enables dielectric modes for better radiation performance especially enhanced gain.

With the enhanced element gain, DRA shows potential to minimize the element scale for simplified, low-profile, and high-gain antenna array. Figure 1.2 depicts the number of research articles on DRAs, showing a consistent upward trend since 1989, highlighting the increasing interest in this area.

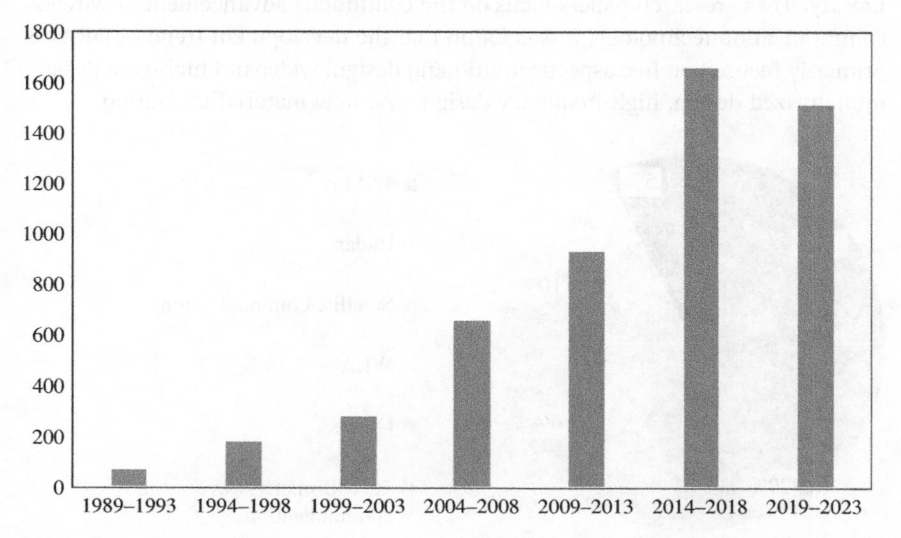

Figure 1.2 The number of articles retrieved from the IEEE website from 1989 to 2023. *Source:* Adapted from survey based on 258 papers in IEEE Xplore Digital Library.

1.2 Background

DRAs have gained significant attention in recent years due to their ability to efficiently radiate and confine electromagnetic energy. DRAs are constructed using high dielectric constant materials, which can support specific resonant modes based on their shape and size. When an electromagnetic wave interacts with a dielectric resonator, it creates a resonant response in the dielectric material, leading to the interaction between the electric and magnetic fields. At a specific frequency, this interaction can result in multiple resonant modes and high radiation efficiency of the DRA. As a result, compared to other types of antennas, DRAs offer wider bandwidths or cover multiple frequency bands with proper design on their resonant modes. Additionally, DRAs are cost-effective due to their use of inexpensive materials and simple fabrication processes, which also makes them easy to integrate and assemble especially on higher frequency bands.

Figure 1.3 shows a pie chart based on the investigation on 258 research papers related to DRA that are available on IEEE Xplore Digital Library. It reveals that DRA is highly attractive for various applications such as mobile communications, radar, satellite communications, and WLAN. Overall, DRA has proven to be a promising and versatile candidate with significant potential for future advancements.

Figures 1.4 and 1.5 survey the frequency range and research characteristics of DRA based on the 258 research papers carried out in the IEEE Xplore Digital Library. These research papers focus on the continuous advancement of wireless communication technology. It was found that the development trend of DRAs is primarily focused on five aspects: multi-band design, wideband high-gain design, miniaturized design, high-frequency design, and new material utilization.

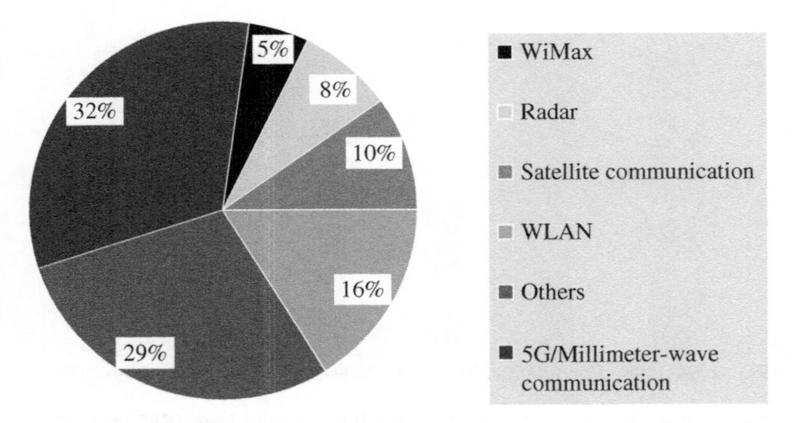

Figure 1.3 Application field of DRAs. *Source:* Adapted from survey based on 258 papers in IEEE Xplore Digital Library.

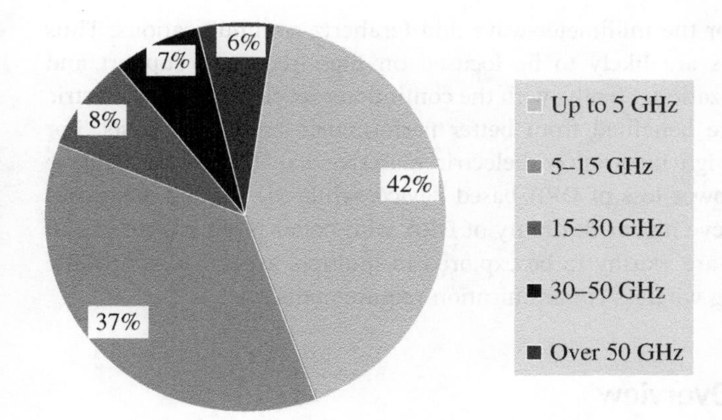

Figure 1.4 Frequency ranges of DRAs. *Source:* Adapted from survey based on 258 papers in IEEE Xplore Digital Library.

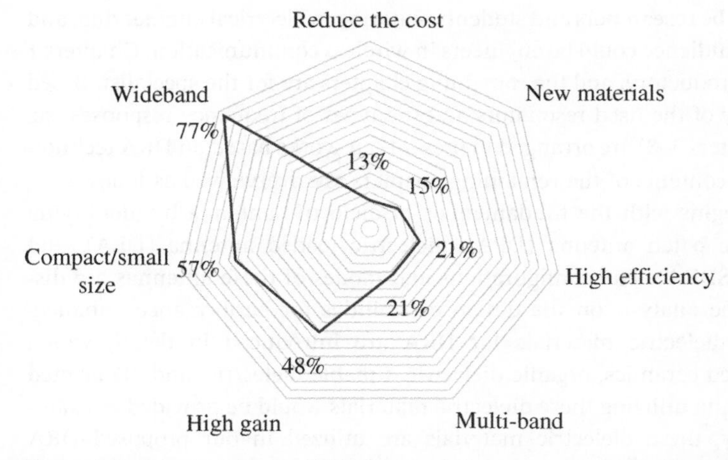

Figure 1.5 Research on the DRA characteristics. *Source:* Adapted from survey based on 258 papers in IEEE Xplore Digital Library.

First, DRA designs increasingly emphasize multi-band support to meet the communication needs of various frequency bands since more and more devices require multi-band communication. Second, DRA with high-efficiency and multi-mode characteristics are utilized to enhance the antenna gain and bandwidth. Third, DRA with high dielectric constant is employed to miniaturize the antenna size, which is essential for implementing the compact structure and high integration of the electronic devices. Fourth, by overcoming the metal loss on high frequency,

DRA is preferred for the millimeter-wave and terahertz communications. Thus future DRA designs are likely to be focused on high-frequency support and performance optimization. Finally, with the continuous development of dielectric materials, DRAs are benefited from better performance and lower costs. For instance, low-loss high integration dielectric materials can be used to achieve compact size and lower loss of DRA-based device, while 3D printed dielectrics can be used to achieve higher flexibility of DRA with better performance. It can be seen that DRAs are worthy to be explored in multiple aspects to contribute to the ever-changing wireless communication requirements.

1.3 Chapter Overview

This book offers a unique and comprehensive treatment of microwave and millimeter-wave DRAs designed and experimented by our group in the past few years. This book is relevant for readers in academia, industry, and education: the primary audience could be researchers and students majoring in electrical engineering, and the secondary audience could be engineers in wireless communication. Chapters 1 and 2 act as introductory, and the remaining chapters are for the specialist. Based on the category of the used resonators and the types of frequency responses, six chapters (Chapters 3–8) are arranged to present our works based on DRA technology. The main contents of the remaining chapters are overviewed as follows.

Chapter 2 begins with the fundamentals of dielectric antenna by identifying DRA, dielectric patch antenna (DPA), dielectric loaded antenna (DLA), and stacked DRA (SDRA). The configurations and modes of these antennas are discussed, with the analysis on the design capabilities for performance enhancements. Then, dielectric materials for DRA are introduced in detail, which includes sintered ceramics, organic dielectric, tunable dielectric, and 3D printed material. Antenna utilizing these dielectric materials would be provided as examples. Especially, these dielectric materials are utilized in our proposed DRA designs, and these DRA designs would be specified in the following chapters.

Chapter 3 introduces the fundamentals of SDRA by classifying its analytical methods. Aiming for the indoor base station antenna applications, a series of SDRA structures are proposed to enable full exploitation of the IEEE 802.11ac WLAN and the long-term evolution (LTE) services. And a wideband circularly polarized antenna with passive beam steering function is proposed based on the SDRA for indoor base station.

Chapter 4 classifies the pattern diverse DRAs according to their working principles and design approaches. As an example, a pattern diverse DRA is analyzed for their modes and operations. Then, the uniform linear antenna array is studied for its limitation in scanning ability. The method of using pattern diverse DRA to widen the array scanning angle is proposed. After that, the schematic of using pattern diverse DRA array is explored for shaped beam synthesis and compared with that of conventional antenna array. Optimization algorithms are compared and applied for both schematic. The results validate the feasibility of using pattern diversity DRA as an energy-efficient means for the base station antenna array to handle with users and interferences.

Chapter 5 overviews the requirements of multiple-in multiple-out (MIMO) antenna in the 5G environment at first, and then classifies the decoupling methods of MIMO antenna based on their principles. This is followed with several DRA MIMO antenna examples operating with different isolation enhancement methods. It can be seen that DRA has great potential to implement MIMO antennas due to the compact size, high radiation efficiency, and versatility in shape and feeding mechanism. Especially, the 3D structure of the DRA offers additional degrees of freedom in exciting various modes in one antenna structure. After that, a MIMO DRA with enhanced isolation and symmetrical pattern is proposed for the future 5G mmWave applications.

Chapter 6 discusses the feasibility of 3D printing technologies in dielectric antenna and antenna with dielectric substrate. 3D printing technology is given in detail with its definition, advantages, application fields, classification, and characteristics. Then, a 3D-printed dielectric antenna and a 3D-printed metal antenna are proposed with complementary structure. Comparisons are made between them in terms of the size, weight, fabrication tolerance, and performance. After that, 3D-printed finger nail antennas for 5G application are introduced to show their fabrication process and fabrication tolerance.

In Chapter 7, DRA and DRA array are proposed for 5G mmWave applications and their advantages have been demonstrated. First, a dual-band dual circularly polarized DRA has been verified for unmanned aerial vehicle (UAV) satellite communication. Then, SIW technology is introduced to provide the merits of lightweight, high gain, and high efficiency of the antenna array in mmWave band. SIW power dividers and waveguide to SIW transition are designed for large-scale antenna array. After that, an SIW-fed DRA array is designed with wideband, high gain, and enlarged dimensions for improved fabrication tolerance, which is suitable for 5G base station antenna. The results and comparison are discussed, showing DRA a promising candidate for 5G mmWave applications.

Chapter 8 focuses on the filtering antenna and diplex/duplex antennas. Classifications have been made according to design methodologies. The comparison between conventional antenna designs and DRA designs features great potential for DRA realizing filtering antennas and diplex/duplex antennas. Recent advanced designs have been given with a wideband and high-gain filtering DRA, and a differentially fed duplex filtering DRA is demonstrated. The differentially fed duplex filtering DRA features high isolation between channels, common mode signal suppression, low cross-polarization level, and good symmetry in radiation patterns. It provides a set of useful references for developing multifunctional antennas that integrate the functions of multiple devices such as duplexer, filter, and antenna into one compact unit.

Chapter 9 summarizes the contributions presented in this book and recommends some future research directions for DRA.

2

Classifications on Dielectric Resonator Antenna

CHAPTER MENU

2.1 Overview

Dielectric resonator antenna (DRA) was first proposed by Long in 1983 for the original goal to avoid the conduction losses of the metal mmWave radiating structure [1]. Following this work, dielectric antennas have been investigated for many advantages and benefits over metal antennas. For instance, dielectric patch antenna (DPA) has been studied by different authors to avoid the severe metal loss of the mmWave patch antenna [2]. Dielectric filled/loaded antennas provide an effective means for developing compact, high-performance mmWave antenna [3]. Metal antennas can reduce their weight by coating the metal to the 3D printed dielectric [4]. They benefited from the dielectric due to low cost and lightweight, but they still belong to the metal antenna because of no penetration into the dielectric.

Dielectric Resonator Antennas: Materials, Designs and Applications, First Edition.
Zhijiao Chen, Jing-Ya Deng, and Haiwen Liu.

Published 2024 by John Wiley & Sons, Inc.

Furthermore, to the best of our knowledge, there is currently no comprehensive study on the classification of these dielectric antennas. The concept of dielectric antenna might tend to be abused. For instance, the antenna models were named differently as dielectric dense patch (DD patch) and DRA but having the same working principle. Some antenna models have the name of DRA but without exciting or specifying the DRA modes. Publications related to the dielectric antenna have been reviewed and summarized [5]. But there is no unified rule for classifying dielectric antenna with identical modes and configurations.

This chapter is organized as follows. Section 2.2 classifies the configurations and modes of several kinds of dielectric antenna. Their operations and fundamentals will be explained in detail. In Section 2.3, dielectric materials for DRA are introduced including sintered ceramics, organic dielectric, tunable dielectric, and 3D printed material. Antennas utilizing these dielectric materials are given as examples, along with the corresponding examples that are specified in the following chapters. Finally, Section 2.4 gives remarks and summary of this chapter.

2.2 Dielectric Antenna Classifications

This section focuses on the printed circuit board (PCB)-based low-profile dielectric antenna designs that function as a concentrator of incoming electromagnetic field. More particularly, we discuss the operations of dielectric antenna and array design on their bandwidth, gain, and efficiency for 5G mmWave communications. Starting from the basic substrate integrated waveguide (SIW)-fed slot antenna, three different dielectric antennas have been designed with their configurations and modes listed in Figure 2.1. It includes a DPA, a dielectric loaded antenna (DLA) or DRA, and a stacked dielectric resonator antenna (SDRA). All these designs take advantage of the dielectric structure but operate differently. The rectangular-shaped dielectric is utilized in all these designs because of the attractive properties of easy fabrication and mode control.

SIW-fed slot antenna uses dielectric substrate to reduce the volume of the conventional rectangular waveguide. The miniaturization and the antenna loss depend primarily on the dielectric constant of the dielectric loading. In general, slot antenna suffers from the high back radiation and narrow bandwidth especially when using small ground plane and high permittivity dielectric. Alternatively, patch antenna is another widely used antenna type with simple structure and easy fabrication. However, the frequency-dependent metal loss from the finite conductivity of metal and the skin effect is severe on the mmWave band.

An effective means for overcoming the metal loss of the patch antenna is using the dielectric patch. Figure 2.1 shows the configuration of the DPA, where a thin

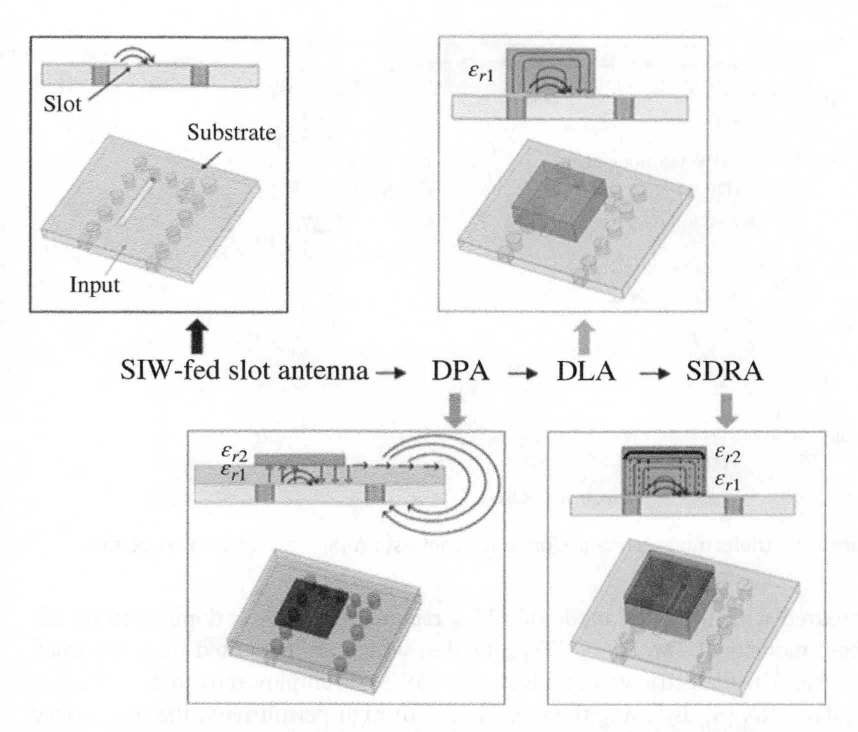

Figure 2.1 Configurations and modes of dielectric antennas. Each configuration has cross-sectional view on the top and perspective at the bottom; the propagating fields are indicated in arrows.

dielectric patch is used to replace the metallic patch. DPA is usually designed with a thin dielectric patch (<0.5λ) with extremely high dielectric constant to ensure the resonance of TM_{10} mode as dominate mode. However, because the short wavelength of the mmWave makes the substrate very thick, a considerable portion of energy is wasted due to the surface wave. The surface wave is indicated in the DPA of Figure 2.1 as the black arrow. It deteriorates the array performance, such as degrades the antenna gain, increases the array sidelobe, and distorts the antenna radiation patterns. Parasitic structures like the electromagnetic band gap (EBG) have been introduced to suppress the surface-wave effectively. But precisely realizing small dimensions of these periodic structures in mmWave band is a tough challenge, especially when applying the conventional PCB processing.

Dielectric loading offers a simple method to suppress the surface wave on mmWave band. As demonstrated by the DLA configuration in Figure 2.1, a dielectric block is loaded over the SIW-backed slot antenna to replace the dielectric substrate. It is named as DRA or DLA depending on the dielectric permittivity.

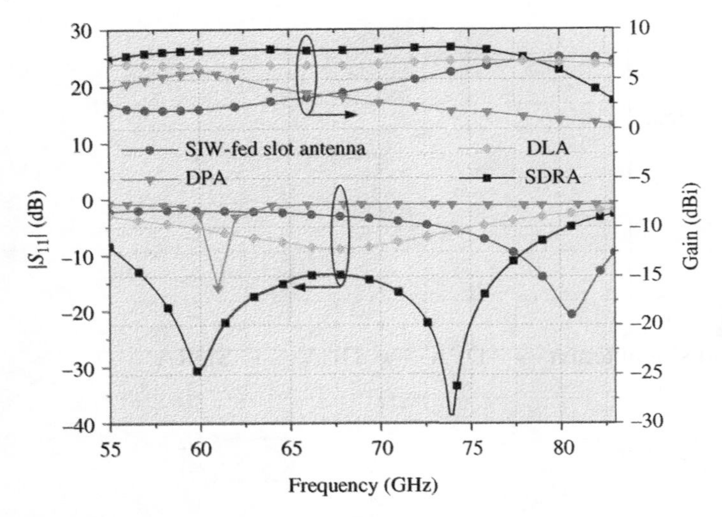

Figure 2.2 Dielectric antenna performance contrast on $|S_{11}|$ and gain in simulation.

Compared with the DPA mode of TM_{10}, rectangular DRA is dominated by the three-dimensional TM_{nml} and TE_{nml} modes, which are benefited from the thick dielectric. The dielectric waveguide model (DWM) is employed to analyze the rectangular DRA [6]. By using the dielectric with high permittivity, the top surface and four sidewalls of the DRA are assumed to be perfect magnetic walls, whereas the bottom ground plane is represented as an electric wall. This creates an exact solution for the resonance frequency of the DRA but results in a narrow operating bandwidth. In comparison, DLA with the low permittivity dielectric assumes the top surface and four sidewalls to be imperfect magnetic walls. In this case, the quality factor of the DLA is reduced, and its resonance frequency is extended to form a wideband operation. Another purpose of using dielectric loading is generating slow-wave effects for antenna dimension minimization, while exciting extra antenna modes for performance enhancement. A number of DLA designs including the dielectric-loaded waveguide antenna and dielectric-loaded horn are proposed for a specific gain, sidelobe requirement, or low cross-polarization. Some of them are recently published, providing attractive candidates for the 5G mmWave communications.

In Figure 2.2, dielectric antennas pictured in Figure 2.1 are compared in terms of the reflection coefficients ($|S_{11}|$) and antenna gains. For each configuration, the feeding structure was optimized to maximize the impedance bandwidth and gain performance, whereas the dimensions of the dielectric loading remained unchanged. The SIW-fed slot antenna using the dielectric constant of $\varepsilon_r = 2.2$

suffers from narrow bandwidth. Its resonant mode is located at 80.0 GHz with the peak gain of 7.0 dBi. By loading the dielectric block with $\varepsilon_{r1} = 2.2$, the operating frequency of the antenna is downshifted to 67.0 GHz and the bandwidth is broadened. The antenna gain is degraded due to the use of the low-permittivity loading material with low quality factor. For the DPA loading with the dielectric of $\varepsilon_{r2} = 10.2$, the operating frequency of the antenna is further downshifted to 60 GHz. The resonance of the TM_{10} mode increases the quality factor but narrows down the bandwidth again. This is expected due to the limited resonance mode of the DPA.

The above three dielectric antennas operate at diverse bands due to different operations. This gives rise to a new design by merging the antenna modes to obtain wideband operation. An SDRA is proposed in this approach as shown in Figure 2.1. The modes are merged from the fact that the modes of slot, cavity, patch, and dielectric resonator (DR) are excited at the same time. As observed in the $|S_{11}|$ of Figure 2.2, two spaced resonances on 60 and 74 GHz are combined into a flat response to achieve the −10 dB bandwidth of 55.6–78.1 GHz or fractional bandwidth of 33.6%. Due to the hybrid mode of the SDRA, an enhanced gain up to 8.1 dBi is achieved over the operating band, as presented in the gain plot in Figure 2.2.

The wideband performance of the SDRA was first investigated by Petosa et al. in 2000 [7], proposing an approximate equivalent method to analyze the dielectric constant of multilayer DRs. This method equivalents the multilayer dielectric constant to a single value, but results in a large deviation especially for the design with different feeding methods. To solve this problem, numerous theories have been proposed for the composite DRA, considering the influence factors of the feeding method and the merged modes. This includes the theory of hybrid mode [8], where a compact wideband mmWave antenna is developed by exciting hybrid modes of slot, dielectric, and cavity resonators.

2.3 Dielectric Material Classifications

Driven by the demand for miniaturized and low-cost solutions for the mmWave communications, the dielectric-based fabrication techniques have been investigated by many research groups. This includes the investigation on novel dielectric materials and fabrication processes to bring the fundamental improvements for mmWave antennas. Functional dielectrics such as sintered ceramics, organic dielectric materials, tunable dielectric materials, and 3D printed materials have expanded the antenna applications to many priority research areas like on-chip implementations.

2.3.1 Sintered Ceramics

Sintered ceramics have a wide range of dielectric constant, low dielectric loss, high *Q* value, and near-zero temperature at resonance frequency. These characteristics are essential for the millimeter-wave antenna. For example, coefficient of resonance frequency (τf) reflects resonant frequency drift with temperature and near-zero temperature can obtain high frequency stability. Dielectric with around $\varepsilon_r = 30$ can improve the bandwidth and gain of millimeter-wave antenna arrays [9]. Dielectric antennas using high ε_r sintered ceramic have high *Q* value and narrow bandwidth resonance characteristics.

Sintered ceramics are usually produced at high temperature, which follows the process of ceramics surface area reduction, porosity reduction, compactness and strength improvement, and recrystallization. According to sintering temperature, sintered ceramics are divided into three classifications: high-temperature co-fired ceramics (HTCCs, >1000°C), low-temperature co-fired ceramics (LTCCs, 700–1000°C), and ultralow-temperature co-fired ceramics (ULTCCs, <700°C). The manufacturing of HTCC is energy-intensive and cannot be co-fired with base metal or low melting point electrode (such as Ag, Cu, Al). Conventional HTCC used in antenna include alumina ceramics [10], nitride ceramics [11], and quartz ceramics [12]. The miniaturization and integration of microwave devices promote the development of LTCC and ULTCC, which can co-fired with the base metal electrode without melting and interaction. Manufacturing process of lower temperature ULTCC is energy-saving and cost-saving, and allows the co-firing with a wider range of conductor materials.

2.3.1.1 High-Temperature Co-fired Ceramics

Alumina ceramic is the most widely used HTCC material in the electronic industry due to the low dielectric loss, low dielectric constant, and high stability (Table 2.1). It is primarily composed of Al_2O_3 with lightweight properties (density = 3.5 g/cm^3), making it an ideal choice for antenna applications. In Figure 2.3a, parallel tetragonal alumina ceramics of different densities and their corresponding

Table 2.1 Parameters for HTCC.

Component	Dielectric constant ε_r	Dielectric loss tan δ
Al_2O_3 (95% density)	9	0.0003
Al_2O_3 (99% density)	9.8	0.0001
AlN	8.8	0.0002
Quartz ceramic (SiO)	3.42	0.004

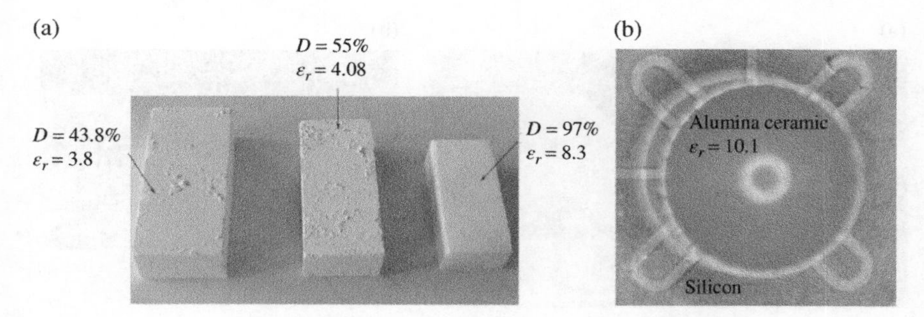

Figure 2.3 Alumina ceramic [10, 13]. (a) Parallel tetragonal alumina ceramic of different density. (b) Dual-line polarizer resonator of alumina ceramic.

dielectric properties are presented [13]. Figure 2.3b shows a millimeter-wave radiator designed on a dual-line polarizer based on a DR. The DR is a spherical structure made of alumina ceramic, which has the dielectric constant of $\varepsilon_r = 10.1$ and loss tangent of $\tan\delta = 0.0005$. The pattern reconfigurable DRA proposed in Section 4.2 and the stacked DRA with passive beam steering proposed in Section 3.4 also use alumina ceramic as DRA dielectrics.

Aluminum nitride (AlN) ceramics have low dielectric constant and low dielectric loss. In comparison to alumina ceramics, they are non-toxic and have thermal expansion coefficient that is well-matched with silicon. Additionally, they exhibit good thermal shock resistance and stable performance at high temperatures. Figure 2.4a illustrates the optical micrographs of the ceramic [14]. Figure 2.4b shows an aluminum nitride ceramic dielectric rod antenna proposed for neurological tissue pulse simulation [11].

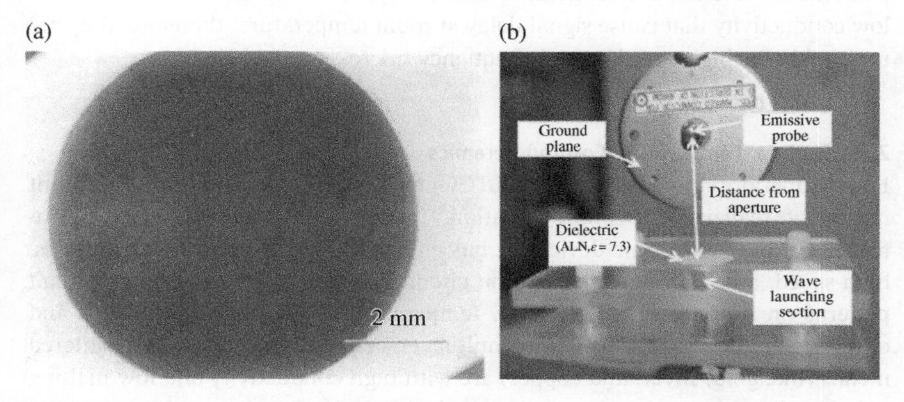

Figure 2.4 Aluminum nitride ceramic [11, 14]. (a) Optical micrographs of AlN ceramic. (b) Neurological tissue pulse simulation rod AlN antenna.

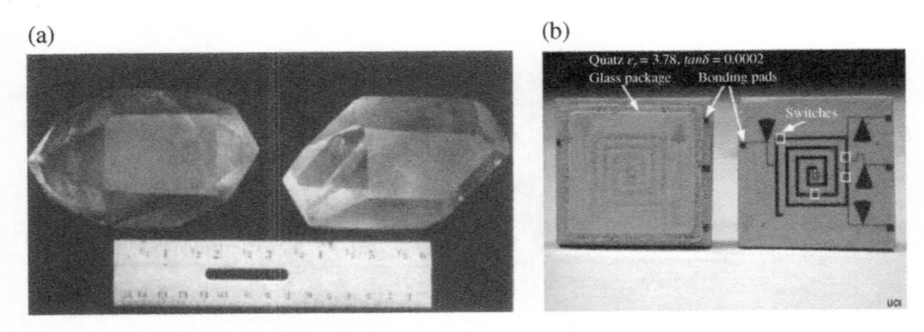

Figure 2.5 Quartz crystals [12, 15]. (a) Photos of original quartz crystals. (b) Quartz substrate used in spiral antenna.

Quartz ceramic is a single oxide ceramic comprised of silicon dioxide (SiO_2). This opaque ceramic is produced through slurry casting and sintering. Quartz ceramics inherit from the properties of quartz, including low dielectric constant, low loss tangent, and low expansion coefficient. They also exhibit remarkable thermal shock resistance and a dielectric constant that is stable with respect to both frequency and temperature. Figure 2.5a illustrates the original quartz crystals [15]. In [12], quartz substrate of $\varepsilon_r = 3.78$, $\tan\delta = 0.0002$ is used in reconfigurable scan-beam single-arm spiral antenna as shown in Figure 2.5b. The use of a quartz glass substrate with cavity structures offers a cost-effective mass production solution with better electronic compatibility and enhanced radiation efficiency and gain.

The above mentioned HTCCs have to be co-fired with refractory metals such as tungsten, molybdenum, and manganese. They cannot be co-fired with low melting point metal materials such as gold, silver, and copper. These refractory metals have low conductivity that cause signal delay at room temperature; therefore, they are unsuitable for high-speed or high-frequency micro-assembly circuits.

2.3.1.2 Low-Temperature Co-fired Ceramics

Low-temperature co-fired ceramic (LTCC) technology is a reliable and efficient means of realizing the miniaturization, integration, lightweight, and multi-functionality of electronic devices. It has emerged as a response to the low-loss, high-speed, and high-density trend on circuit packaging. By creating the circuit patterns on sintered ceramics at low temperatures, LTCC buries antennas and other electronic components into multilayer ceramic substrates. The sintered metals (like gold, silver, and copper) are with high conductivity and low melting points and are sintered at temperatures below 900°C to form a stacking geometry. The low energy consumption of the low-temperature sintering process reduces the complexity and cost of sintering. Compared with thick film printing, LTCC does not require multiple print and sinter runs, and the number of layers has no limitation. Due to the inherent characteristics such as high speed of transmission, high frequency, low loss, high temperature resistance, and high current resistance, LTCC technology has found widespread application in various fields, including mobile phones, base stations, automotive electronics, radar, and aerospace.

The dielectric constant of LTCC materials varies depending on the composition. The use of high-conductivity metal materials as conductors improves the quality factor of the circuit system and increases the flexibility of circuit design. There has been a significant investment in research and development of LTCC by both academic and business communities worldwide. The three leading LTCC manufacturers are Murata (https://www.murata.com), Kyocera (www.kyocera.co.jp), and TDK Corporation of Japan, with a combined market share of over 55%. Other prominent Japanese manufacturers include TAIYO YUDEN, KOA, and Yokowod. The second largest market is located in China, with major manufacturers including Huaxinke (https://www.huaxinke.cn), Guoju, Qilixin, and Jingde, accounting for nearly 25% of the market share. Other manufacturers include Bosch in Europe; apitech and Thales in France; Samsung Motor in Korea; and LG Innotek, Ferro (https://www.ferro.com), and Dupont in the United States. Table 2.2 lists some commercial LTCC products for antennas.

The notable attributes of LTCC are dielectric loss ($1/Q$, Q = quality factor), dielectric constant (<10), sintering temperature, and co-firing characteristics with

Table 2.2 Commercial LTCC products.

Product	tan δ	ε_r	Applicable band
Ferro-A6m	0.002	5.5–5.9	60–140 GHz
Dupont-GreenTape9K7	0.001	6.9–7.3	100 GHz
Dupont-951	0.006	7.8	2–5 GHz
ESL41110	—	4.3–4.7	1–30 GHz
Heraeus-HL2000	0.0026	6.7–7.6	—
NEC	—	3.9	—
Dupont-943	0.002	7.5	30 GHz
TDK	0.006	7.28	—
CeramTape GC	0.0033	7.3–8.5	Microwave
KEKO SK47	0.0019	6.9–7.1	Microwave
ESL41020	0.0033	7–7.5	Microwave

metal electrodes. These properties are determined by the proportion of glass phase in ceramics, and classified into three categories: glass-ceramic (50%–80%), ceramic-glass composite (20%–50%), and ceramic materials (<20%).

Glass-ceramics exhibit both glassy and ceramic properties as they are composed of crystals with regular arrangements of atoms. Unlike sintered ceramics, glass-ceramics can maintain high dielectric constant when adjusting the temperature coefficient τf at the resonance frequency close to zero by mixing crystal phases with opposite signs. This system can be divided into two categories: magnesium aluminosilicate (MAS) and calcium borosilicate (CBS). In MAS, the $MgO\text{-}Al_2O_3\text{-}SiO_2$ (cordierite $Mg_2Al_4Si_5O_{18}$) glass-ceramic system has desirable properties, including stability, low expansion coefficient ($112\text{–}119 \times 10^{-6}$/K), and low dielectric constant ($\varepsilon_r = 5.1$) when co-fired with copper conductor in a nitrogen atmosphere.

In CBS, borosilicate glass is produced by melting the glass raw material at 1400°C and quenching it at high temperature. After crystallization treatment, a mixed crystalline phase of $CaSiO_3$ and CaB_2O_4 is formed. The resulting sample has a low dielectric constant ($\varepsilon_r = 5.9$) and $\tan \delta = 0.002$. Ferro's A6 with CBS glass ($CaO\text{–}B_2O_3\text{–}SiO_2$) is a commonly used CBS material for LTCC antenna applications [16, 17]. In [18], an integrated DRA array for 60 GHz communication is proposed. DRA element is constructed on an 18-layer LTCC substrate of Ferro A6-M to form a 16-element DRA array as shown in Figure 2.6.

Ceramic-glass composite materials have a volume fraction of glass phase of 20%–50%. DuPont 951 is a classic example for antenna applications by offering material parameter controls. Its main components are $Al_2O_3CaZrO_3$ glass, with a dielectric constant of 7.8 and a dielectric loss of 0.006. However, ceramic-glass composite materials suffer from the issue of high losses and require glass melting during

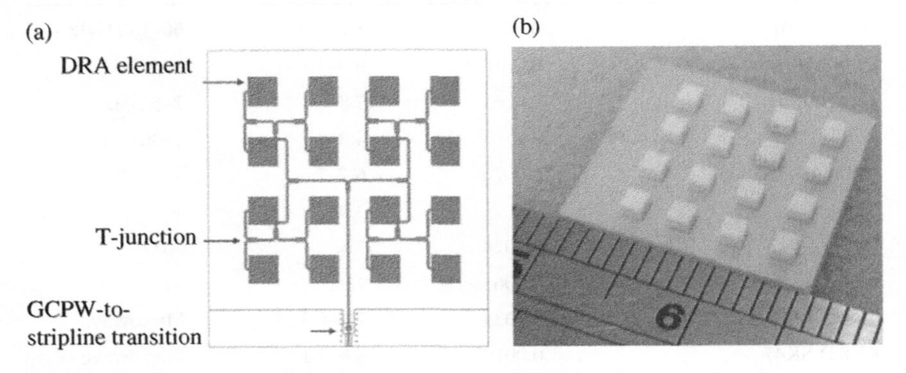

Figure 2.6 16-element DRA array based on LTCC technology. *Source:* Guo and Chu [18]/IEEE. (a) Configuration. (b) Fabrication prototype.

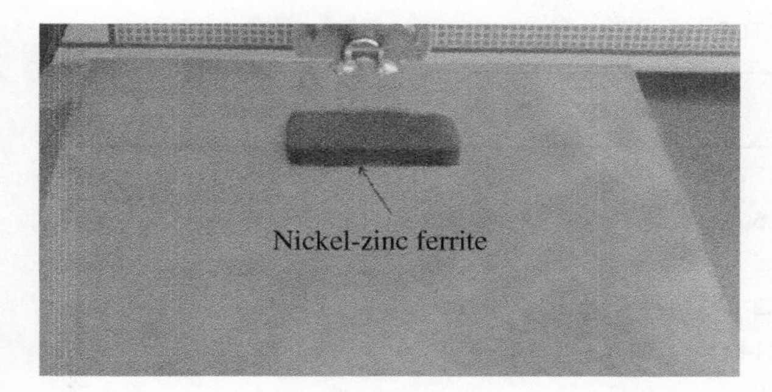

Figure 2.7 Manufactured $Ni_{0.5}Zn_{0.5}Fe_2O_4$ ceramic antenna [19].

the manufacturing process. In the co-firing process, the glass phase in LTCC substrates needs crystallization process. This process is sensitive to temperature and atmosphere, and the final crystalline phase obtained may differ significantly depending on the specific temperature and atmospheric conditions, making it challenging to ensure consistent performance of LTCC substrates. As a result, the high energy consumption and expensive equipment limit the application of glass-ceramic composite materials.

Ceramic materials have received extensive research attention. Their structure is similar to that of traditionally sintered ceramics, with a glass content of 10%–20%. In [19], researchers designed a compact DRA built on a microstrip substrate for microwave applications. As shown in Figure 2.7, the DR is made of nickel-zinc ferrite ($Ni_{0.5}Zn_{0.5}Fe_2O_4$) with a dielectric constant of 10.6 and a dielectric loss of 0.0002. The characteristics of wide bandwidth have been obtained when properly designed.

2.3.1.3 Ultralow-Temperature Co-fired Ceramics

Ultralow-temperature co-fired ceramics (ULTCC) can be sintered at very low temperatures ranging from 400 to 700°C, providing an energy-saving means for manufacturing. The low sintering temperature enables the use of a wider range of conductor materials for functionalization, allowing integration with other technologies such as semiconductor technology and polymer-based microcircuit manufacturing. Circuits and packages can be also embedded and sintered into ceramics, thus expanding the application range of ULTCC in low costs. ULTCC modules are well-suited for shell and packaging technology or used as substrates for high-frequency technology applications such as antennas, filters, and circulators.

The composition of ULTCC includes molybdates, tungstates, phosphates, and borates. In [20], Li_3BO_3 ceramic is synthesized with two low-melting materials

Table 2.3 Properties of ULTCC materials.

Material	ε_r	tan(δ)	$Q \times F$/GHz	τf ppm/°C	T/°C
Li_3BO_3	5	—	37 200	3.1	600
Li_2WO_4	5.5	—	62 000	−146	640
Li_2MoO_4	5.5	—	46 000	−160	540
$Li_2Mg_2(WO_4)_3$	8.2	—	90 000	−52.4	800
$Bi_6B_{10}O_4$	12.14	0.003	—	−72	625
Li_2O-Al_2O_3-B_2O_3	3.9	0.03	—	−64	75
ZnO-$3B_2O_3$	7.5	0.014	—	110	650
$(K_{0.5}La_{0.5})MoO_4$	10.3	0.0017	59 000	—	660
$(AgBi)_{0.5}WO_4$	35.9	0.0058	13 000	—	580

Li_2CO_3 and B_2O_3 at 580–640°C. It was found that Li_3BO_3 sintered at 600°C has the best microwave dielectric properties with $\varepsilon_r = 5$, $Q \times F = 37\,200$ GHz, and $\tau f = 3.1$ ppm/°C. $Li_6B_4O_9$ also has low sintering temperature, low dielectric constant, and high $Q \times F$ value. Both Li_2WO_4 and Li_2MoO_4 can be sintered at ultralow temperature (640 and 540°C) to achieve good chemical compatibility with Ag and Al electrodes in the co-firing process.

Li_2WO_4 ceramic has dielectric properties of $\varepsilon_r = 5.5$, $Q \times F = 62\,000$ GHz, and $\tau f = -146$ ppm/°C [21], whereas Li_2MoO_4 ceramic has dielectric properties of $\varepsilon_r = 5.5$, $Q \times F = 46\,000$ GHz, and $\tau f = -160$ ppm/°C [22]. Other commonly used ULTCC and their properties are concluded in Table 2.3.

Compared with other ULTCC materials, molybdates have four major advantages: high-frequency characteristics ($Q \times F > 45\,000$ GHz), low dielectric loss (tan $\delta < 2.8 \times 10^{-4}$), low K value (as low as 5.5), and high temperature stability (TCF $\rightarrow$ 0). These characteristics are derived from its special tetrahedral structure, as shown in Figure 2.8.

The transfer of electrons in tetrahedral molybdenum-based ceramic crystals is easier, resulting in low dielectric loss in the microwave and millimeter wave frequency band. Additionally, molybdenum-based raw materials have high porosity and the relative density of the ceramic can be controlled through doping and sintering ρ, allowing for control of the relative dielectric constant of the ceramic. The K value of molybdenum-based ceramics can be as low as 5.5 and as high as 81. The stable resonant frequency for molybdenum-based microwave components under varying operating temperatures is achieved by adjusting the doping ratio of components that makes the temperature coefficient of resonant frequency (TCF) close to zero [23]. Additionally, the cold-sintered molybdenum-based ceramics

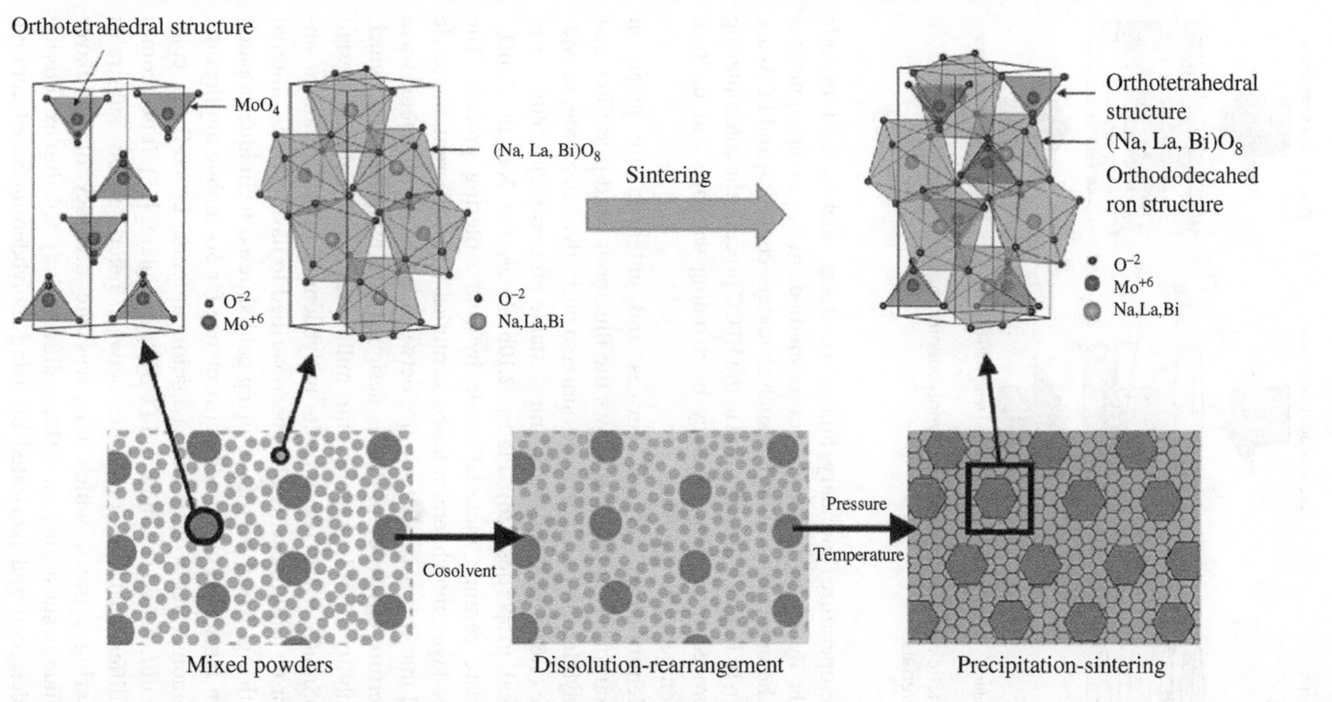

Figure 2.8 Tetrahedral structure and formation principle of molybdenum-based ceramics.

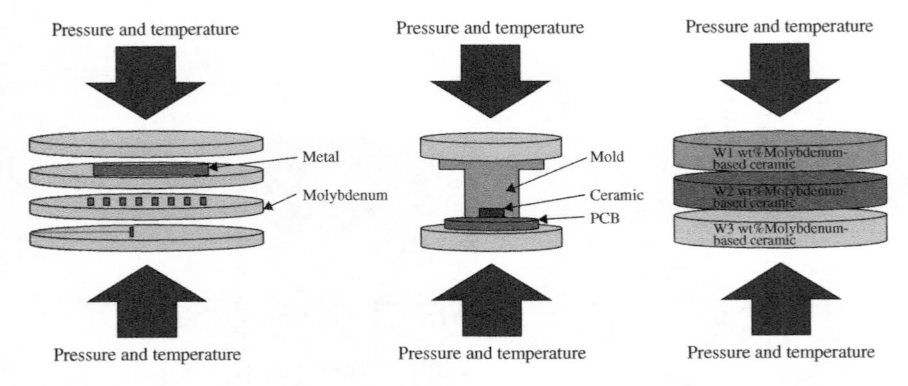

Figure 2.9 Schematic diagram of co-firing of molybdenum-based ceramics and heterogeneous materials.

overcome the temperature boundary among ceramics, metals, and organic polymer materials. By adding pressure or other methods in the sintering process [24], the co-fired heterogeneous materials such as metals, ceramics, and PCBs are obtained as shown in Figure 2.9. Compared to the LTCC process, the cold sintering process has improved processing accuracy by avoiding warpage and fracture caused by high-temperature sintering.

Some molybdenum-based ceramic devices and antennas are given in Figure 2.10 as examples. Figure 2.10a shows the Bluetooth band-pass filter that is made from molybdenum-based ceramics sintered with silver electrode, providing small volume (1.6 mm × 0.8 mm × 0.7 mm), stable process performance, and excellent electrical properties [26]. Figure 2.10b shows the $Na_{0.5}Bi_{0.5}MoO_4$-Li_2MoO_4 composite ceramic material made by cold sintering process. The co-firing of three-layer molybdenum-based ceramics with different dielectric constants proved that it is possible to co-fire between layers. A millimeter-wave lens with an aperture efficiency of 78% is designed and manufactured, which proved the feasibility of this material for millimeter wave antenna design [27]. In [28], $CaSnSiO_5$-K_2MoO_4 composite molybdenum matrix ceramics sintered at 180°C for 60 minutes, which has been verified to have good co-sintered compatibility with Ag. Based on this co-firing cold-sintered technique, a patch antenna has been fabricated with good performance for 5G mobile application. Figure 2.10c demonstrates a satellite navigation antenna by co-firing cold-sintered molybdenum-based ceramics and PCB organic plates [25]. It overcomes the problem of differential shrinkage rates between heterogeneous substrates during the sintering process, which improves the accuracy of multilayer alignment of different substrates. In 2022, a differentially fed duplex filtering DRA has been designed and fabricated by using molybdenum-based ceramic

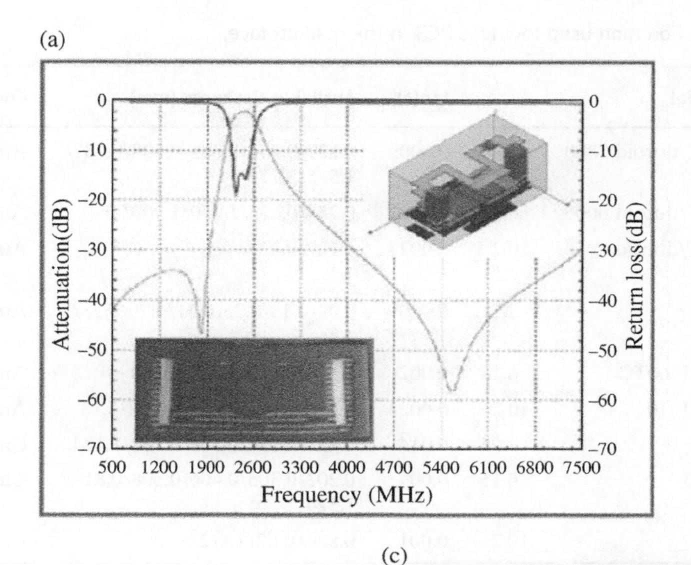

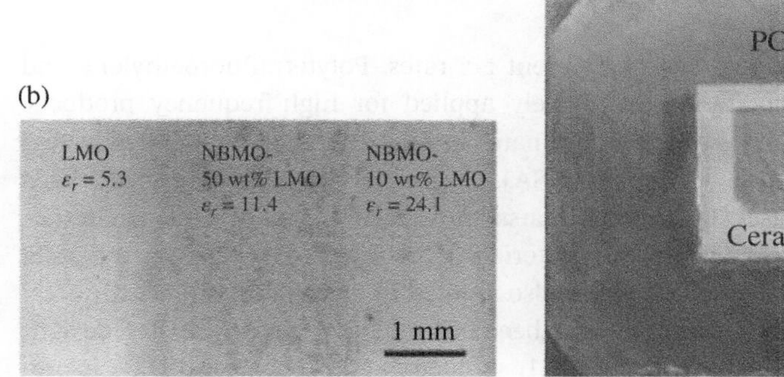

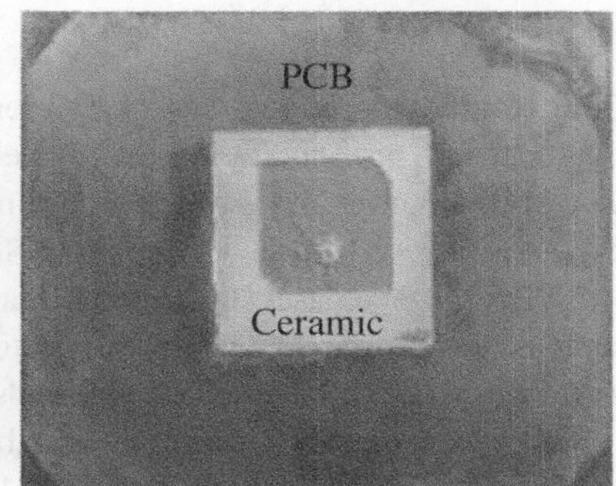

Figure 2.10 Fabrication prototypes of molybdenum-based ceramic devices and antennas [25–27]. (a) 645°C co-fired six-layer heterogenous miniaturized band-pass filter. (b) 150°C co-fired three-layer heterogenous millimeter wave dielectric lens. (c) 150°C co-fired molybdenum-based ceramic and PCB-integrated satellite navigation antenna.

($Bi_2Mo_2O_9$) [29]. The verifications on the dielectric material and DRA performance have been elaborated in detail in Section 8.4.2.

2.3.2 Organic Dielectric Material

In the commercial marketplace, the organic dielectric materials with a wide range of dielectric permittivity are available as listed in Table 2.4. Their prices are lower than LTCC and complementary metal oxide semiconductor (CMOS), but vary

Table 2.4 Common used low-loss PCB in the marketplace.

PCB Material	ε_r	tan(δ)	Available thickness (mm)	Country
Rogers RT/duroid 5880	2.20	0.0009	0.127/0.254/0.381/0.508/0.787/1.575/3.175	America
Rogers RT/duroid 6006	6.15	0.0027	0.254/0.635/1.270/1.900/2.50	America
Rogers RT/duroid 6010LM	10.2	0.0023	0.127/0.239	America
Taconic TLY-5A	2.17	0.0009	0.09/0.13/0.25/0.51/0.79/1.58/2.36	America
Taconic RF-60TC	6.15	0.002	0.130/0.250/0.510/0.760	America
Taconic RF-10	10.2	0.0025	0.25/0.64/0.76/1.19/1.58/3.18	America
TFA-2	2.94	0.014	0.127/0.254/0.508/0.762/1.524	China
WL-CT615	6.15	0.004	0.203/0.305/0.406/0.508/0.813/1.524	China
TF-1/2	10.2	0.001	0.8/1.0/1.2/1.5/2.0	China

from country to country due to different tax rates. Polytetrafluoroethylene and hydrocarbon resin materials are widely applied for high-frequency products. 90% of the market share of PTFE laminates in high-frequency substrates is held by manufacturers such as Rogers (USA), Taconic (USA), Parker (USA), Isola (USA), Nakagata Kasei (Japan), and Matsushita Electric (Japan). Other materials such as new inorganic and organic materials composite TPH-1/2, Slygard PDMS, and liquid crystal polymer (LCP) are also applied in antenna manufacturing.

The organic dielectric materials have been utilized in our proposed DRA designs in the following chapters. In Section 3.1, a wideband dual-polarized DRA is proposed as shown in Figure 2.11a. The loaded white dielectric cube is made of Rogers

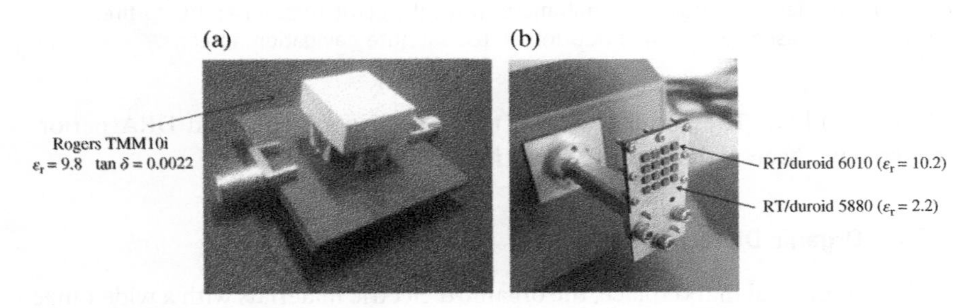

Figure 2.11 Fabrication prototypes of organic dielectric material. (a) Wide-band dual-polarized DRA. (b) Stacked DRA array for enlarged dimensions.

TMM 10i with dielectric constant of $\varepsilon_r = 9.8$ and loss tangent of $\tan\delta = 0.0022$. In Section 7.4, stacked DRA array for enlarged dimensions is proposed as shown in Figure 2.11b. The stacked DRA is made of RT/duroid 6010 ($\varepsilon_r = 10.2$) as top layer, which is the PTFE with ceramic. The DRA bottom layer and the SIW feeding network are designed based on RT/duroid 5880 ($\varepsilon_r = 2.2$), which is random microfiber glass.

2.3.3 Tunable Dielectric Material

2.3.3.1 Liquid

Liquid antennas are proposed by using liquid as the metal and dielectric materials of the antenna. Compared with the traditional solid-state antennas with fixed performance, the liquid antennas have the advantages of high plasticity and strong reconfigurability. When the liquid antenna is not working, its liquid can be emptied to reduce the radar cross section (RCS). The antenna mode can be changed by controlling the shape and amount of the liquid, thus realizing the reconfigurable characteristic of the antennas [30, 31].

According to materials, liquid antennas can be divided into liquid metal antennas [32, 33], water antennas [34–37], and other liquid antennas. Liquid metal antennas already have some applications in communication systems, but there are some disadvantages. For example, the chemical properties of the metal are harmful to human health and communication equipment. The existing liquid metal materials are expensive and available only in very few types. Compared with liquid metal antennas, water antennas are cheap, easy to obtain, non-toxic and harmless, and can be converted into conductive liquids by adding electrolytes. Water can be used to make dielectric antennas, including water DRAs [34], water DLAs [35], and water DPAs [36]. Conductive liquids (such as seawater) can be used for metal antennas [37].

Figure 2.12 shows a hybrid water DRA with broadband characteristics [34], which can be useful in many very high-frequency-band wireless applications. In this design, a simple seawater monopole is loaded with two distilled water cylinders working as DRs. By employing the loading technique, multiple close resonances are introduced to broaden the impedance bandwidth of the seawater antenna. It achieves a wide bandwidth from 69 to 171 MHz for $|S_{11}| < -10$ dB. Meanwhile, the radiation patterns are stable in the desired band.

The liquid in water antennas could be seawater or pure water (distilled water). The difference is that seawater contains a variety of inorganic salts, mainly sodium chloride, calcium chloride, potassium chloride, etc. Most of them exist in the ocean in the form of ions. The vast majority of the main ions are sodium ions, magnesium ions, chloride ions, potassium ions, etc. These ions can move freely in seawater, enabling seawater with good conductivity. The dielectric constant of seawater

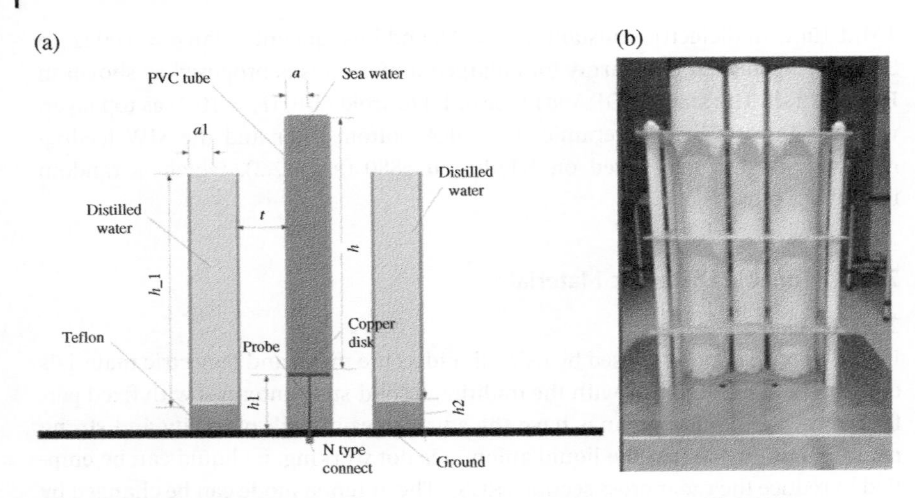

Figure 2.12 Hybrid water monopole antenna. *Source:* Qian and Chu [34]/IEEE. (a) Geometry. (b) Fabrication prototype.

and pure water is ranged from 78 to 81 at room temperature, whereas the conductivity is ranged from 0.0005 to 4 S/m [37].

The dielectric constant and conductivity have the greatest impact on the antenna performance. It was found that the permittivity and conductivity of water and its salt solution are not constant, which would change greatly with the working frequency, ambient temperature, salinity, and others. Therefore, it is very important to quantitatively analyze and obtain the variation functions of permittivity and conductivity with respect to these considerations.

The complex permittivity ε of any material can be expressed as follows:

$$\varepsilon = \varepsilon_r - j\varepsilon' \tag{2.1}$$

$$\tan(\delta) = \frac{\varepsilon'}{\varepsilon_r} \tag{2.2}$$

where ε_r is the relative permittivity, ε' is the imaginary part of the complex relative permittivity, and $\tan(\delta)$ is the loss tangent.

In 1977, Klein and Swift proposed the Single-Debye model, which describes the variation curve of the complex relative permittivity of water and its salt solution with temperature and salinity [38].

$$\varepsilon(S,T) = \varepsilon_\infty + \frac{\varepsilon_s(S,T) - \varepsilon_\infty}{1 + [j\omega\tau(S,T)]^{1-\alpha}} - j\frac{\sigma(S,T)}{\omega\varepsilon_0} \tag{2.3}$$

where S is the solution salinity, T is the solution temperature, ε_∞ is the dielectric constant of water when the frequency approaches infinity, ε_s is the dielectric constant of water when the frequency is 0, ω is the angular frequency, τ is the relaxation time, σ is the electrical conductivity, and ε_0 is the vacuum dielectric constant.

This model is accurate in the low-frequency band <1 GHz, but the error increases with the increase in frequency. In 2004, Meissner et al. proposed the Double-Debye model [39] with better accuracy. It describes the relative permittivity and conductivity of liquid water and its salt solution from −20 to 40°C without considering the change of dielectric properties caused by water freezing. The formula is shown as follows:

$$\varepsilon(S,T) = \frac{\varepsilon_s(S,T) - \varepsilon_1(S,T)}{1 + j\nu/\nu_1(S,T)} + \frac{\varepsilon_1(S,T) - \varepsilon_\infty(S,T)}{1 + j\nu/\nu_2(S,T)} + \varepsilon_\infty(S,T) - j\frac{\sigma(S,T)}{(2\pi\varepsilon_0)\nu} \tag{2.4}$$

where ν is the radiation frequency, and ν_1 and ν_2 are the first and second Debye frequencies, respectively. The dielectric constant ε_s of water at zero frequency can be expressed as,

$$\varepsilon_s(S,T) = \frac{3.70886 \cdot 10^4 - 8.2168 \cdot 10^1 T}{4.21854 \cdot 10^2 + T} \tag{2.5}$$

For pure water, let $S = 0$, we obtain,

$$\varepsilon_1(T) = a_0 + a_1 T + a_2 T^2 \tag{2.6}$$

$$\nu_1(T) = \frac{45 + T}{a_3 + a_4 T + a_5 T^2} \tag{2.7}$$

$$\varepsilon_\infty(T) = a_6 + a_7 T \tag{2.8}$$

$$\nu_2(T) = \frac{45 + T}{a_8 + a_9 T + a_{10} T^2} \tag{2.9}$$

For salt solution, the conductivity and permittivity can be calculated by Double-Debye function model; in that case the conductivity of salt solution with salinity 35 ($S = 35$) is measured by experiment.

$$\sigma(S,T) = \sigma(T, S = 35) \cdot R_{15}(S) \cdot \frac{R_T(S)}{R_{15}(S)} \tag{2.10}$$

$$\begin{aligned}\sigma(T, S = 35) = {} & 2.903602 + 8.607 \cdot 10^{-2} \cdot T + 4.738817 \cdot 10^{-4} \cdot T^2 \\ & - 2.991 \cdot 10^{-6} \cdot T^3 + 4.3047 \cdot 10^{-9} \cdot T^4\end{aligned} \tag{2.11}$$

$$R_{15}(S) = S \cdot \frac{35.5109 + 5.45216 \cdot S + 1.4409 \cdot 10^{-2} \cdot S^2}{1004.75 + 182.283 \cdot S + S^2} \tag{2.12}$$

$$\frac{R_T(S)}{R_{15}(S)} = 1 + \frac{\alpha_0(T - 15)}{\alpha_1 + T} \tag{2.13}$$

where

$$\alpha_0 = \frac{6.9431 + 3.2841 \cdot S - 9.9486 \cdot 10^{-2} \cdot S^2}{84.850 + 69.024 \cdot S + S^2} \quad (2.14)$$

$$\alpha_1 = 49.843 - 0.2276 \cdot S + 0.198 \cdot 10^{-2} \cdot S^2 \quad (2.15)$$

As a result, the parameters of Double-Debye function model for brine with different temperatures and salinity can be deduced as,

$$\varepsilon_s(S,T) = \varepsilon_s(S=0,T) \cdot \exp\left[b_0 S + b_1 S^2 + b_2 TS\right] \quad (2.16)$$

$$v_1(S,T) = v_1(S=0,T) \cdot \left[1 + S\left(b_3 + b_4 T + b_5 T^2\right)\right] \quad (2.17)$$

$$\varepsilon_1(S,T) = \varepsilon_1(S=0,T) \cdot \exp\left[b_6 S + b_7 S^2 + b_8 TS\right] \quad (2.18)$$

$$v_2(S,T) = v_2(S=0,T) \cdot [1 + S(b_9 + b_{10} T] \quad (2.19)$$

$$\varepsilon_\infty(S,T) = \varepsilon_\infty(S=0,T) \cdot [1 + S(b_{11} + b_{12} T)] \quad (2.20)$$

The values of parameters a_0~a_{10}, b_0~b_{12} are given in Table 2.5.

2.3.3.2 Piezodielectric Materials

Piezoelectric materials are anisotropic and their properties are related to the propagation direction. Different properties can be observed in three directions, which

Table 2.5 Parameters for a_0~a_{10}, b_0~b_{12}.

i	a_i	j	b_j
0	5.7230	0	-3.5642×10^{-3}
1	2.2379×10^{-2}	1	4.4787×10^{-6}
2	-7.1237×10^{-4}	2	1.1557×10^{-5}
3	5.0478	3	2.3936×10^{-3}
4	-7.0315×10^{-2}	4	-3.1353×10^{-5}
5	6.0059×10^{-4}	5	2.5247×10^{-7}
6	3.6143	6	-6.2891×10^{-3}
7	2.8841×10^{-2}	7	1.7603×10^{-4}
8	1.3652×10^{-1}	8	-9.2214×10^{-5}
9	1.4825×10^{-3}	9	-1.9972×10^{-2}
10	2.4166×10^{-4}	10	1.8118×10^{-4}
		11	-2.0427×10^{-3}
		12	1.5788×10^{-4}

means three different tensile elastic modulus, shear modulus, Poisson's ratio, and dielectric constant. Piezoelectric materials can be divided into soft piezoelectric ceramics, hard ceramics, piezoelectric single crystals, textured ceramics, piezoelectric metamaterials, high-temperature piezoelectric materials, and lead-free piezoelectric ceramics [40].

Soft piezoelectric ceramics have high piezoelectric properties and electromechanical coupling coefficient. Textured ceramics possess higher piezoelectric properties and electromechanical coupling coefficients than soft piezoelectric ceramics. Hard ceramics include hard doping process by adding K^+, Na^+, $Fe_3{}^+$, $Mn_4{}^+$, $Co_3{}^+$, etc. to soft piezoelectric ceramics. Their dielectric loss is low, but their piezoelectric properties are not as high as soft piezoelectric ceramics. Single crystals are widely used in electronic information industry and can be used to manufacture electronic components that select and control frequencies, such as crystal oscillators. Piezoelectric metamaterial is an artificial structural material that can excite or control the piezoelectric properties by designing and aligning key size piezoelectric elements. High-temperature piezoelectric material can be used in high temperature environment. Lead-free piezoelectric ceramics with perovskite structure are developed to reduce the environmental pollution of lead-based piezoelectric ceramics. Compared with lead-based piezoelectric ceramics, lead-free piezoelectric ceramics still have problems such as poor stability of electrical properties and high temperature sensitivity. All of these piezoelectric materials can be applied for antenna manufacturing but should be carefully selected based on their applications.

The properties of piezoelectric materials are concluded in Table 2.6, including the relative permittivity ε_r, the loss tangent $\tan(\delta)$, the curie temperature T_c, the mechanical quality factor Q_m, and the piezoelectric coefficient d_{33}. The characteristics of these piezoelectric materials are introduced as follows.

Figure 2.13 shows acoustically actuated nanomechanical magnetoelectric (ME) antennas with a suspended ferromagnetic/piezoelectric thin-film heterostructure [41]. A 50-nm-thick Pt film was sputter-deposited and patterned by lift-off on top of the high resistivity silicon (Si) wafers (>10 000 Ω cm) to define the bottom electrodes. The 500 nm AlN film was sputter-deposited, and the via holes were formed by H_3PO_4 etching to access the bottom electrodes. The AlN film was etched by inductively coupled plasma etching in Cl_2-based chemistry to define the shape of the resonant nanoplate. A 100-nm-thick gold (Au) film was evaporated and patterned to form the top ground. A 500-nm-thick $FeGaB/Al_2O_3$ multilayer layer was deposited by a magnetron sputtering and patterned by lift-off process. A 100 Oe in situ magnetic field bias was applied during the magnetron deposition along the width direction of the device to pre-orient the magnetic domains.

It has been demonstrated that ME antennas can achieve miniaturization by a factor of 2–3 compared to the state-of-the-art compact antennas without any performance degradation. These antennas have sizes as small

Table 2.6 Performance characterization of different piezoelectric materials.

Type	Material	ε_r	tan(δ)	T_c [°C]	Q_m	d_{33} [pC/N]
Soft ceramics	PMN-0.29 PT	4500	0.05	120	—	650
	2.5Sm-PMN-29PT	13 000	0.025	89	—	1510
	Sr-0.1PZN-0.1PNN-0.8PZT	4081	0.04	176	—	800
	0.1 wt% Li_2CO_3-modified PZN–PNN–PZT	4700	—	167	—	950
	1Cu-65PZT-35PNN	3750	0.003	176	70	605
	55PNN-45PZT	9000	0.03	120	36	1753
	4Ta-PMS-49.5PNN-49.5PZ	6838	0.014	119		805
Textured ceramics	PMN-PZT	1723	0.003	234	403	720
	PMN-PYN-PT	2230	—	205	—	1100
	PIN-PSN-PT	2300	0.012	147	—	1100
	PIN-PSN-PT	2250	0.013	141	—	1060
	Eu-PMN-PT	—	—	—	—	1950
Hard ceramics	CuO-PMMnN-PZT	—	0.004	325	1003	218
	BNT-PMN-PT	1512	0.004	336	2010	355
	BMZ-BS-PT	1464	0.0058	449	61	360
Single crystals	PMN-30PT	5000	0.005	138	—	1600
	PIN-PMN-PT (MPB)	7240	—	197	120	2740
	PMN-PZT	5000	0.005	210	150	1750
	Mn:PIN-PMN-PT	3700	—	193	810	1120
	Mn:PMN-PZT	3410	—	203	1050	1140
	Sm: PMN-PT	12 000	0.006	121		4100
Piezoelectric metamaterials	PZT-5H bimorph	—	—	—	—	—
	PZT	—	—	—	—	540
High-temperature piezoelectric materials	BS-64PT	1000	0.05	450	—	460
	BS-63.5PT	—	—	446	—	700
	BF-PT-BT	546	0.013	—	—	222
Lead-free ceramics	KNNS-BNKZ-Fe-As	5000	0.04	140	—	650
	(1-*x*)BNKT-BNT	1815	0.02	125		683

Source: Jin et al. [40]/IEEE.

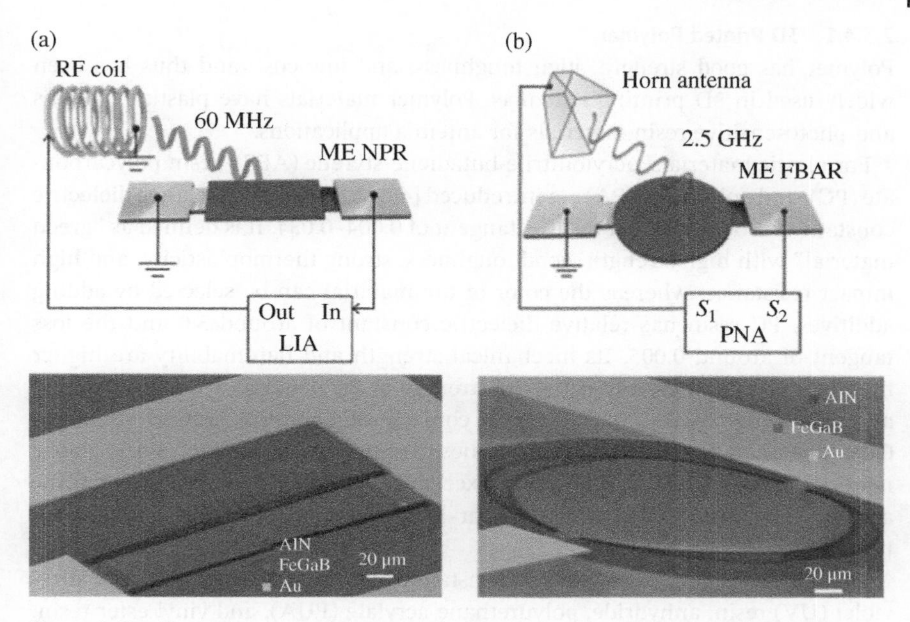

Figure 2.13 Two structures of the acoustically actuated nanomechanical ME antennas. *Source:* Nan et al. [41]/Springer Nature/CCBY 4.0/Public domain. (a) Nanoplate resonators (NPR). (b) Thin-film bulk acoustic wave resonators (FBAR).

as one-thousandth of a wavelength and are designed to resonate in different modes, enabling them to operate at very-high frequencies (60 MHz) and ultra-high frequencies (2.525 GHz). Overall, ME antennas have the potential implications for our future antennas and communication systems, including Internet of Things, wearable technology, bio-implantable and bio-injectable devices, smartphones, and wireless communication systems.

2.3.4 3D Printing Material

3D printing technology has received significant attention for antenna application because it provides additional design freedom and reduces manufacturing costs [42]. In Chapter 6, the feasibility of 3D printing technologies will be discussed for dielectric antenna and antenna with dielectric substrate.

3D printing materials can be divided into polymer, ceramic, and metal. To facilitate the selection of suitable materials for 3D printing antennas, this section provides a detailed overview of these materials.

2.3.4.1 3D Printed Polymer

Polymer has good strength, high toughness, and low cost, and thus has been widely used in 3D printing antennas. Polymer materials have plastic materials and photosensitive resin materials for antenna applications.

For plastic materials, acrylonitrile-butadiene-styrene (ABS), resin polycarbonate (PC), and polyamide (PA) are introduced [43]. ABS resin has relative dielectric constant of around 3.0 and the loss tangent of 0.004–0.034. It is defined as "green material" with high strength, good toughness, strong thermoplasticity, and high impact resistance, whereas the color of the material can be selected by adding additives. PC resin has relative dielectric constant of around 3.0 and the loss tangent of around 0.005. Its mechanical strength and flammability are higher than that of ABS, thus enhancing the strength of the material to avoid shrinking and deforming. PA resin has dielectric constant of 4 and loss tangent of around 0.01. It has superior mechanical properties than that of ABS and PC. PA66 plastic material in the PA resin family shows excellent properties in toughness, ductility, and wear resistance, but suffers from high melting point and difficulty in processing.

Photosensitive resin has dielectric constant of around 4.0, which includes ultraviolet (UV) resin, anhydride, polyurethane acrylate (PUA), and vinyl ester resin. UV resin is liquid and has high mechanical strength, non-volatile odor, easy storage, short preparation process, easy curing, high forming precision, and good surface effect, which is suitable for stereo lithography apparatus (SLA). The anhydride has moderate viscosity but has low mechanical strength. So the printed part with anhydride has low forming degree and easy to shrink. PUA has good optical properties, good wear resistance, and toughness, but suffers from difficulties with control and adjusting the coloring degree in polymerization process. Vinyl ester resin has good chemical stability, has high mechanical strength, and is not easy to shrink. However, the viscosity of the material is high, the fluidity is poor, and it takes a long time in the process of polymerization, which brings difficulties to the 3D printing process and affects the product forming and processing.

Figure 2.14 shows a 3D printed ABS antenna that is embeddable in a plastic CubeSat [44]. The antenna size is miniaturized to $71.5 \times 71.5 \times 13$ mm^3, and its weight is reduced to 51 g. The ABS material ($\varepsilon_r = 2.3$, $\tan\delta = 0.01$ at 2.15 GHz) is printed in hollow structure to achieve desired antenna performance.

2.3.4.2 3D Printed Ceramic

Ceramic material has relative dielectric constant of 8.5–10. It is suitable for 3D printing material due to the characteristics of high mechanical strength, compression and wear resistance, high hardness, high temperature resistance and melting resistance, poor electrical conductivity, and thermal conductivity [45]. However, it

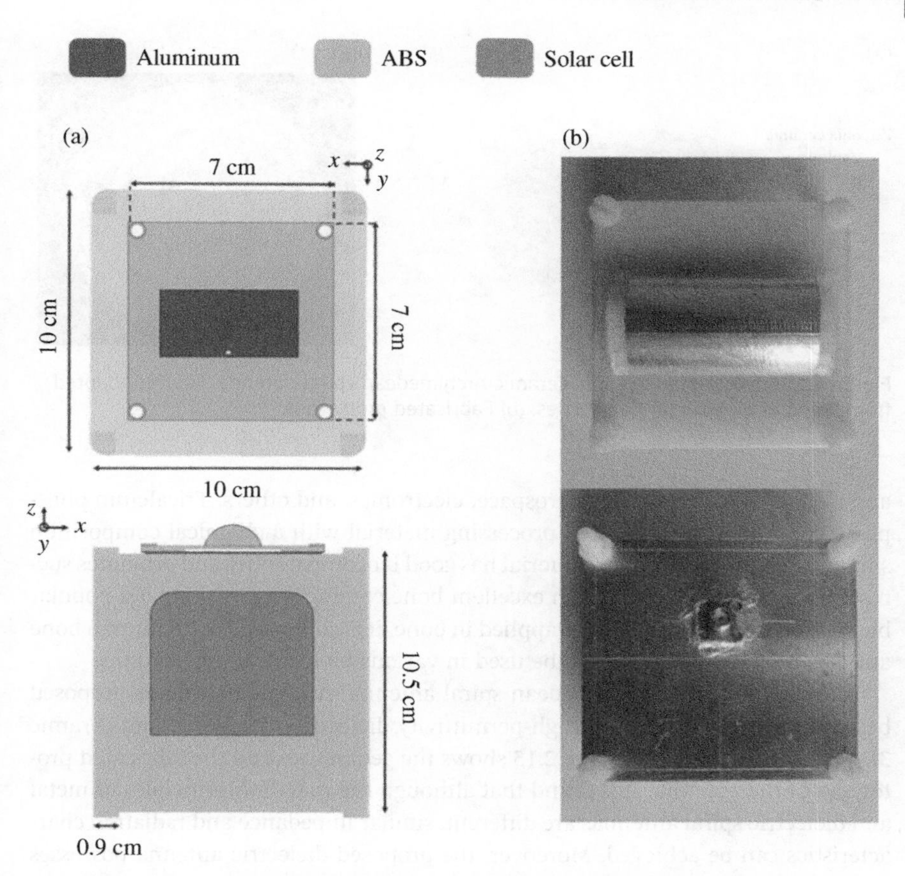

Figure 2.14 3D printed ABS antenna. *Source:* Adapted from Muntoni et al. [44]. (a) Antenna configuration. (b) Antenna prototypes.

suffers from some problems such as poor mechanical properties, low density, and low rate of finished products. At present, the commonly used ceramic materials in 3D printing are Al_2O_3, Si_3N_4, $Ca_3(PO_4)_2$, and so on.

Alumina ceramic material is a widely used material due to its advantages such as wide source, low cost, large output, and versatile applications. It has relative dielectric constant of 9–10 and the dielectric loss tangent of about 0.0004. By adding organic matter and alloy powder and sintering, alumina ceramics can be synthesized and modified for specific purposes. For 3D printed ceramics, modified ceramic powders are used for convenient processing, low cost, and strong maneuverability, resulting in shorter production time. This technology has broad

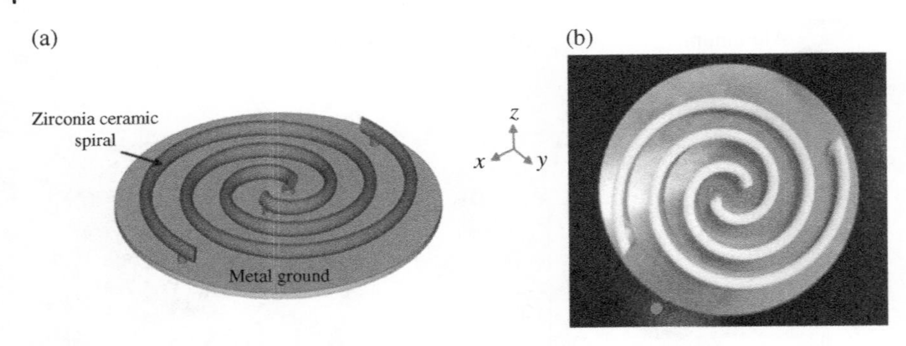

Figure 2.15 3D printed zirconia ceramic Archimedean spiral antenna. *Source:* Adapted from Wang et al. [46]. (a) Geometries. (b) Fabricated prototypes.

applications in construction, aerospace, electronics, and others. Tricalcium phosphate ceramic is a synthesized processing material with a chemical composition similar to human bone. This material has good biocompatibility and promotes successful human metabolism with excellent bone conductivity, making it a popular biomedical material. When it is applied in bone, it is compatible with human bone and has no variability and can be used in various biomedical applications.

In [46], a dielectric Archimedean spiral antenna made of zirconia is proposed based on the radiation from high-permittivity dielectric waveguide and ceramic 3D printing technology. Figure 2.15 shows the geometries and the fabricated prototypes of the antenna. It is found that although the radiation principles of metal and dielectric spiral antennas are different, similar impedance and radiation characteristics can be achieved. Moreover, the proposed dielectric antenna possesses lower profile, lower RCS at high-frequency spectrum (9–13 GHz), and higher radiation efficiency compared with the conventional metal spiral antenna.

2.3.4.3 Metal

3D printed metal materials can be classified into various types based on their composition, including iron-based alloys, titanium and titanium-based alloys, nickel-based alloys, cobalt-chromium alloys, aluminum alloys, copper alloys, and noble metals. 3D printing metal antennas [47] are proposed to address the significant waste of metal materials in traditional casting or forging methods. This technology eliminates the need for excessive material processing and significantly reduces the cost of production, thereby minimizing material wastage.

An axial corrugated horn covering the full Ka-band (26.5–40 GHz) is proposed in [48]. Figure 2.16 shows the fabricated prototypes of the 3D metal printed horn antenna. The measured radiation pattern demonstrates good beam symmetry

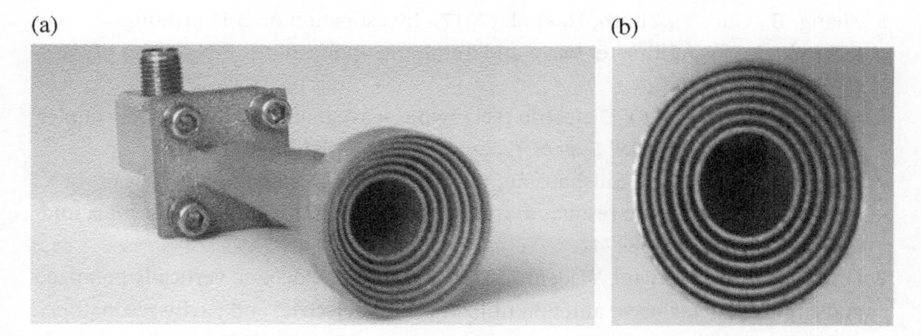

Figure 2.16 Photograph of 3D metal printed horn antenna [48]. (a) Isometric. (b) Front view.

and achieves a cross-polarization level better than 29 dB in the principal planes over the entire frequency band.

2.4 Summary

This chapter begins with the fundamentals of the dielectric antenna by identifying DRA, DPA, DLA, and SDRA. The configurations and modes of these antennas are discussed, with analysis of the design capabilities for performance enhancements. Then, dielectric materials for DRA are introduced in detail, which includes sintered ceramics, organic dielectric, tunable dielectric, and 3D printed material. Antenna utilizing these dielectric materials are provided as examples. Especially, these dielectric materials are utilized in our proposed DRA designs, which are specified in the following chapters.

References

1 Long, S., McAllister, M., and Shen, L. (1983). The resonant cylindrical dielectric cavity antenna. *IEEE Transactions on Antennas and Propagation* 31 (3): 406–412.

2 Lai, H., Luk, K.M., and Leung, K.W. (2013). Dense dielectric patch antenna—a new kind of low-profile antenna element for wireless communications. *IEEE Transactions on Antennas and Propagation* 61 (8): 4239–4245.

3 Shafai, L. (2015). *Dielectric Loaded Antennas*. Wiley.

4 Luk, K.M. and Leung, K.W. (2003). *Dielectric Resonator Antennas*. Wiley.

5 Zhang, B., Guo, Y., Zirath, H. et al. (2017). Investigation on 3-D-printing technologies for millimeter-wave and Terahertz applications. *Proceedings of the IEEE* 105 (4): 1–14.
6 Marcatili, E.A.J. (1969). Dielectric rectangular waveguide and directional coupler for integrated optics. *Bell System Technical Journal* 48 (7): 2071–2102.
7 Petosa, A., Simons, N., Siushansian, R. et al. (2000). Design and analysis of multisegment dielectric resonator antennas. *IEEE Transactions on Antennas and Propagation* 48 (5): 738–742.
8 Omar, A., Park, J., Kwon, W. et al. (2021). A compact wideband vertically polarized end-fire millimeter-wave antenna utilizing slot, dielectric, and cavity resonators. *IEEE Transactions on Antennas and Propagation* 69: 5234–5243.
9 Chen, Z., Shen, C., Liu, H. et al. (2020). Millimeter-wave rectangular dielectric resonator antenna array with enlarged DRA dimensions, wideband capability, and high-gain performance. *IEEE Transactions on Antennas and Propagation* 68 (4): 3271–3276.
10 Ahmad, Z. and Hesselbarth, J. (2018). On-chip dual-polarized dielectric resonator antenna for millimeter-wave applications. *IEEE Antennas and Wireless Propagation Letters* 17 (10): 1769–1772. https://doi.org/10.1109/LAWP.2018.2865453.
11 Petrella, R.A., Schoenbach, K.H., and Xiao, S. (2016). A dielectric rod antenna for picosecond pulse stimulation of neurological tissue. *IEEE Transactions on Plasma Science* 44 (4): 708–714. https://doi.org/10.1109/TPS.2016.2537213.
12 Jung, C.W., Lee, M.-J., Li, G.P. et al. (2006). Reconfigurable scan-beam single-arm spiral antenna integrated with RF-MEMS switches. *IEEE Transactions on Antennas and Propagation* 54 (2): 455–463. https://doi.org/10.1109/TAP.2005.863407.
13 Ghorbel, I., Ganster, P., Moulin, N. et al. (2023). Direct microwave heating of alumina for different densities: experimental and numerical thermal analysis. *Journal of the American Ceramic Society* 1–13. https://doi.org/10.1111/jace.18971.
14 Lee, R.-R. Development of high thermal conductivity aluminum nitride ceramic. *Journal of the American Ceramic Society* 74: 2242–2249. https://doi.org/10.1111/j.1151-2916.1991.tb08291.x.
15 Walker, A.C. Hydrothermal synthesis of quartz crystals. *Journal of the American Ceramic Society* 36: 250–256. https://doi.org/10.1111/j.1151-2916.1953.tb12877.x.
16 Zhang, B. and Zhang, Y.P. (2012). Grid array antennas with subarrays and multiple feeds for 60-GHz radios. *IEEE Transactions on Antennas and Propagation* 60 (5): 2270–2275. https://doi.org/10.1109/TAP.2012.2189733.
17 Sun, H., Guo, Y.-X., and Wang, Z. (2013). 60-GHz circularly polarized U-slot patch antenna array on LTCC. *IEEE Transactions on Antennas and Propagation* 61 (1): 430–435. https://doi.org/10.1109/TAP.2012.2214018.
18 Guo, Y.-X. and Chu, H. (2013). 60-GHz LTCC dielectric resonator antenna array. *2013 IEEE Antennas and Propagation Society International Symposium (APSURSI)*, Orlando, FL, USA. pp. 1874–1875. 10.1109/APS.2013.6711595.

19 de Oliveira, E.E.C., de Araujo, W.C., de Assis, P.C. et al. (2015). Small-size compact nickel-zinc ferrite dieletric resonator antenna with high dielectric constant. *2015 SBMO/IEEE MTT-S International Microwave and Optoelectronics Conference (IMOC)*, Porto de Galinhas, Brazil. pp. 1–4. 10.1109/IMOC.2015.7369134.

20 Chang, S.Y., Pai, H.F., Tseng, C.F. et al. (2017). Microwave dielectric properties of ultra-low temperature fired Li_3BO_3 ceramics. *Journal of Alloys and Compounds* 698: 814–818.

21 Zhou, D.I., Randall, C.A., Pang, L.-X. et al. (2011). Microwave dielectric properties of Li_2WO_4 ceramic with ultra-low sintering temperature. *Journal of the American Ceramic Society* 94 (2): 348–350.

22 Zhou, D., Randall, C.A., Wang, H. et al. (2010). Microwave dielectric ceramics in Li_2O-Bi_2O_3-MoO_3 system with ultra-low sintering temperatures. *Journal of the American Ceramic Society* 93 (4): 1096–1100.

23 Wang, D., Zhang, S., Zhou, D. et al. Temperature stable cold sintered ($Bi_{0.95}Li_{0.05}$) ($V_{0.9}Mo_{0.1}$) O_4-$Na_2Mo_2O_7$ microwave dielectric composites. *Materials* 12 (9): 1370.

24 Wang, D., Zhang, S., Wang, G. et al. (2020). Cold sintered $CaTiO_3$-K_2MoO_4 microwave dielectric ceramics for integrated microstrip patch antennas. *Applied Materials Today* 18: 100519.

25 Wang, D., Siame, B., Zhang, S. et al. (2020). Direct integration of cold sintered, temperature-stable $Bi_2Mo_2O_9$-K_2MoO_4 ceramics on printed circuit boards for satellite navigation antennas. *Journal of the European Ceramic Society* 40 (12): 4029–4034.

26 Wang, S., Luo, W., Li, L. et al. (2020). Improved tri-layer microwave dielectric ceramic for 5G applications. *Journal of the European Ceramic Society* 41 (1): 418–423.

27 Wang, D., Zhou, D., Zhang, S. et al. (2017). Cold sintered temperature stable $Na_{0.5}Bi_{0.5}MoO_4$-Li_2MoO_4 microwave composite ceramics. *ACS Sustainable Chemistry & Engineering* 6 (2): 2438–2444.

28 Ji, Y., Song, K., Zhang, S. et al. (2020). Cold sintered, temperature-stable $CaSnSiO_5$-K_2MoO_4 composite microwave ceramics and its prototype microstrip patch antenna. *Journal of the European Ceramic Society* 41 (1): 424–429.

29 Tian, H., Chen, Z., Chang, L. et al. (2021). Differentially fed duplex filtering dielectric resonator antenna with high isolation and CM suppression. *IEEE Transactions on Circuits and Systems II: Express Briefs* doi: 10.1109/TCSII.2021.3120726.

30 Xing, L., Xu, Q., Zhu, J. et al. (2021). A high-efficiency wideband frequency-reconfigurable water antenna with a liquid control system: usage for VHF and UHF applications. *IEEE Antennas and Propagation Magazine* 63 (1): 61–70.

31 Martínez, J.O., Rodríguez, J.R., Shen, Y. et al. (2022). Toward liquid reconfigurable antenna arrays for wireless communications. *IEEE Communications Magazine* 60 (12): 145–151.

32 Cheng, S., Wu, Z., Hallbjorner, P. et al. (2009). Foldable and stretchable liquid metal planar inverted cone antenna. *IEEE Transactions on Antennas and Propagation* 57 (12): 3765–3771.

33 Wang, M., Khan, M.R., Dickey, M.D. et al. (2017). A compound frequency- and polarization- reconfigurable crossed dipole using multidirectional spreading of liquid metal. *IEEE Antennas and Wireless Propagation Letters* 16: 79–82.

34 Qian, Y.-H. and Chu, Q.-X. (2017). A broadband hybrid monopole-dielectric resonator water antenna. *IEEE Antennas and Wireless Propagation Letters* 16: 360–363.

35 Xing, L., Huang, Y., Xu, Q. et al. (2016). A transparent dielectric-loaded reconfigurable antenna with a wide tuning range. *IEEE Antennas and Wireless Propagation Letters* 15: 1630–1633.

36 Sun, J. and Luk, K.-M. (2017). A wideband low cost and optically transparent water patch antenna with omnidirectional conical beam radiation patterns. *IEEE Transactions on Antennas and Propagation* 65 (9): 4478–4485.

37 Wang, M. and Chu, Q.-X. (2019). A wideband polarization-reconfigurable water dielectric resonator antenna. *IEEE Antennas and Wireless Propagation Letters* 18 (2): 402–406.

38 Klein, L. and Swift, C. (1977). An improved model for the dielectric constant of sea water at microwave frequencies. *IEEE Transactions on Antennas and Propagation* 25 (1): 104–111.

39 Meissner, T. and Wentz, F.J. (2004). The complex dielectric constant of pure and sea water from microwave satellite observations. *IEEE Transactions on Geoscience and Remote Sensing* 42 (9): 1836–1849.

40 Jin, H., Gao, X., Ren, K. et al. (2022). Review on piezoelectric actuators based on high-performance piezoelectric materials. *IEEE Transactions on Ultrasonics, Ferroelectrics, and Frequency Control* 69 (11): 3057–3069.

41 Nan, T., Lin, H., Gao, Y. et al. (2017). Acoustically actuated ultra-compact NEMS magnetoelectric antennas. *Nature Communications* 8: 296.

42 Wei, C., Zhang, Z., Cheng, D. et al. (2021). An overview of laser-based multiple metallic material additive manufacturing: from macro- to micro-scales. *International Journal of Extreme Manufacturing* 3 (1): 46–69.

43 Budhu, J. and Rahmat-Samii, Y. (2020). 3D-printed inhomogeneous dielectric lens antenna diagnostics: a tool for assessing lenses misprinted due to fabrication tolerances. *IEEE Antennas and Propagation Magazine* 62 (4): 49–61.

44 Muntoni, G., Montisci, G., Melis, A. et al. (2022). A curved 3D-printed S-band patch antenna for plastic CubeSat. *IEEE Open Journal of Antennas and Propagation* 3: 1351–1363.

45 Hua, L. and Bowen, P. (2021). Application of ceramic 3D printing technology in modern ceramic manufacturing. *7th International Symposium on Mechatronics and Industrial Informatics (ISMII)*, Zhuhai, China. pp. 160–163.

46 Wang, S., Fan, F., Xu, Y. et al. (2022). 3-D printed zirconia ceramic Archimedean spiral antenna: theory and performance in comparison with its metal counterpart. *IEEE Antennas and Wireless Propagation Letters* 21 (6): 1173–1177.

47 Gordon, J.A., Novotny, D.R., Francis, M.H. et al. (2017). An all-metal, 3-D-printed CubeSat feed horn: an assessment of performance conducted at 118.7503 GHz using a robotic antenna range. *IEEE Antennas and Propagation Magazine* 59 (2): 96–102.

48 Agnihotri, I. and Sharma, S.K. Design of a 3D metal printed axial corrugated horn antenna covering full Ka-band. *IEEE Antennas and Wireless Propagation Letters* 19 (4): 522–526.

3

Stacked Dielectric Resonator Antenna

CHAPTER MENU

3.1 Overview

Stacked dielectric resonator antenna (DRA), which is also known as multilayer DRA or composite DRA, is composed of multiple irregular-shaped dielectrics or different dielectric materials. Compared with single-layered DRA, the stacked DRA has higher degrees of design freedom and numerous resonance modes. The broadband stacked DRA was first proposed by Ahmed A. Kishk in 1989 [1]. After that, stacked DRA has been thoroughly investigated to increase the bandwidth of the antenna [2]. Nowadays, stacked DRA has attracted attention for its advantages such as wideband, high gain, and wide beamwidth.

Stacked DRA with different permittivity dielectric layers can reduce the radiation Q-factor of the system and achieves fusion of multiple modes. In past decades,

Dielectric Resonator Antennas: Materials, Designs and Applications, First Edition.
Zhijiao Chen, Jing-Ya Deng, and Haiwen Liu.

Published 2024 by John Wiley & Sons, Inc.

stacked DRAs have been validated to be an effective approach to broaden the impedance bandwidth, increase axial ratio bandwidth, increase the beam bandwidth, improve gain, and increase beam angle scanning.

Figure 3.1 shows two examples of wide impendence bandwidth stacked DRA [3, 4]. The stacked DRA in Figure 3.1a consists of four concentric dielectric rings

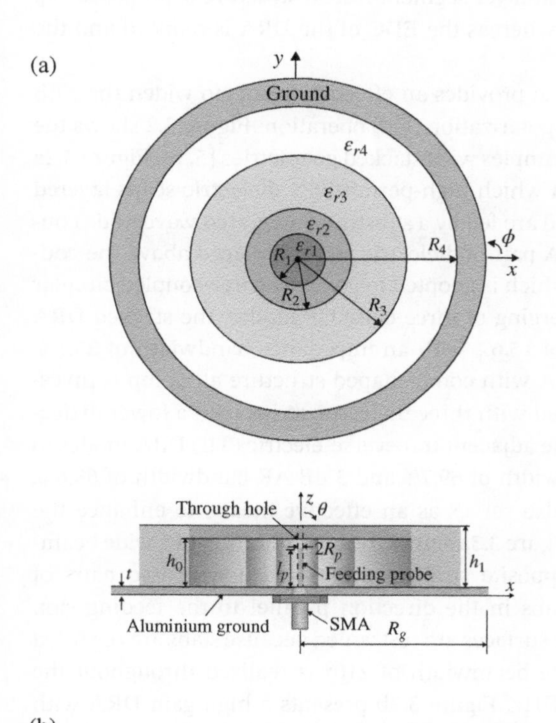

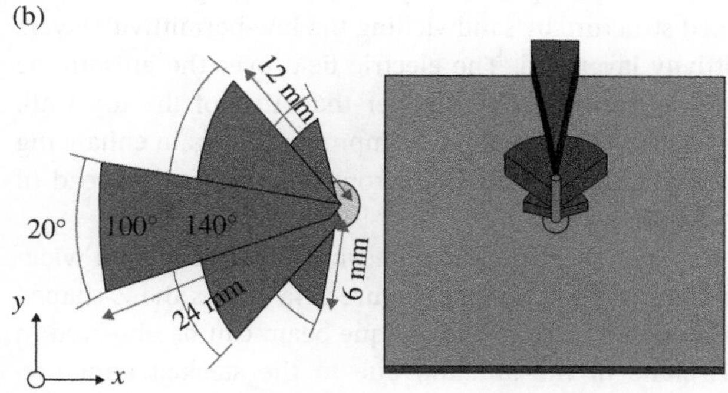

Figure 3.1 The wide impendence bandwidth of stacked DRAs. (a) Multi-ring stacked DRA. *Source:* Xia et al. [3]/IEEE; (b) multi-layer segment fractal DRA. *Source:* Gupta et al. [4]/IEEE.

with decreasing effective dielectric constants (EDCs) in the radial direction. By merging three transverse magnetic (TM) modes of $TM_{01\delta}$, $TM_{02\delta}$, and $TM_{03\delta}$, the DRA achieves a wide 10-dB impendence bandwidth of 60.2% and a high average antenna efficiency of 89%. The multi-ring antenna is fabricated by 3D-pritinng technology, which is suitable for making complex-shaped DRAs. In Figure 3.1b, a wideband stacked DRA with multilayer segment fractal structure is proposed [4]. This gives a compact structure, whereas the EDC of the DRA is reduced and the bandwidth is increased.

The stacked DRA structure also provides an effective means to widen the 3 dB axis ratio bandwidth in circular polarization (CP) operation. Figure 3.2 shows the two circularly polarized DRA examples with stacked geometries [5, 6]. Figure 3.2a illustrates a stacked structure in which high-permittivity dielectric strips layered on low-permittivity substrate and are fed by a substrate integrated waveguide coupling slot from the bottom [6]. A pair of dielectric strips is placed above the coupling slot with 45° inclination, which is adopted to generate three coupled circular polarized modes. Due to the merging of three close CP modes, the stacked DRA exhibits a 1 dB AR bandwidth of 15.6% with an impedance bandwidth of 25.1%. In Figure 3.2b, a multilayer DRA with comb-shaped structure at its top is investigated [5]. The DRA is embedded with three dielectric strips with a lower dielectric constant, generating multiple adjacent transverse-electric (TE) DRA modes to achieve 10-dB impedance bandwidth of 69.7% and 3-dB AR bandwidth of 68.6%.

The stacked DRA structure also serves as an effective means to enhance the antenna beamwidth and gain. Figure 3.3a shows the configuration of a wide beamwidth rectangular DRA [7]. Opposite surfaces are attached with two pairs of higher-permittivity dielectric slabs in the direction parallel to the feeding slot. The radiations from the opposite surfaces are restrained because slabs are regarded as magnetic walls. Wide E-plane beamwidth of 210° is realized throughout the impendence band of 3.02–3.26 GHz. Figure 3.3b presents a high-gain DRA with uniaxial anisotropic stacked structure by sandwiching the low-permittivity layers between the high-permittivity layers [8]. The electric field over the anisotropic cylindrical DRA side walls is considerably stronger than that of the top wall. Consequently, the boresight gain of the antenna is improved. The gain enhancing method of increasing the radiations of the DRA from the side wall instead of the top walls can be also found in [9].

The stacked geometry can also be utilized to achieve oblique beam and wide-beam scanning characteristics of DRA designs. Figure 3.4a shows the Z-shaped DRA with oblique beam proposed in [10]. The oblique beam can be observed in both the E-plane and H-plane of the antenna due to the stacked structure. Figure 3.4b shows a novel passive beam-steering liquid DRA [11]. This stacked DRA is made of two different layers of dielectric liquids with extremely low electrical conductivity, which are filled in polytetrafluoroethylene container. The

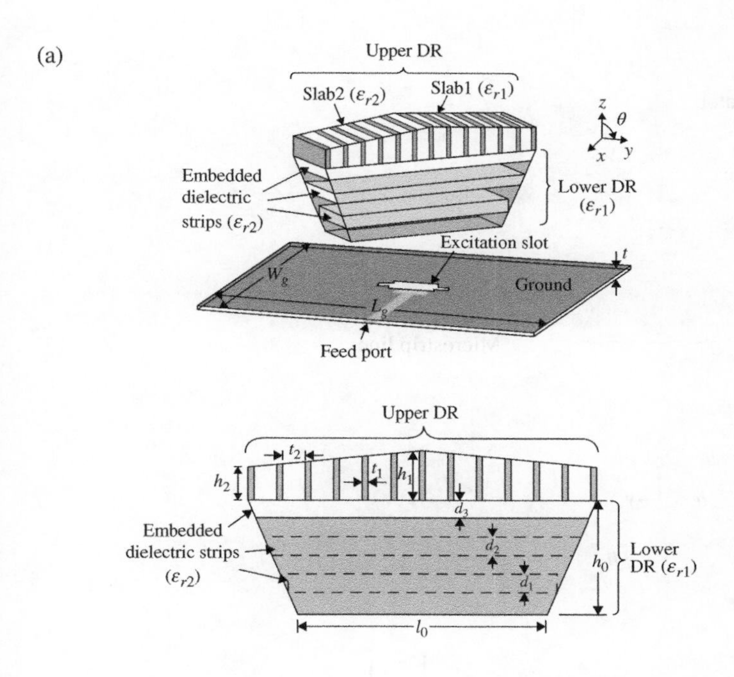

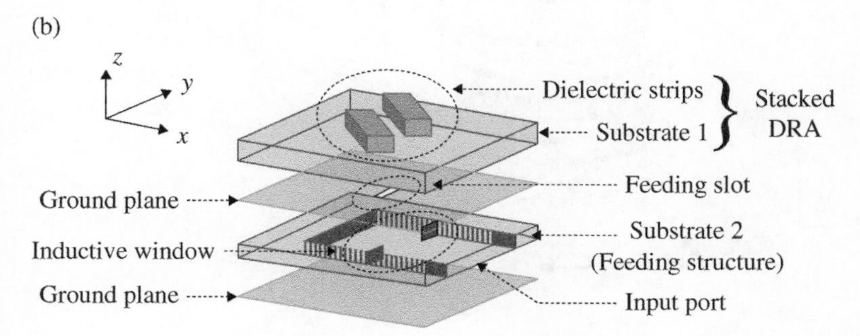

Figure 3.2 The wide axial ratio bandwidth of stacked DRAs. (a) 3D-printed DRA with two printing materials. *Source:* Yang et al. [5]/IEEE; (b) rectangular stacked DRA. *Source:* Xia and Leung [6]/IEEE.

beam-steering feature is achieved by controlling the variation of the antenna radiation mode originating from the biphasic liquid fluidity behavior due to gravity. The radiation beam of the proposed antenna can be automatically steered/tilted to the target direction upon dynamic physical movement of the antenna.

The rest of this chapter is organized as follows. Section 3.2 illustrates the analysis methods for stacked DRA, which includes equivalent circuit methods, approximate

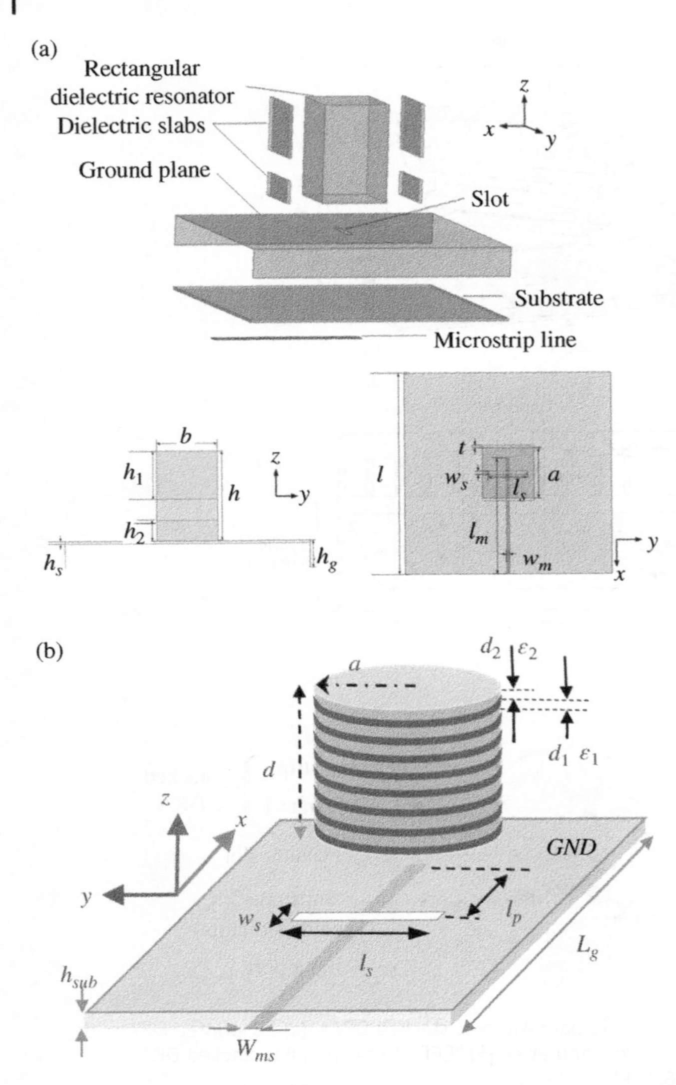

Figure 3.3 The widebeam and high-gain stacked DRA. (a) The widebeam DRA. *Source:* Sun et al. [7]/IEEE; (b) the high-gain DRA. *Source:* Fakhte et al. [8]/IEEE.

magnetic boundary, EDC, and characteristic mode (CM) theory. Section 3.3 gives four examples of stacked DRAs with wideband design. The modes of four proposed DRA are studied in detail and the performance of DRAs is given. Section 3.4 gives a wideband stacked DRA example with passive beam steering. Finally, Section 3.5 concludes the summary of this chapter.

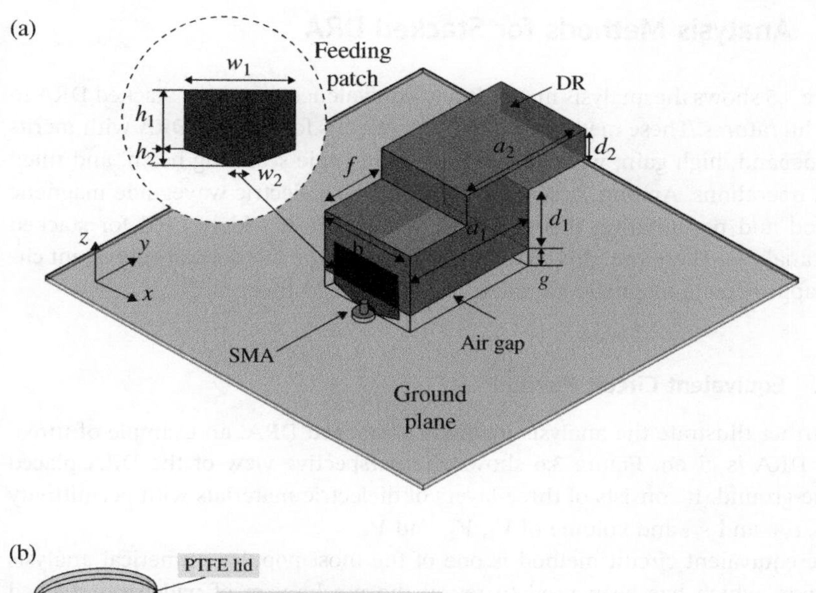

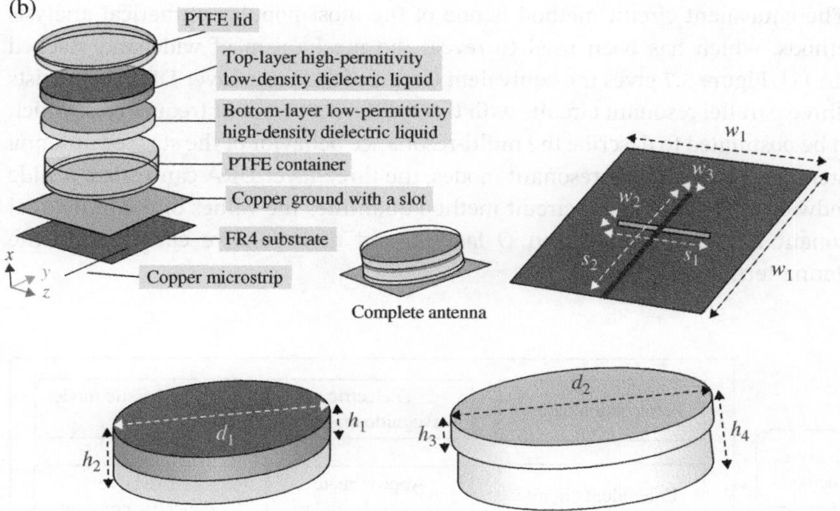

Figure 3.4 The oblique beam and beam-steering stacked DRA. (a) The oblique beam DRA. *Source:* Denidni et al. [10]/IEEE; (b) the beam-steering DRA. *Source:* Song et al. [11]/IEEE.

3.2 Analysis Methods for Stacked DRA

Figure 3.5 shows the analysis methods and equivalence theory for stacked DRA in open literatures. These methods explain the reasons for stacked DRA with merits of wideband, high gain, wide beamwidth, wide-angle scanning beam, and tilted beam operations. Among these analysis methods, dielectric waveguide magnetic method and perturbation theory in [12, 13] are most widely used for stacked DRA analysis. Therefore, this chapter will focus on the methods of equivalent circuit, approximate magnetic boundary, EDC, and CM theory.

3.2.1 Equivalent Circuit Method

To further illustrate the analysis methods of stacked DRA, an example of three-layer DRA is given. Figure 3.6 shows the perspective view of the DRA placed on the ground. It consists of three layers of dielectric materials with permittivity of ε_{r1}, ε_{r2}, and ε_{r3} and volume of V_1, V_2, and V_3.

The equivalent circuit method is one of the most popular numerical analysis methods, which has been used to reveal the mechanism of wideband stacked DRA [2]. Figure 3.7 gives the equivalent circuit of the three-layer DRA. It consists of three parallel resonant circuits with three different resonant frequencies, which can be postulated to describe the multi-resonance behavior of the stacked antenna system. Due to the multi-resonant modes, the three-layer DRA can realize a wide bandwidth. The equivalent circuit method quantifies the values of the individual resonant frequencies, radiation Q factors, and their relative couplings to the antenna terminals.

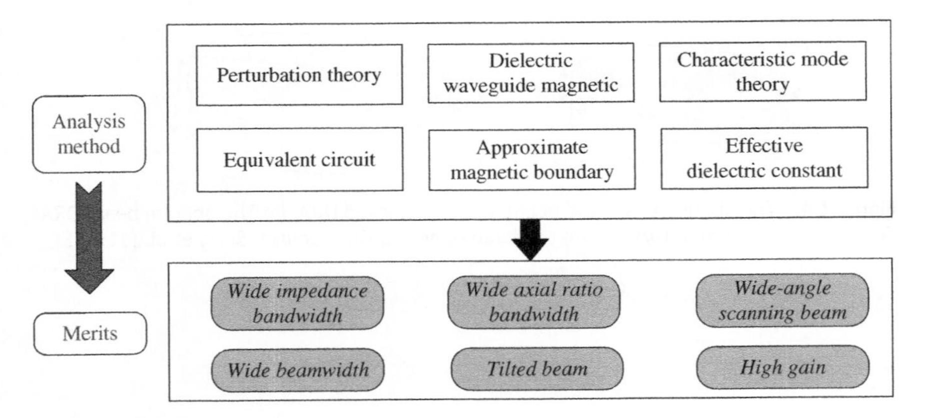

Figure 3.5 The analysis methods and merits of stacked DRA.

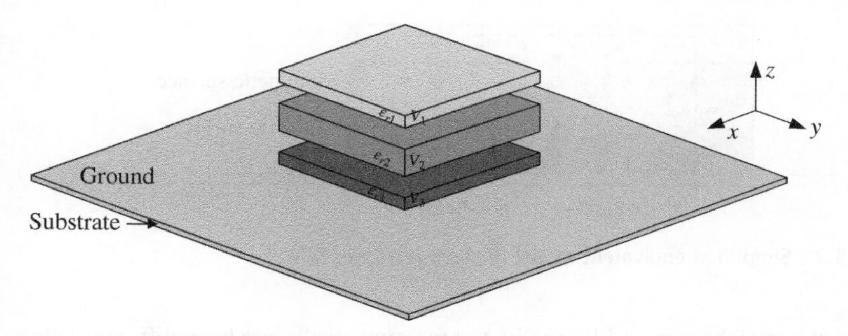

Figure 3.6 The perspective view of a three-layer DRA.

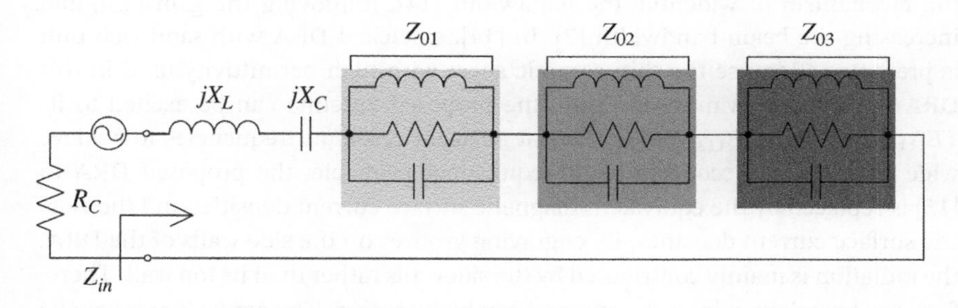

Figure 3.7 The equivalent circuit of the three-layer DRA.

3.2.2 Approximate Magnetic Boundary

If the permittivity of the thin dielectric resonator approaches to extremely large values, the surface of dielectric film can be considered as a perfect magnetic conductor (PMC) wall. Therefore, it can be equivalent to an open-circuit boundary when looking from the dielectric film to the air, which causes total internal reflections. In stacked DRA design, the thin dielectric resonator with high permittivity can be approximated as a magnetic wall boundary. For the three-layer DRA, if the dielectric resonator with low permittivity (ε_{r2}) is sandwiched between the other two dielectric sheets with high permittivity (ε_{r1}, ε_{r3}), the simplified equivalent model is represented in Figure 3.8. As shown in Figure 3.8, the ground is considered as electric surface and the interfaces between high and low permittivity dielectric sheets in the *yoz* plane can be approximately considered as magnetic surface if the permittivity contrast ($\varepsilon_{r1}/\varepsilon_{r2}$, $\varepsilon_{r3}/\varepsilon_{r2}$) is high enough. Owing to the equivalent magnetic wall boundary, the interface in the *yoz* plane will have high reflection, which increases the radiations from the sidewalls of the DRA, leading to an

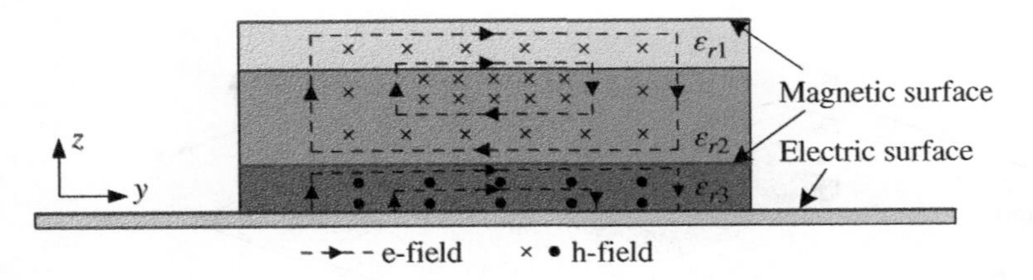

Figure 3.8 Simplified equivalent model of the three-layer DRA.

improved boresight gain. Additionally, the magnetic walls can keep different DRs independent, which allows the DRs to be designed and utilized separately.

Technique of approximate magnetic boundary has been widely used to analyze the mechanism of widening the bandwidth [14], improving the gain [15], and increasing the beam bandwidth [7]. In [14], a stacked DRA with sandwich unit is presented. Because the thin ceramic sheet with high permittivity used in this DRA is regarded as magnetic wall, the proposed antenna can be excited to its TE_{111} (half) and TE_{113} (full) modes at adjacent resonant frequencies to achieve wide bandwidth. According to the equivalent principle, the proposed DRA in [15] is replaced by the equivalent magnetic surface current densities and the electric surface current densities. By engraving grooves on the side walls of the DRA, the radiation is mainly contributed by the sidewalls rather than its top wall. Therefore, the boresight gain of the antenna can be increased. The equivalent magnetic currents field distribution of the DRA can be adjusted by changing the geometry of the antenna, which can be utilized to enhance the gain or widen the beam bandwidth.

3.2.3 Effective Dielectric Constant

The effective dielectric constant (EDC) method homogenizes and simplifies the stacked DRA equivalent to a conventional DRA filled with a uniform electric material. The equivalent dielectric constant can be obtained according to the simple quasi-static capacitance model (SCM). The SCM can be classified into vertically series capacitance model and transversely parallel capacitance model, corresponding to volume weighted average formula and volume weighted arithmetic average formula. For the three-layer DRA shown in Figure 3.9, the EDC ε_{eff} for volume weighted geometric can be calculated with formula (3.1).

$$\varepsilon_{eff} = \frac{\varepsilon_{r1}*V_1 + \varepsilon_{r2}*V_2 + \varepsilon_{r3}*V_3}{V_1 + V_2 + V_3} \tag{3.1}$$

where ε_{r1}, ε_{r2}, and ε_{r3} are dielectric constants, V_1, V_2, and V_3 are the volumes of dielectric layer.

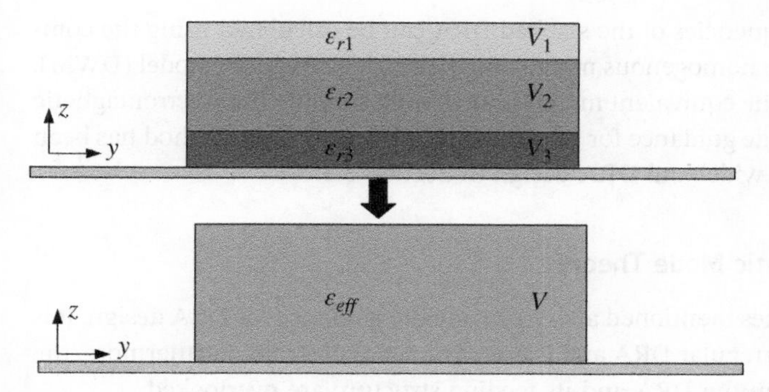

Figure 3.9 The effective homogenous model of three-layer DRA.

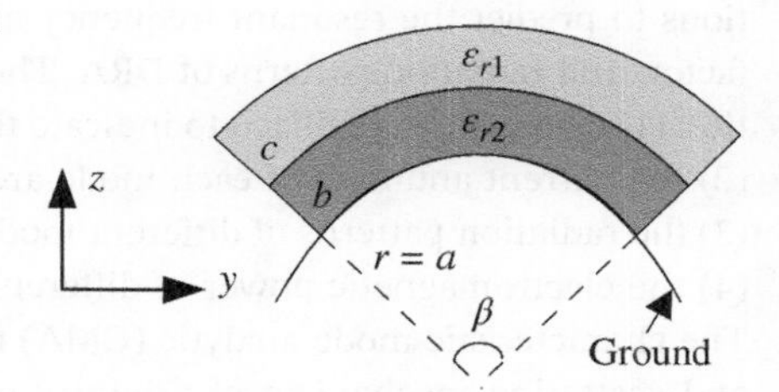

Figure 3.10 The configuration of two-layer conformal stacked DRA placed on curved ground plane.

For the conformal stacked DRAs placed on curved ground plane, the calculation formula (3.1) of equivalent dielectric constant is no longer applicable and the quasi-static capacitance model for DRA needs to be modified. Figure 3.10 presents the configuration of two-layer conformal stacked DRA. For this stacked DRA, the modified capacitance models have been studied based on the capacitance for cylindrical structures [16], which contains R-directionally series capacitance model (RSC), Z-directionally parallel capacitance model (ZPC), and Φ-directionally parallel capacitance model (PPC). According to the RCS, ZPC, and PPC models for the conformal DRA, the corresponding calculation formulas of equivalent permittivity are given in formulas (3.2)–(3.4), respectively.

$$\frac{1}{C} = \frac{1}{C_1} + \frac{1}{C_2} \Rightarrow \frac{\ln\left(\frac{c}{a}\right)}{\beta d \varepsilon_{eff}} = \frac{\ln\left(\frac{b}{a}\right)}{\beta d \varepsilon_1} + \frac{\ln\left(\frac{c}{b}\right)}{\beta d \varepsilon_2} \tag{3.2}$$

$$C = C_1 + C_1 \Rightarrow \frac{\varepsilon_{eff}\beta(c^2 - a^2)}{2d} = \frac{\varepsilon_1\beta(b^2 - a^2)}{2d} + \frac{\varepsilon_2\beta(c^2 - b^2)}{2d} \tag{3.3}$$

$$C = C_1 + C_1 \Rightarrow \frac{d_c\varepsilon_{eff}}{\beta_c}\ln\left(\frac{c}{a}\right) = \frac{d_c\varepsilon_1}{\beta_c}\ln\left(\frac{b}{a}\right) + \frac{d_c\varepsilon_2}{\beta_c}\ln\left(\frac{c}{b}\right) \tag{3.4}$$

The resonant frequencies of the stacked DRA can be calculated using the combination of effective homogenous model and dielectric waveguide model (DWM). It can be seen that the equivalent method can greatly simplify the electromagnetic problems and provide guidance for stacked antenna design. This method has been applied for stacked wideband DRA design in Section 3.3.

3.2.4 Characteristic Mode Theory

Although the theories mentioned above provide the guidance for DRA design, it is difficult to handle irregular DRA and inhomogeneous substrate. Furthermore, the interactions between the DRA and its feeding structure are overlooked.

Characteristic mode (CM) theory can reveal physical insights of resonance behavior and guide the design of irregular DRA. It provides a set of simple equations to predict the resonant frequency and determines the input impedance, *Q* factor, and radiation patterns of DRA. The CM theory has many superior properties: (1) eigenvalue is utilized to indicate the resonant characteristics of the mode, (2) the current and field of each mode are co-phased on the surface of structure, (3) the radiation patterns of different modes are orthogonal in the far region, and (4) the electromagnetic power of different modes is orthogonal with each other. The characteristic mode analytic (CMA) method can clearly reveal the radiation and scattering mechanism of arbitrary electromagnetic structure, avoiding the redundant process of blind simulation and optimization in analytic methods.

In the perspective of CM theory, there are two prevalent strategies to design the wideband antennas: strategy A, the excitation of wideband mode(s) [17], or strategy B, a combination of multiple modal resonances [18]. By following these two strategies, the proper excitation of wideband modes and the optimization of the feeding mechanism to combine resonance modes have been investigated to achieve wideband antenna performance. An example of using CM theory to guide DRA design is given below.

Figure 3.11a shows the rectangular DRA placed on the ground with four intersected slots fed by a microstrip ring with series feeding lines [19]. The CM theory is applied to reveal the intrinsic behavior of the proposed DR and guide the mode excitations for wideband operation. Figure 3.11b gives the eight modal significances (MSs) of the DR CMs. In CMA, the MS indicates the intrinsic radiation characteristics of a structure. Particularly, a value of MS $= 1$ indicates that the mode resonates and radiates most efficiently. Therefore, observing the MSs of these modes in Figure 3.11b, the mode 1/2 has wide MS bandwidth and covers the bandwidth of mode 5/6.

The far-field radiation patterns of the DR of the first eight CMs at resonant frequencies are shown in Figure 3.12. Although the far-field radiation in the endfire direction is slightly stronger than that of the broadside direction, the broadside

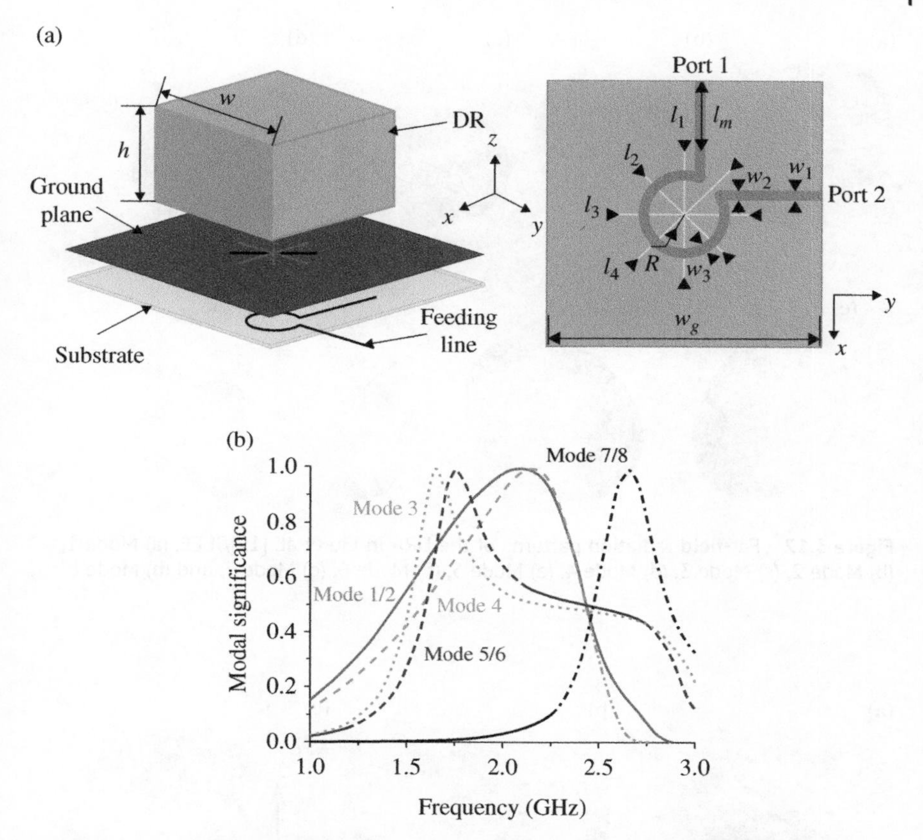

Figure 3.11 The configuration and MSs of the DRA. (a) The configuration of the DRA, (b) the modal significances of the DR characteristic modes. *Source:* Liu et al. [19]/IEEE.

radiation of the mode 1/2 and mode 5/6 can be enhanced and the front-to-back ratio of mode 7/8 will be improved when a metallic ground is used. Hence, modes 1/2, 5/6, and 7/8 play the key role in achieving broadside CP radiation.

Figure 3.13 shows the magnetic field of the DR in selected CMs at resonant frequencies. In order to obtain wide bandwidth and broadside radiation, the modes 1/2 and 7/8 are selected. To guarantee the magnetic field of the DR and the magnetic current source of the slots have the same direction, four crossing slots are divided into two pairs of crisscross slots. A pair of crisscross slots is used to excite H_1/H_2, and the other part of crisscross slots that are along the *x*- and *y*-axes is used to excite H_7/H_8. As can be seen from the example, the CMA provides a design guidance for proper excitation of wideband modes.

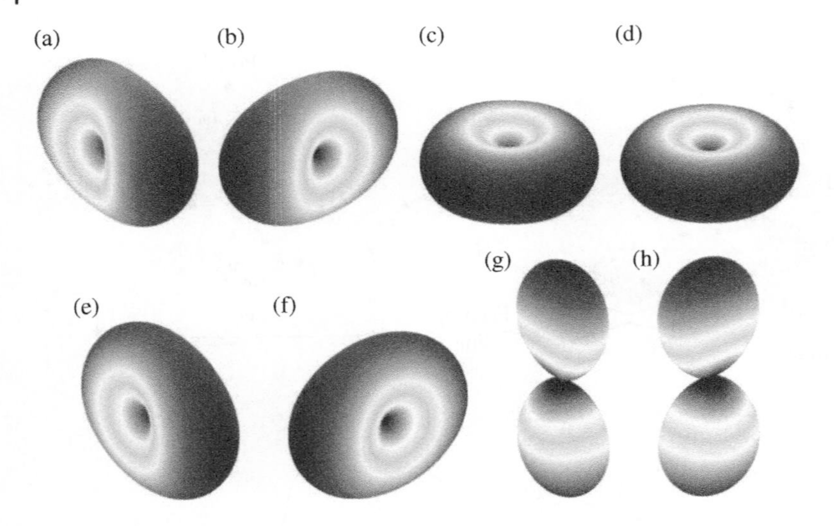

Figure 3.12 Far-field radiation patterns of the DRA in Liu et al. [19]/IEEE. (a) Mode 1, (b) Mode 2, (c) Mode 3, (d) Mode 4, (e) Mode 5, (f) Mode 6, (g) Mode 7, and (h) Mode 8.

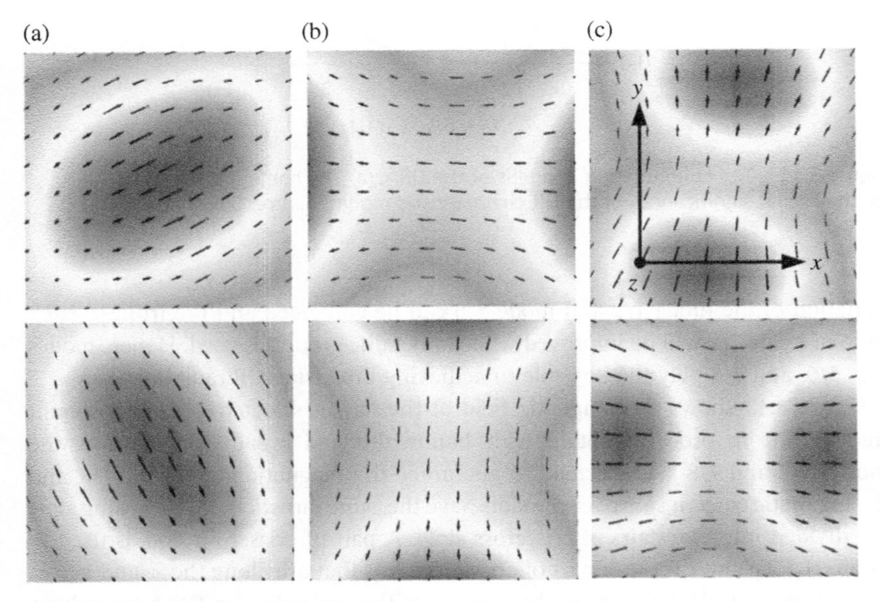

Figure 3.13 Corresponding characteristic magnetic field of the DR. (a) H1/2, (b) H5/6, and (c) H7/8. (Dark represents the maximum value and light represents the minimum value).

The CM theory can be utilized in analyzing broadband operation of basic-shaped DRAs such as rectangular, cylindrical, and hemispherical DRAs and the corresponding analysis will be given in Appendices A, B, and C, respectively. For stacked DRA and other irregular DRAs, the CMA also provides an available method to reveal the intrinsic wideband mechanism of the antenna, which will be illustrated in Appendices D and E, respectively.

3.3 Stacked Wideband DRA Designs

3.3.1 Antenna Design

In order to investigate stacked DRA geometry and identify wideband DRA modes, four DRA examples offering wide-/dual-band, dual-/circular-polarized, and pattern reconfigurable DRAs were proposed based on the EDC method [20]. In this design, the monopole supports are employed to form an air gap between the dielectric resonator and the ground plane and we have termed this configuration a stacked DRA. The four stacked DRAs and their modes and performances have been studied in this chapter. Good agreement between the simulated and measured results is observed, which proves the feasibility of DRA structures for the IEEE 802.11ac wireless local area network (WLAN) antennas.

3.3.1.1 Wideband Dual-Polarized DRA (WBDP-DRA) Design

The geometry of a wideband dual-polarized DRA (WBDP-DRA) and its fabricated prototype are shown in Figure 3.14. In general, a dielectric loaded structure is centrally placed on sandwiched substrates. The dielectric loaded structure is comprised of four thin dielectric plates vertically placed on the substrate that support the dielectric cube. Both the dielectric plates and the loaded cube are made of low-cost, commercial composite Rogers TMM 10i microwave dielectric material ($\varepsilon_r = 9.8$ and $\tan\delta = 0.002$), which is machined by a diamond saw and glued together to form a novel dielectric loaded structure. A set of inverted-trapezoid monopoles is printed on the outer side of the thin dielectric plates to excite the dielectric cube. The sandwiched substrate below the dielectric loaded structure is made of two stacked double side printed Taconic TLC-30 substrates ($\varepsilon_r = 3.0$ and $\tan\delta = 0.003$). The ground plane is located between the substrates. The monopoles on the dielectric plates are fed by the feed network under the substrate by utilizing vias that are inserted into the double-layer structure. Four holes are cut into the ground plane to avoid the vias shorting to the ground plane. The feed network printed under the substrate contains two T-junction microstrip power dividers to provide polarization diversity. A $\lambda/2$ transmission line, equivalent to

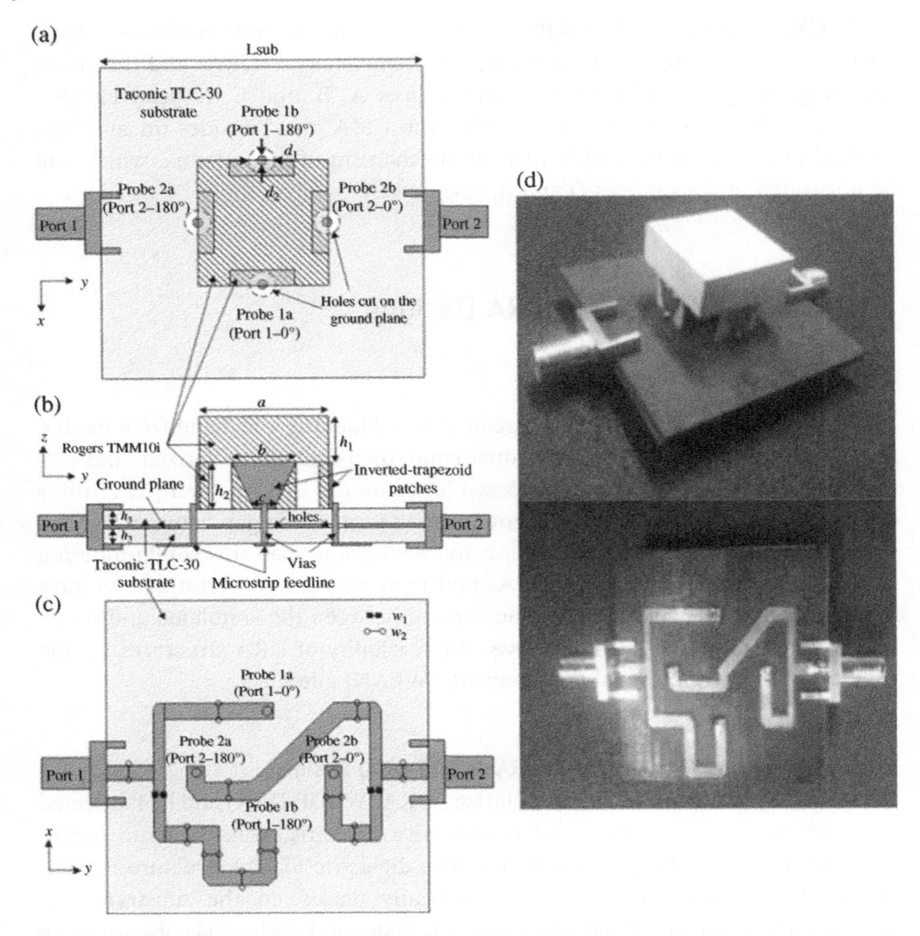

Figure 3.14 Configuration and photos of the WBDP-DRA. (a–c) Configuration on the top view, side view, and back view: L_{sub} = 40 mm, a = 17, b = 8 mm, c = 2 mm, d_1 = 3 mm, d_2 = 1 mm, h_1 = 6.35 mm, h_2 = 5.5 mm, h_3 = 0.8 mm, W_1 = 1.1 mm, and W_2 = mm; (d) photos on the side view and back view.

a 180° phase shift, is added to one arm of each power divider to provide a differential feed on Probe 1a, 1b and Probe 2a, 2b, respectively.

3.3.1.2 Dual-Band Dual-Polarized DRA (DBDP-DRA) Design

The configuration and photos of the DBDP-DRA are presented in Figure 3.15. Compared with WBDP-DRA, the first layer of the double layer substrate in Figure 3.14 is removed to avoid the fabrication problems caused by the gap

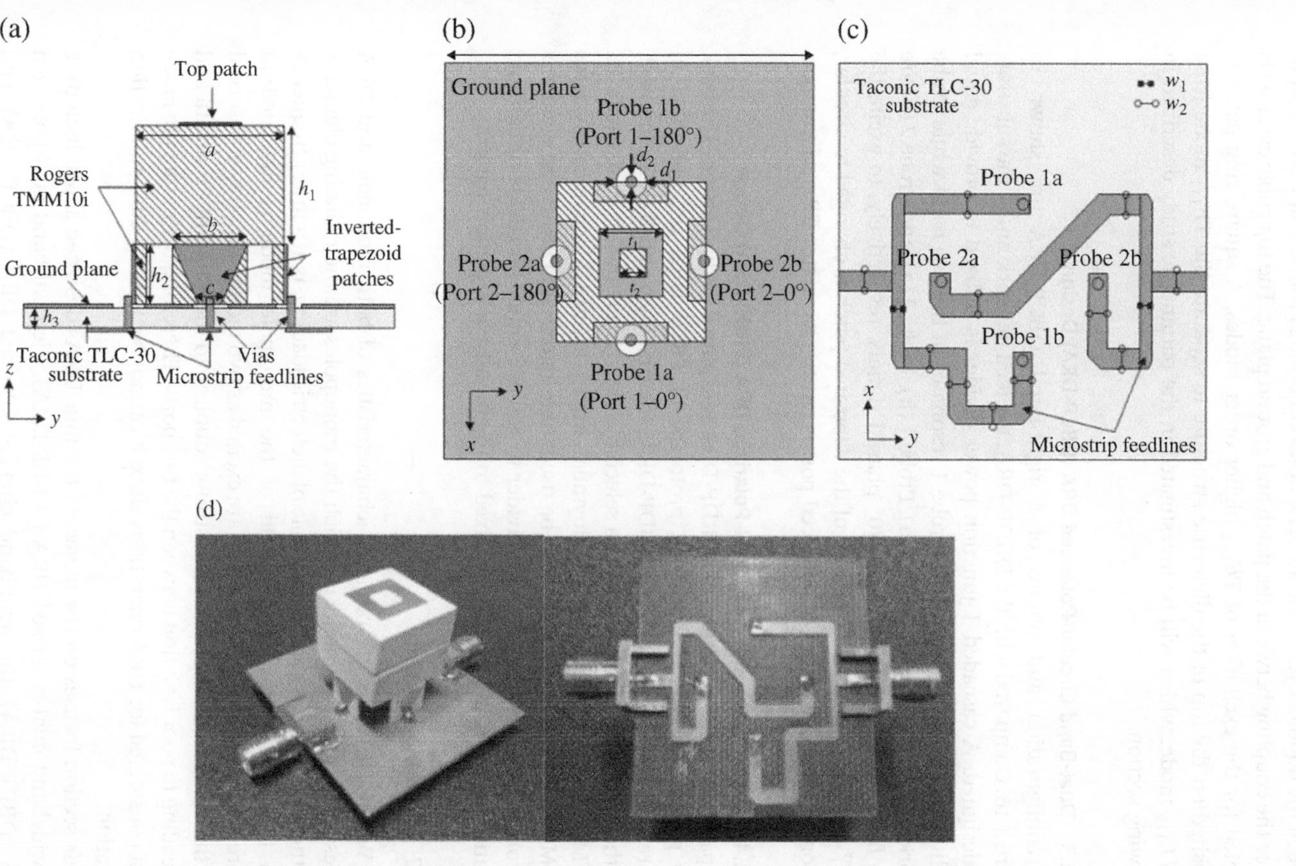

Figure 3.15 Configuration and photos of the DBDP-DRA. (a–c) Configuration on the top view, side view, and back view: a = 16.2 mm, b = 7.2 mm, c = 1.8 mm, h_1 = 12.7 mm, h_2 = 5.85 mm, h_3 = 0.762 mm, t_1 = 10 mm, t_2 = 4 mm, L_{sub} = 36 mm, W_1 = 1 mm, W_2 = 1.8 mm, d_1 = 3.6 mm, d_2 = 1 mm; (d) photos on the side view and back view.

between the double layer substrate. The holes cut on the ground plane are larger to reduce the coupling between the patch and ground plane. The top cubic ceramic is doubled for the excitation of TE_{112} higher order modes. A square ring patch is deposited on the top of the dielectric structure to separate the TE_{111} mode from the TE_{112} mode, which will be investigated in the parametric study described in following section.

3.3.1.3 Dual-Band Circular-Polarized DRA (DBCP-DRA) Design

The configuration and photos of a right-hand DBCP-DRA are shown in Figure 3.16. Compared with the DBDP-DRA, the feed network under the substrate is redesigned. A cascaded T-junction power divider is utilized to provide equal amplitude with a 90° phase shift. Probe 1 resonates in the low band while Probe 2 works for the high band. Probe 1 (a), Probe 1 (b), Probe 2 (a), and Probe 2 (b) are each fed with 0°, 90°, 180°, and 270° phased signals, respectively, to excite the $TE_{111}^{x \sim y}$ and $TE_{112}^{x \sim y}$ modes. The shape of the deposited patch is changed to circular, and one corner of each monopole feed post is cut to give better CP performance.

3.3.1.4 Pattern Reconfigurable Dual-Polarized DRA (PRDP-DRA) Design

The pattern re-configurability of a PRDP-DRA is explored in this section. The back view geometry and the fabricated prototype of PRDP-DRA are presented in Figure 3.17. Compared with the WBDP-DRA, the first layer substrate is removed, and the power divider of Port 1 has a selectable path ($\Delta\varphi = 180^{\circ}$ or 0°) to excite TE or TM modes. In a practical deployment Micro Electro Mechanical Systems (MEMs) or PIN diode switches can be used; here hard-wired switching elements are soldered on the Port 1 power divider to achieve either broadside or cardioid radiation patterns, while Port 2 is fixed to the broadside radiation pattern.

3.3.2 Working Principle

The WBDP-DRA has the radiating characteristics of both monopole and DRA modes. The monopole mode, in which the monopoles are the radiating elements and the dielectric cube has just the role of dielectric loading, is at a lower frequency than the DRA modes. The E-field of the monopole mode is indicated in Figure 3.18, where Probes 2a and 2b are excited as two monopoles. It was observed that these two excited probes have strong coupling to the adjacent Probes 1a and 1b, leading to poor isolation between the two ports. Therefore, the monopole mode is not considered for DRA operations since it cannot be used for dual-polarization operation.

This section focuses on the modes of the four DRAs designed above including the wideband dual-polarized DRA (WBDP-DRA), the dual-band dual-polarized DRA (DBDP-DRA), the dual-band circular-polarized DRA (DBCP-DRA), and

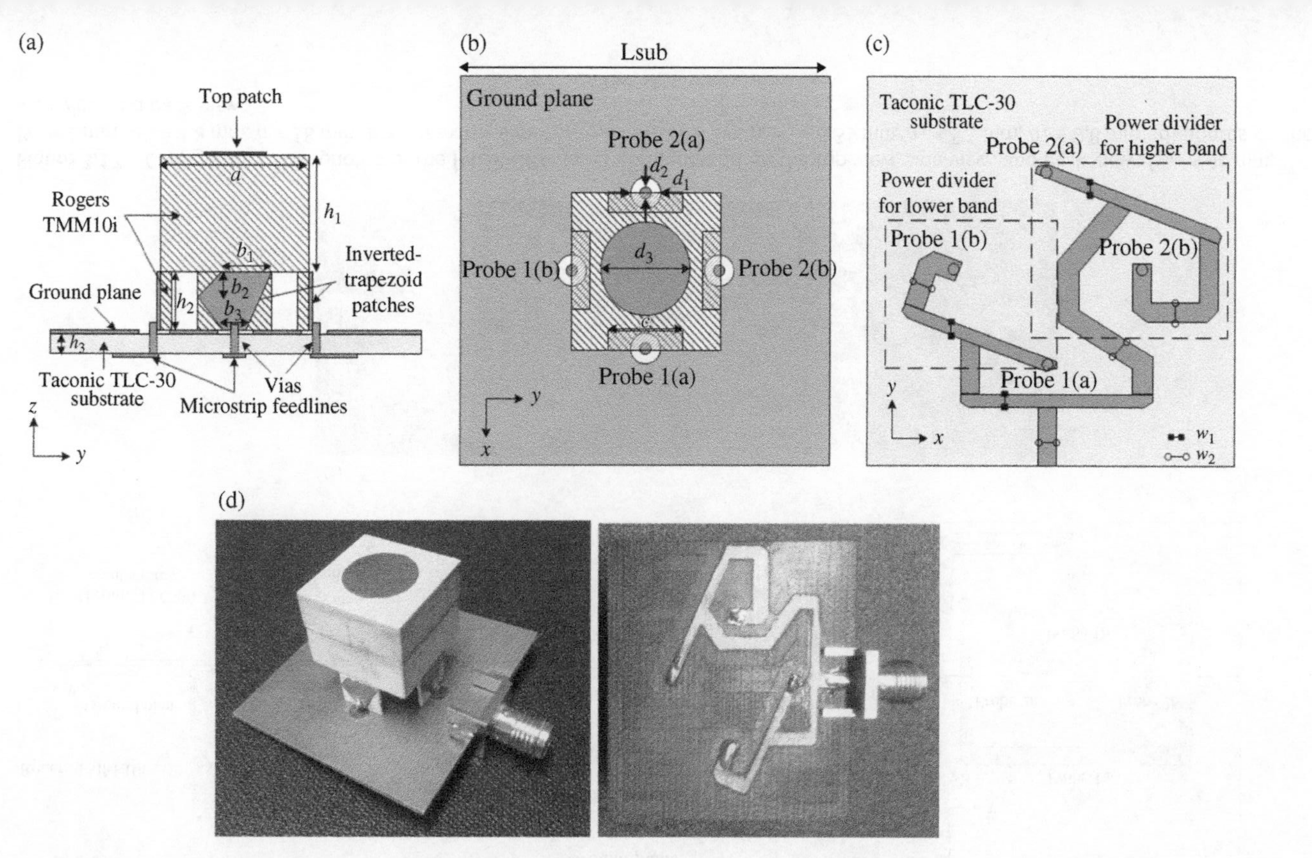

Figure 3.16 Configuration and photos of the DBCP-DRA. (a–c) Configuration on the top view, side view, and back view: L_{sub} = 39.6 mm, W_1 = 1 mm, W_2 = 1.8 mm, b1 = 5.4 mm, b2 = 2.25 mm, b3 = 1.8 mm, d_3 = 11 mm; a = 16 mm, c = 5.4 mm, h_1 = 12.7 mm, h_2 = 5.85 mm, h_3 = 0.8 mm, d_1 = 3.6 mm, d_2 = 1 mm; (d) photos on the side view and back view.

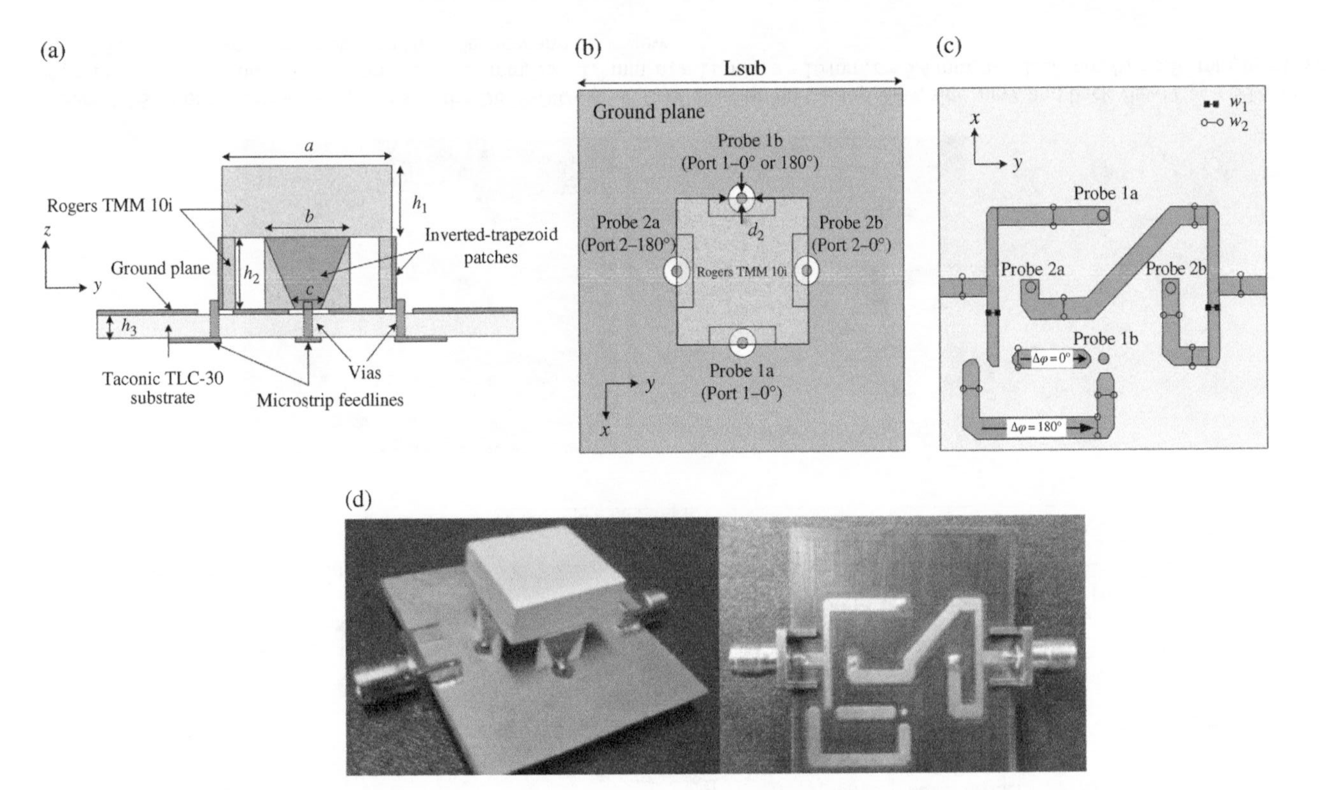

Figure 3.17 Configuration and photos of the PRDP-DRA. (a–c) Configuration on the top view, side view, and back view: L_{sub} = 40 mm, W_1 = 1 mm, W_2 = 1.8 mm, a = 18 mm, b = 8 mm, c = 4 mm, d_1 = 4 mm, d_2 = 1 mm, h_1 = 6.35 mm, h_2 = 5.5 mm, h_3 = 0.8 mm; (d) photos on the side view and back view.

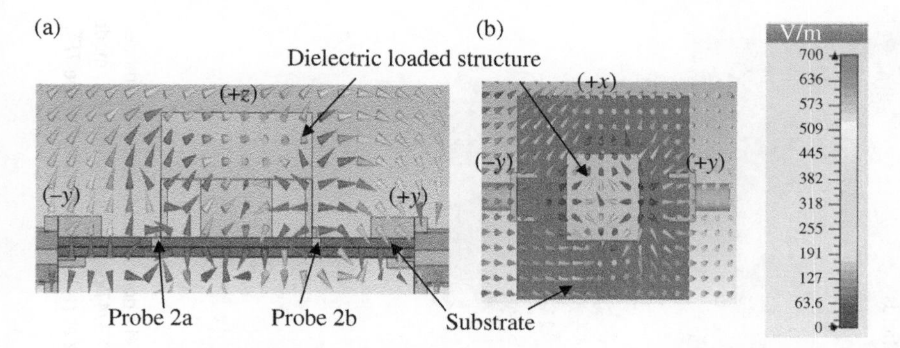

Figure 3.18 Simulated resonant *E*-field of monopole mode of Port 2 on 4.4 GHz using CST Microwave Studio. (a) *YZ* plane; (b) *XY* plane.

Table 3.1 Modes for each DRA.

DRAs	Port	Modes		
		LODR mode	FDR mode	HODR mode
WBDP-DRA	1	$TE^{y}_{1\delta 1}$	TE^{y}_{111}	TE^{y}_{122}
	2	$TE^{x}_{\delta 11}$	TE^{x}_{111}	TE^{x}_{212}
DBDP-DRA	1	$TE^{y}_{1\delta 1}$	TE^{y}_{111}	TE^{y}_{112}
	2	$TE^{x}_{\delta 11}$	TE^{x}_{111}	TE^{x}_{112}
DBCP-DRA	—	$TE^{x\sim y}_{111}$	$TE^{x\sim y}_{111}$	$TE^{x\sim y}_{112}$
PRDP-DRA	1	$TE^{y}_{1\delta 1}$	TE^{y}_{111}	TE^{y}_{122}
		TM_{111}		—
	2	$TE^{x}_{\delta 11}$	TE^{x}_{111}	TE^{x}_{212}

the pattern reconfigurable dual-polarized DRA (PRDP-DRA). Table 3.1 lists the corresponding modes of the four DRAs. To enable understanding of the operation of DRA modes in this class of device, we defined the following classification of modes, which are lowest order dielectric resonator (LODR) mode, fundamental dielectric resonator (FDR) mode, and higher order dielectric resonator (HODR) mode. The resonant frequencies of these modes can be observed in the return losses of each DRA, as indicated in Figures 3.21, 3.25, 3.29, and 3.33, respectively.

The *E*-fields and *H*-fields in these DRAs are simulated in CST microwave studio. As examples, the *E*-field and *H*-field at various mode resonances are shown in Figures 3.19 and 3.20: 5.2, 6.5, and 7.4 GHz in the WBDP-DRA; 5.2 GHz in the DBDP-DRA and 5.7 GHz in the DBCP-DRA; and 5.5 GHz in the PRDP-DRA.

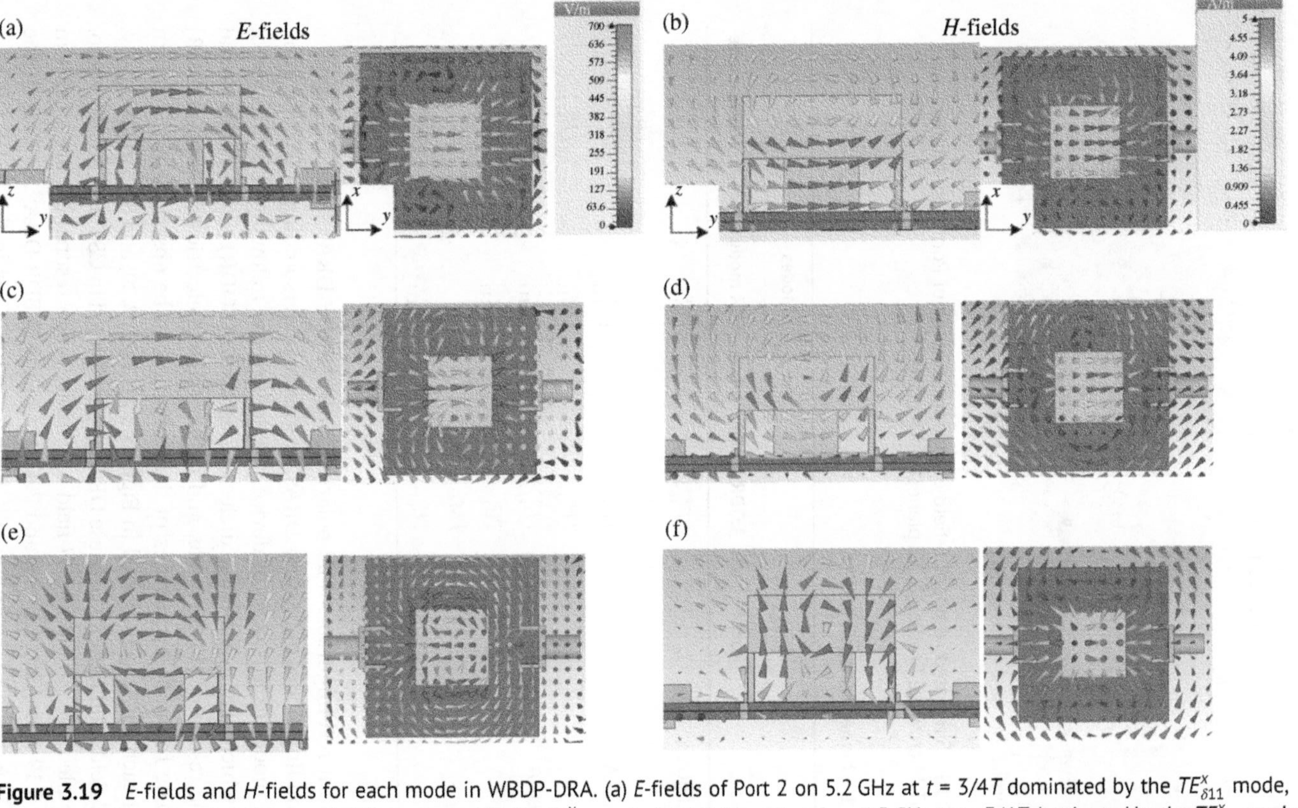

Figure 3.19 *E*-fields and *H*-fields for each mode in WBDP-DRA. (a) *E*-fields of Port 2 on 5.2 GHz at $t = 3/4T$ dominated by the $TE^x_{\delta 11}$ mode, (b) *H*-fields of Port 2 on 5.2 GHz at $t = 3/4T$ dominated by the $TE^y_{\delta 11}$ mode, (c) *E*-fields of Port 2 on 6.5 GHz at $t = 3/4T$ dominated by the TE^x_{111} mode, (d) *H*-fields of Port 2 on 6.5 GHz at $t = 3/4T$ dominated by the TE^y_{111} mode, (e) *E*-fields of Port 2 on 7.4 GHz at $t = 1/4T$ dominated by the TE^x_{212} mode, and (f) *H*-fields of Port 2 on 7.4 GHz at $t = 3/4T$ dominated by the TE^y_{212} mode.

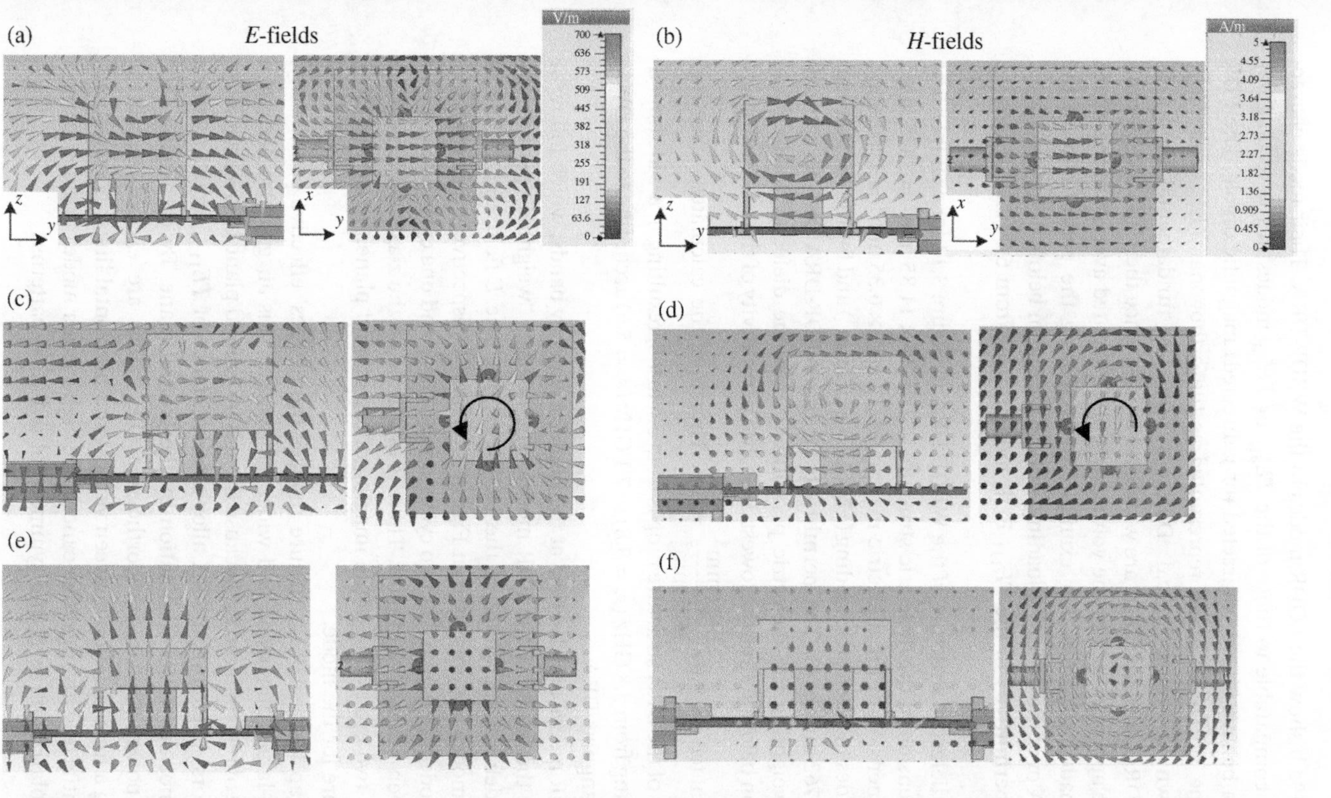

Figure 3.20 *E*-fields and *H*-fields for the modes in DBDP-DRA, DBCP-DRA, and PRDP-DRA. (a) *E*-fields of DBDP-DRA Port 2 on 5.2 GHz at $t = 1/4T$ dominated by the TE^{x}_{112} mode, (b) *H*-fields of DBDP-DRA Port 2 on 5.2 GHz at $t = 1/4T$ dominated by the TE^{y}_{112} mode, (c) *E*-fields of DBCP-DRA on 5.7 GHz at $t = 0$ (side) and $t = 1/4T$ (top) dominated by the $TE^{x \sim y}_{112}$ mode, (d) *H*-fields of DBCP-DRA on 5.7 GHz at $t = 0$ (side) and $t = 1/2T$ (top) dominated by the $TE^{x \sim y}_{112}$ mode, (e) *E*-fields of PRDP-DRA Port 1 ($\Delta\varphi = 0°$) on 5.5 GHz at $t = 0$ dominated by the TM_{111} mode, (f) *H*-fields of PRDP-DRA Port 1 ($\Delta\varphi = 0°$) on 5.5 GHz at $t = 0$ dominated by the TM_{111} mode.

Figure 3.19a,b show the LODR modes in the WBDP-DRA. Their mode characteristics are comparable to those of the $TE^x_{\delta 11}$ or $TE^y_{1\delta 1}$ modes in the rectangular DRA, which behave like an x-directed (or y-directed) magnetic dipole. The FDR modes in the WBDP-DRA are represented as the orthogonal TE^x_{111} and TE^y_{111} modes shown in Figure 3.19c,d. Different from the fundamental modes in DRA, the FDR modes in a DRA are wideband modes, since their effective permittivity can be approximated by the volume-fraction weighted average of the whole dielectric loaded structure. For example, by weighting the permittivity of the WBDP-DRA over the dielectric loading and the air region below it, the calculated "effective" permittivity of the TE_{111} FDR mode ranges from 6.2 (7.1 GHz) to 9.2 (5.8 GHz).

Take the dielectric structure of the WBDP-DRA in Figure 3.14 for example. The overall dimension of the dielectric structure is $17 \times 17 \times 11.85\ \text{mm}^3 = 3425\ \text{mm}^3$. And the dimension of the dielectric material is $17 \times 17 \times 6.35\ \text{mm}^3 = 1969\ \text{mm}^3$, which includes the dielectric loading ($17 \times 17 \times 6.35\ \text{mm}^3$) and four monopole legs ($5.5 \times 8 \times 0.762\ \text{mm}^3 \times 4$). The permittivity of the WBDP-DRA can be approximated by weighting the dielectric permittivity over the dielectric loading and the air region below it, thus the lowest dielectric permittivity of the structure could be approximated by $9.8 * \dfrac{1969\ \text{mm}^3}{3425\ \text{mm}^3} = 5.6$. Therefore, the calculated "effective" permittivity of the DRA structure ranges from 5.6 to 9.8, resulting in a wide impedance matching from 5.8 GHz ($\varepsilon_r = 9.8$) to 7.1 GHz ($\varepsilon_r = 5.6$) according to the DWM for the rectangular DRA.

The HODR modes are observed at the higher frequency bands, which follow or merge with the bands of the FDR modes. The devices' "weighted" permittivity leads to wideband operations of the HODR modes. The E-field and H-field of the HODR modes are presented in Figure 3.19e and f, respectively. For example, the TE^x_{212} modes are denoted as two cycles on the x-axis and one cycle on the y-axis as can be seen in the E-field plots. There is one cycle on the z-axis, which can be regarded as two cycles due to the imaging of the ground plane, hence the TE_{212} nomenclature for the mode.

The air gap in the DRA structure reduces the devices' effective permittivity, resulting in larger dimension and wider bandwidth. This air gap also permits a wider range of resonant modes because there is no ground plane below the dielectric resonators. Therefore, a DRA allows the existence of TE_{112n} modes, while a DRA does not due to its elimination by the ground plane. By employing 180° T-junction power dividers, the orthogonal TE modes are excited across the DRA, giving a high isolation between vertical and horizontal linear polarizations. The differential feed arrangement cancels the higher order mode currents, giving a low cross-polarization level and symmetrical radiation patterns.

As can be seen in Table 3.1, three pairs of degenerate modes of WBDP-DRA are excited for two bands: $TE^x_{\delta 11}$, $TE^y_{1\delta 1}$, TE^x_{111}, TE^y_{111} modes for the lower band (3.7–4.2 GHz) and TE^x_{112}, TE^y_{112} modes for the higher band (5.5–5.9 GHz). Figures 3.20a and b give the *E*-fields and *H*-fields in DBDP-DRA dominated by the TE^x_{112} mode. Similar to the dual-band operation in the DBDP-DRA, the two widebands in the DBCP-DRA are attributed to the resonance of the TE_{111} and TE_{112} modes. Circular polarization operation can be observed from the varying $TE^{x\sim y}_{111}$ and $TE^{x\sim y}_{112}$ modes in Figures 3.20c and d, where the *E*-/*H*-fields are rotating on the *x*–*y* plane around the *z*-axis. Figures 3.20e and f demonstrates the TM_{111} mode (or Quasi-TM_{011} [21]) in PRDP-DRA, which is excited by feeding the monopole pairs with in-phase signals rather than differential ones. The *E*-field is vertically polarized while the *H*-field is rotating around the *z*-axis, giving a cardioid far-field radiation pattern. The pattern re-configurability of the PRDP-DRA can be achieved by having one port resonant with TE modes while exciting the other port with TE or TM modes corresponding to a broadside or cardioid radiation pattern.

3.3.3 Fabrication and Measurement

Based on the mode guidelines in Table 3.1, design examples of the WBDP-DRA, DBDP-DRA, DBCP-DRA, and PRDP-DRA are given in this section for IEEE 802.11ac applications. Their *S* parameters, radiation patterns, axial ratio (CP DRA only), and antenna gains were adjusted to achieve the best performances. The antenna prototypes were fabricated and measured in the antenna laboratory at Queen Mary University of London to verify the simulation results. The return loss was measured by an Agilent N5230C PNA-L microwave network analyzer. A standard linearly polarized (LP) horn was used to measure the total radiation patterns in the horizontal and vertical planes. For the dual-polarized DRAs, the radiation patterns of one port were measured with the other port connected to a 50 Ω load. The CP radiation pattern was derived from the measured LP pattern by employing the method described in [22]. For gain measurement, a broadband standard gain horn (Gain = 10.34 dB at 5.5 GHz) was used for the gain comparison method [23]. Finally, the simulated and measured results are compared for each DRA and their applications for the IEEE 802.11ac standard are proposed.

3.3.3.1 Wideband Dual-Polarized DRA (WBDP-DRA)

The WBDP-DRA in Figure 3.14 is discussed first. Its simulated and measured *S* parameters are shown in Figure 3.21. The measured 10-dB impedance bandwidths are 40.9% (4.87–7.37 GHz) for Port 1 and 39.0% (4.93–7.32 GHz) for Port 2, with more than 25 dB isolation between the two polarizations. The S_{11} and S_{22} are different due to the different feed network layout. Small frequency shifts for the

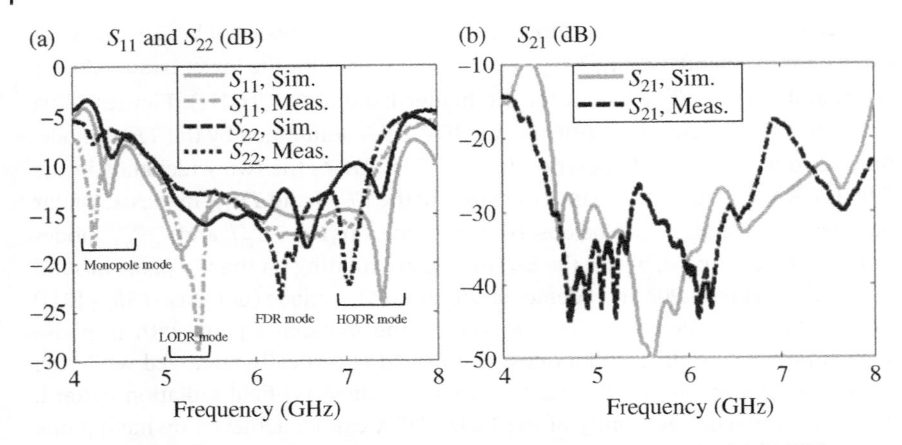

Figure 3.21 Simulated and measured (a) reflection coefficients for the DBCP-DRA, (b) transmission coefficients for the DBCP-DRA.

return loss and lower gains can be observed, which can be accounted by the fabrication tolerances and the unstable permittivity of Rogers TMM 10i material. Figure 3.22 shows the radiation patterns for Port 1 on *XZ*-plane (or *E*-plane). The broadside radiation patterns of $TE^{y}_{1\delta 1}$, TE^{y}_{111} modes at 5.2, 6.0 GHz are presented in Figure 3.22a, b. Figure 3.22c shows the radiating characteristics at 7.0 GHz, which is the dominated by the TE^{y}_{122} modes. The cardioid radiation pattern observed at 7.4 GHz in Figure 3.22d is the combination of the TE^{y}_{122} mode and other HODR modes. The measured broadside gain varies from 5.0 to 6.4 dBi over the 5–6 GHz band compared in Figure 3.23. The proposed DRA is suitable for the two-element MIMO communication system in the IEEE 802.11ac standard.

Parametric studies of the WBDP-DRA were carried out using CST microwave studio, and they demonstrate the mode control that these various parameters offer. Figure 3.24 plots the reflection coefficients at Port 1 (S_{11}) as a function of frequency for different thicknesses of the dielectric cube (h_1), the height of the supports (h_2), and the dielectric constant (ε_r) of the DRA. In each figure, only one parameter is varied with all the other parameters fixed at their optimal values from Figure 3.14.

First, the effect of the dielectric cube size is investigated by increasing the thickness of the dielectric cube (h_1), as shown in Figure 3.24a. As expected, the impedance of the monopole mode is fixed, while the impedances of the DRA modes shift downward because the larger DRA should have a lower resonance frequency. With reference to Figure 3.24b, it was found that the impedance of the LODR ($TE^{y}_{1\delta 1}$) mode remains almost unchanged with the tuning of the height of the supports (h_2). This reveals that the LODR mode is the lowest DRA mode of dielectric loading and only depends on the parameters of the dielectric cube. The impedances of monopole

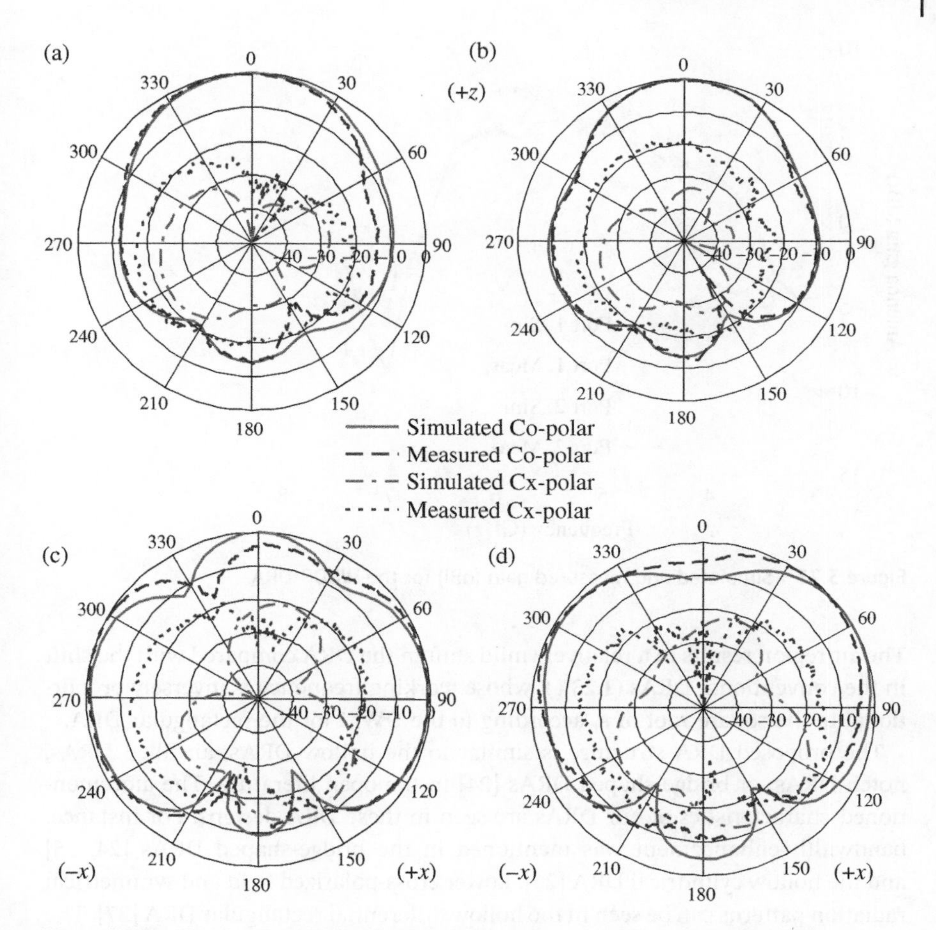

Figure 3.22 Simulated and measured radiation patterns for Port 1 on *XZ* plane: (a) 5.2 GHz, (b) 6.0 GHz, (c) 7.0 GHz, and (d) 7.4 GHz.

mode, FDR mode, and HODR mode shift downward as h_2 increases, which are expected due to the larger monopoles and overall weighted volume. It can be seen from Figure 3.24c when the dielectric constant (ε_r) increased from 7.8 to 11.8 and the 10-dB bandwidth decreased from 51.4% to 40.4%. This result is expected and similar to the effects in a DRA, where a larger permittivity gives a narrower bandwidth. The impedances of all modes shift downward as ε_r increases since their electrical dimensions are increased. It is worth mentioning that the resonant frequencies of the monopole mode, $TE^y_{1\delta 1}$ mode, and TE^y_{122} mode increase 1.05%, 1.10%, and 1.19% respectively, as calculated from the data in Figure 3.24.

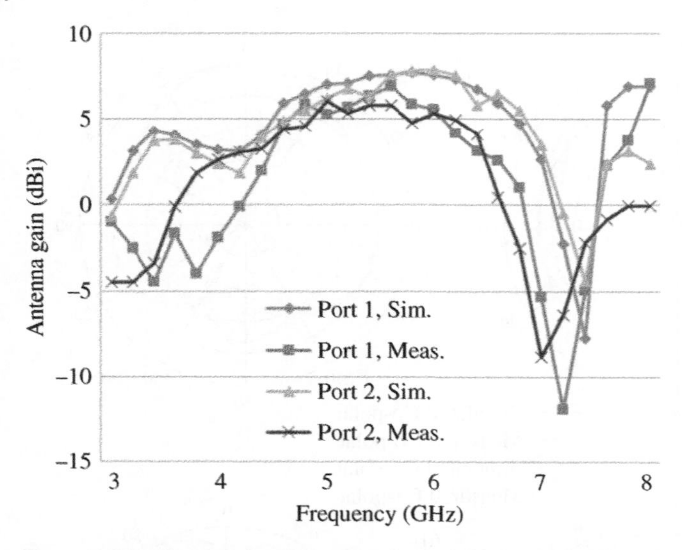

Figure 3.23 Simulated and measured gain (dBi) for the WBDP-DRA.

The air region results in a relatively mild shift in the DRA compared with the shift in the conventional DRAs (1.23%), whose working frequency is inversely proportional to the square root of ε_r according to the DWM for the rectangular DRA.

The proposed DRA structure is similar to the hollow DRAs, air-filled DRAs, notch DRAs, or bridge-shaped DRAs [24] in the open literature. The aforementioned characteristics of our DRAs are seen in these DRA designs. For instance, bandwidth enhancement was mentioned in the bridge-shaped DRAs [24, 25] and the hollow cylindrical DRA [26]. Lower cross-polarized field and symmetrical radiation patterns can be seen in the hollow differential rectangular DRA [27]. The isolation enhancement between the two polarizations was observed in the air-filled DRAs [28] by utilizing two pairs of balanced excitation ports. The TE_{112} mode in the hollow structure was noticed in the hollow hemispherical DRA, but has not been observed in rectangular dielectric antennas to the best of our knowledge.

3.3.3.2 Dual-Band Dual-Polarized DRA (DBDP-DRA)

Figure 3.25 exhibits the simulated and measured S parameters of the proposed DBDP-DRA. Discussion on the monopole modes at 2.6 GHz is not included here as it cannot be used for dual-polarization operation. The measured return loss bandwidths for the DRA modes are 18.3% (3.48–4.18 GHz), 30.1% (4.85–6.57 GHz) and 17.0% (3.55–4.21 GHz), 29.2% (4.77–6.40 GHz) on vertical and horizontal linear polarizations, respectively, with high isolation (>25 dB) between the two

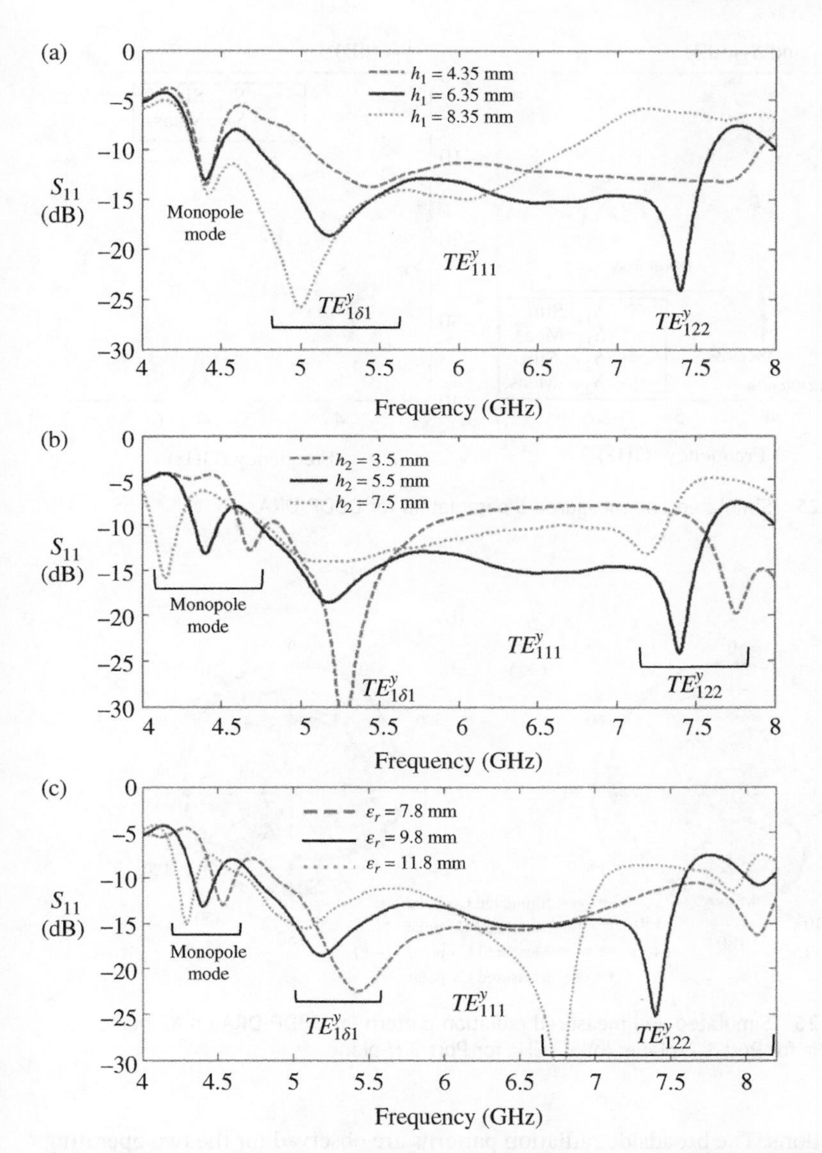

Figure 3.24 Simulated S_{11} of the WBDP-DRA as a function of frequency for different parameters. (a) h_1: the thickness of dielectric loading, (b) h_2: the height of the monopole supporters, and (c) ε_r: the permittivity of the dielectric loaded structure.

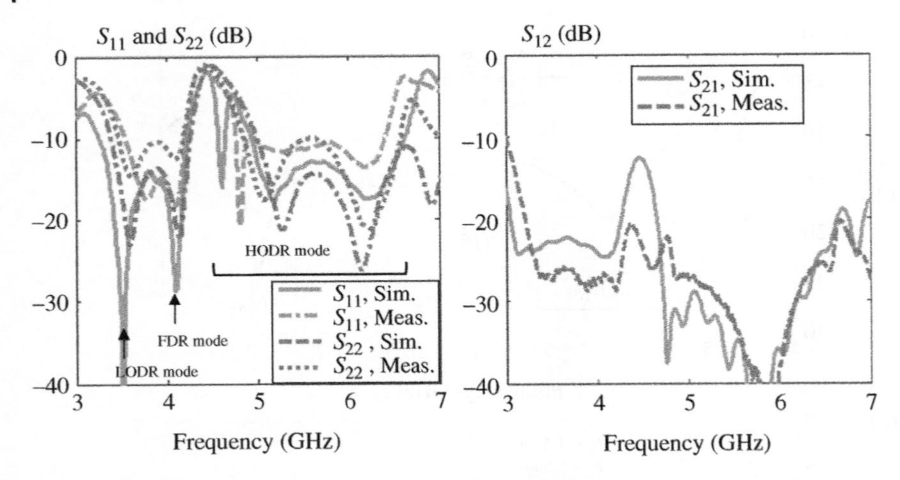

Figure 3.25 Simulated and measured S parameters for DBDP-DRA.

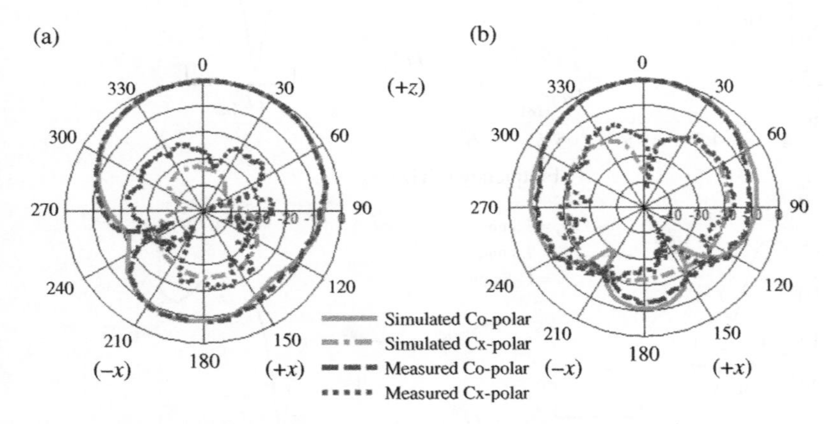

Figure 3.26 Simulated and measured radiation pattern for DBDP-DRA on XZ plane: (a) 3.7 GHz for Port 1 E-plane; (b) 5.5 GHz for Port 2 H-plane.

polarizations. The broadside radiation patterns are observed for the two operating bands. The broadside radiation patterns at 3.7 GHz for E-plane Port 1 and 5.5 GHz for H-plane Port 2 are demonstrated in Figure 3.26. The simulated and measured peak gain is compared in Figure 3.27. The measured gain variation is from 4.7 to 5.0 dBi for the lower band and 5.0 to 6.5 dBi for the upper band.

A parametric study on the size of the square ring patch against the return loss is performed, as shown in Figure 3.28. Without the top patch, the $TE^x_{\delta 11}$, TE^x_{111}, and

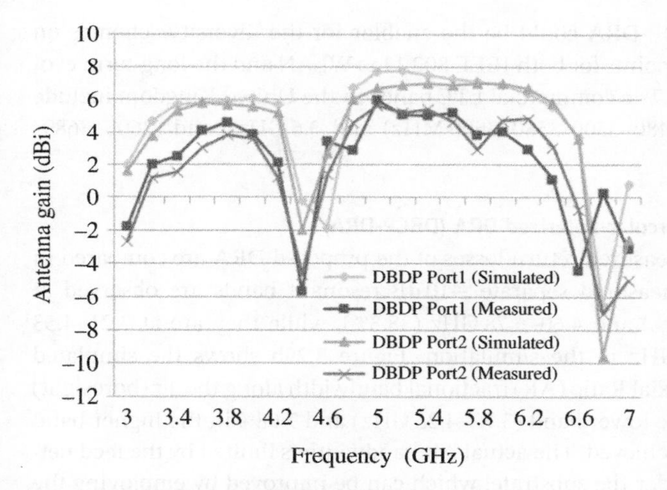

Figure 3.27 Gain comparison for DBDP-DRA.

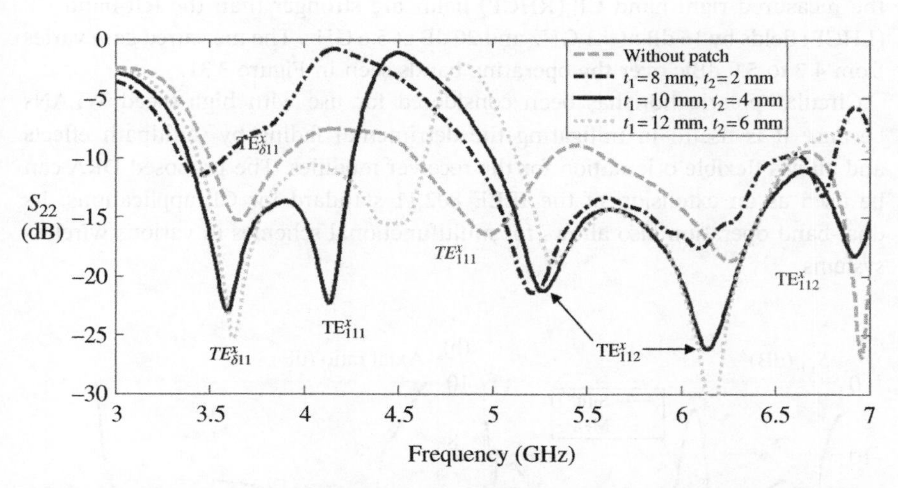

Figure 3.28 Simulated S_{22} of the DBDP-DRA as a function of frequency for the patch size.

TE^x_{112} modes are resonant at 3.6, 5, and 6.3 GHz, respectively. After depositing a rectangular top patch on the dielectric load, a notch band is generated between the TE^x_{111} and TE^x_{112} modes. Meanwhile, the resonances for the two separated wide-bands are enhanced. The size of the top patch, which shifts the generated notch band, is optimized to $t_1 = 10$ mm, $t_2 = 4$ mm to cover the 3.5/3.6 GHz LTE and 5–6 GHz IEEE 802.11ac bands.

The proposed DBDP-DRA could be the enabler for the diversity schemes on indoor MIMO access points for both IEEE 802.11ac WLAN and the long-term evolution (LTE) services. The commercial LTE bands in the United Kingdom include the 3.5 GHz band (3480–3500/3580–3600 MHz) and 3.6 GHz band (3605–3689/3925–4009 MHz).

3.3.3.3 Dual-Band Circular-Polarized DRA (DBCP-DRA)

The simulated and measured return losses of the proposed DRA are compared in Figure 3.29a. Two measured separate −10 dB resonant bands are observed at 3.36–4.47 GHz (28.35%) and 4.60–6.78 GHz (38.3%), while they are at 3.21–4.53 GHz and 4.79–6.80 GHz in the simulation. Figure 3.29b shows the simulated and measured 3-dB Axial Ratio (AR) fractional bandwidth along the +*z* (boresight) direction. 13.2% in the lower band (3.72–4.25 GHz) and 7.08% in the higher band (5.45–5.85 GHz) are achieved. The actual AR bandwidth is limited by the feed network deployment under the substrate, which can be improved by employing the matching network proposed in [29]. The simulated and measured radiation patterns at 4.1 and 5.6 GHz are presented in Figure 3.30. In the boresight direction, the measured right-hand CP (RHCP) fields are stronger than the left-hand CP (LHCP) fields by 15 dB at 4.1 GHz and 20 dB at 5.6 GHz. The measured gain varies from 4.7 to 5.9 dBic over the operating bands seen in Figure 3.31.

Circular polarization has been considered for use with high-speed WLANs because it is useful in mitigating the detrimental fading by multipath effects and allows flexible orientation for the receiver modules. The proposed DRA can be used as an extension of the IEEE 802.11 standard for CP applications. Its dual-band operation also allows for multifunctional schemes in various wireless systems.

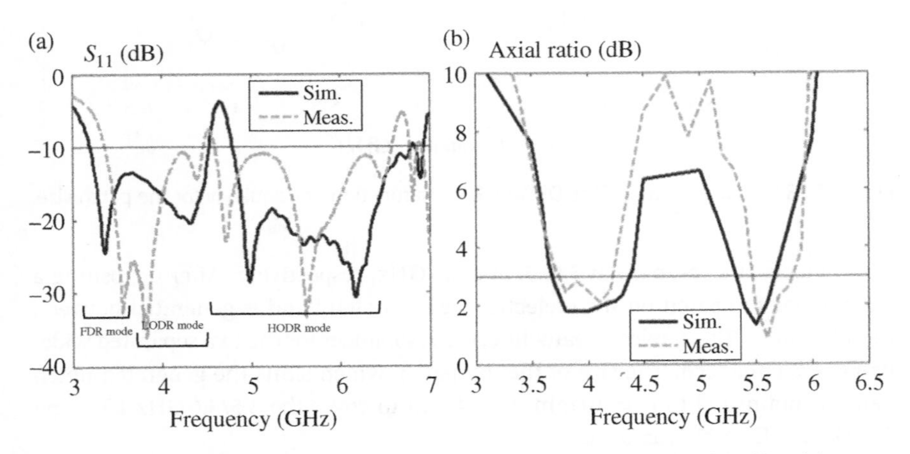

Figure 3.29 Simulated and measured (a) *S* parameter for the DBCP-DRA; (b) axial ratio for the DBCP-DRA.

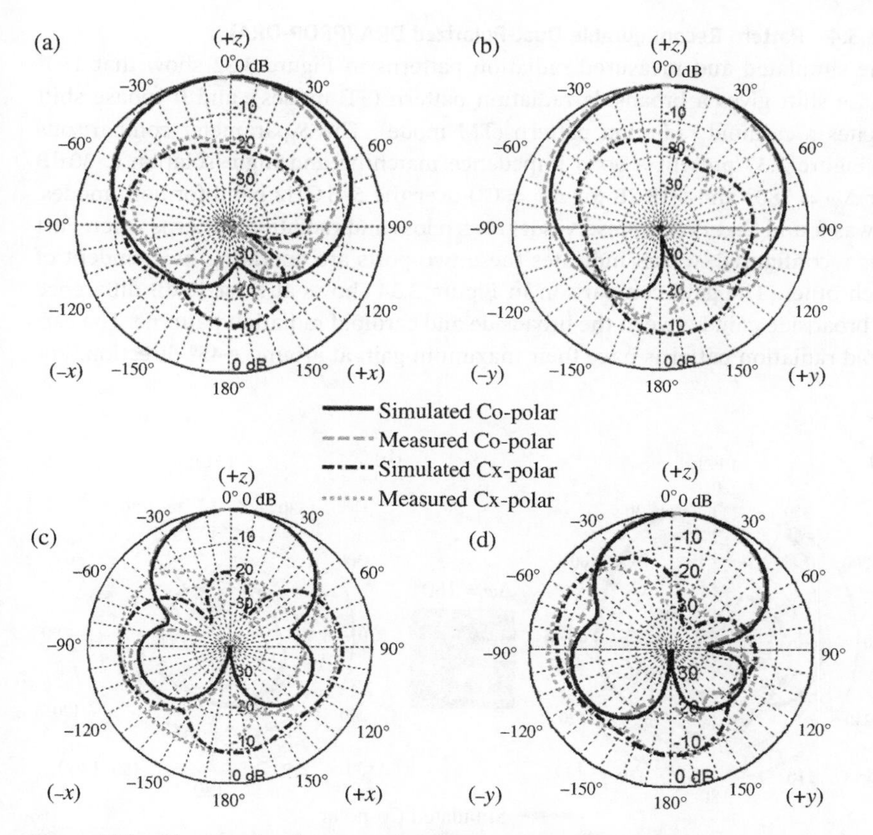

Figure 3.30 Simulated and measured radiation patterns for DBCP-DRA: (a) 4.1 GHz on *XZ* plane, (b) 4.1 GHz on *YZ* plane, (c) 5.6 GHz on *XZ* plane, and (d) 5.6 GHz on *YZ* plane.

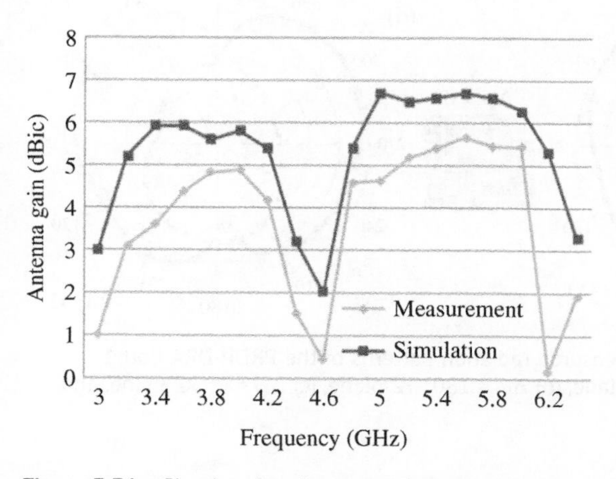

Figure 3.31 Simulated and measured antenna gain for DBCP-DRA.

3.3.3.4 Pattern Reconfigurable Dual-Polarized DRA (PRDP-DRA)

The simulated and measured radiation patterns in Figure 3.32 show that 180° phase shift gives a broadside radiation pattern (TE modes), and 0° phase shift excites a cardioid radiation pattern (TM mode). The S-parameter comparisons in Figure 3.33 present a good impedance matching and high isolation (>30 dB for $\Delta\varphi = 180°$ and >18 dB for $\Delta\varphi = 0°$) over the 5–6 GHz band for both modes. It was found that Port 2 preserves its return loss and radiation patterns when Port 1 is reconfigured, which indicates these two ports are operating independent of each other. The gain comparison in Figure 3.34 shows at least 17 dB difference in broadside gain between the broadside and cardioid radiation patterns. The cardioid radiation patterns have their maximum gain at around ±45° direction. For

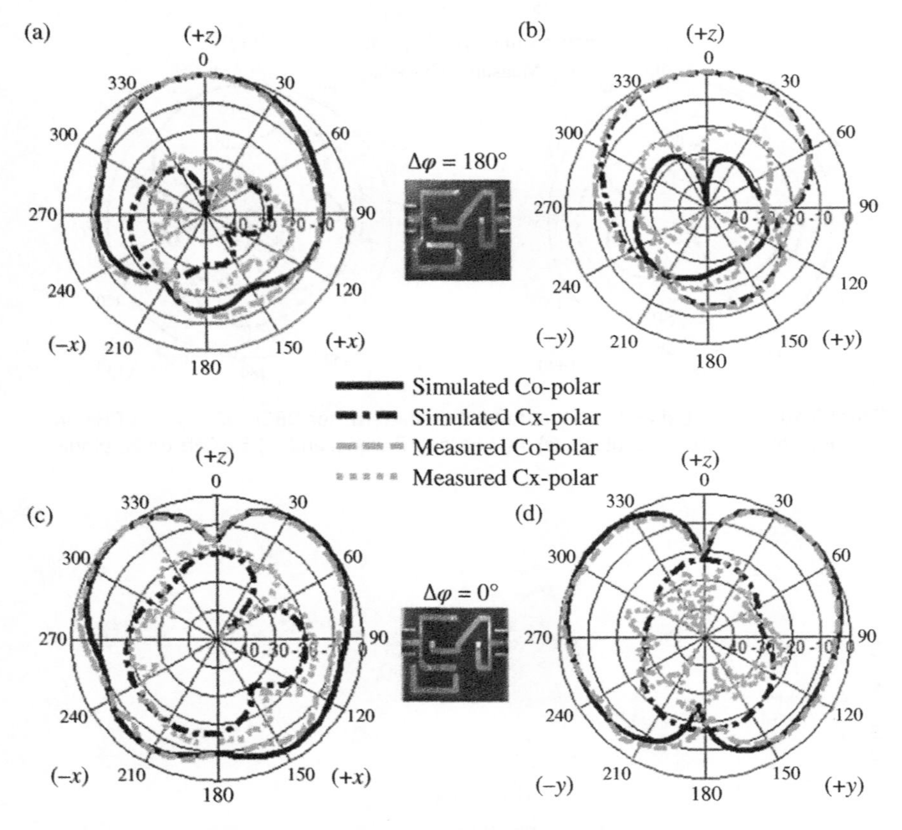

Figure 3.32 Simulated and measured radiation patterns of the PRDP-DRA Port 1 on 5.5 GHz. (a) $\Delta\varphi = 180°$ XZ plane, (b) $\Delta\varphi = 180°$ YZ plane, (c) $\Delta\varphi = 0°$ XZ plane, and (d) $\Delta\varphi = 0°$ YZ plane.

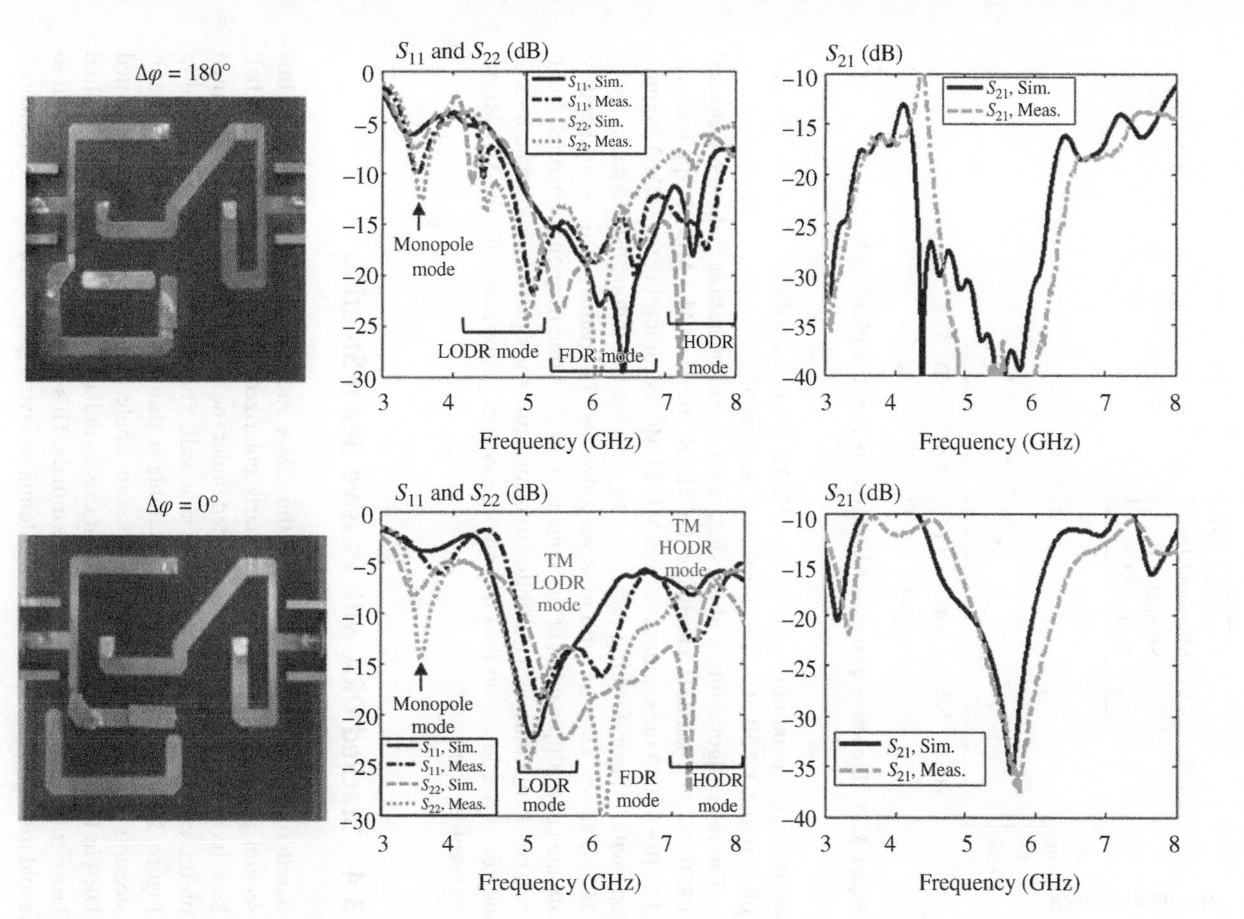

Figure 3.33 Simulated and measured S parameters for the PRDP-DRA.

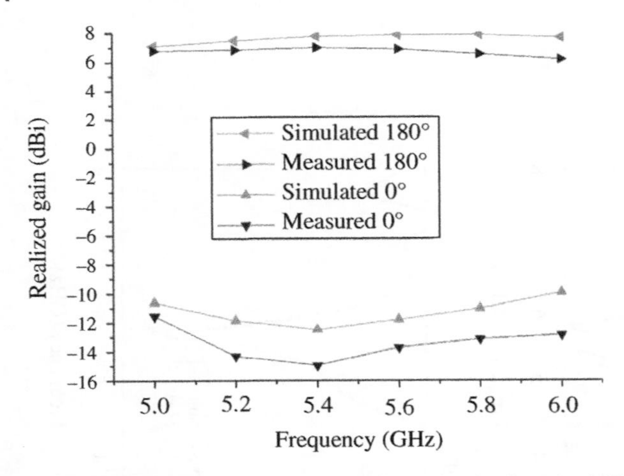

Figure 3.34 Broadside gain comparisons for the Port 1 of PRDP-DRA.

example, the maximum gain on 5.5 GHz *YZ* plane is 2.0 dB at −48.0° for the simulation and 1.2 dB at −51.6° for the measurement.

The re-configurability and the dual-polarization operation make the proposed PRDP-DRA a suitable candidate for IEEE 802.11ac MU-MIMO applications. The IEEE 802.11ac standard defines MU-MIMO as combining the MIMO technology with a beam-forming technique to increase the indoor user capacity. MU-MIMO allows a station with multiple antennas to transmit multiple independent streams to multiple users at the same time in the same frequency channel. Dual-polarized antennas with the ability to reconfigure their patterns add an additional degree of freedom, and thus improve the system performance for the MU-MIMO transceiver front ends.

3.4 Stacked DRA with Passive Beam Steering

Based on the design example in the CMA mentioned in Section 3.2.4, this section proposed a wideband circularly polarized DRA using gravitational ball lens. Its beam direction can be adjusted under the action of gravity when mounted on the celling, oblique structure, or side wall. The main concept is illustrated in Figure 3.35a,b. The ball lens is rotatable within $\theta = \pm 90°$ and enables beam-steering characteristics with ±40° scan angle. Therefore, the wall-mounted DRA in Figure 3.35c automatically adjusts its radiation pattern to a 40° down tilted beam for indoor coverage of user terminals. This pattern adjustment is useful to avoid the significant reduction in antenna receiving gain when WLAN or Picocell

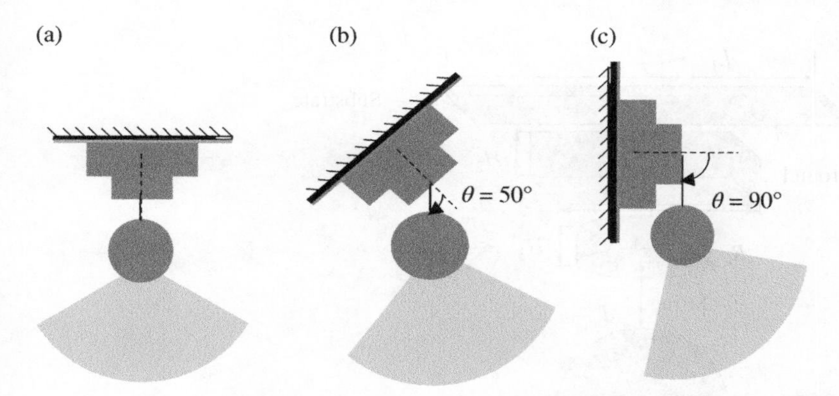

Figure 3.35 Pattern reconfigurability of the mounted DRA by the action of gravity. (a) On the celling, (b) on the oblique structure, and (c) on the side wall.

antennas are installed at an angle on a wall or ceiling. This beam steering can be also controlled by the mechanical movement, providing the flexibility on the steering direction. For all steering angles, DRA has an impedance bandwidth of 42% and axial ratio bandwidth of 39%. It overcomes the difficulties of achieving wideband axial ratio versus the pattern reconfigurability of the existing antennas.

3.4.1 Antenna Design

Figure 3.36a shows the 3D view on the proposed DRA, where dimensions are listed in Table 3.2. The DRA consists of a substrate, a cuboid DRA (DR1), a cylinder DRA (DR2), and a ball lens. The substrate is made of FR4 epoxy, which has relative permittivity of 4.4 and dielectric loss tangent of 0.02. DR1, DR2, and ball lens are made of commercial alumina, which has $\varepsilon_r = 9.2$ and $\tan\delta = 0.008$. The substrate, DR1, and DR2 are stacked as an integrated structure and connects to a ball lens by a string. The string crosses over the antenna structure by drilling small holes (radius = 1.0 mm) on the DRA center. This connection ensures a fixed space of d between the DR2 surface and the ball lens surface. It also permits the continuously rotating of the ball lens over 2D plane under the action of gravity.

Figure 3.36b presents the layout of the double-side printed substrate. The slotted ground plane in yellow is printed below the substrate and a serious feed line in red is printed above the substrate. The ground plane is etched with four intersected slots, whereas a series feed line is crossing over the slots to excite the DRs. The series feed line is terminated by SMA ports, named as Port 1 and Port 2, respectively. The commercially available software ANSYS high-frequency structure simulator (HFSS) is used for simulation and optimization in this work.

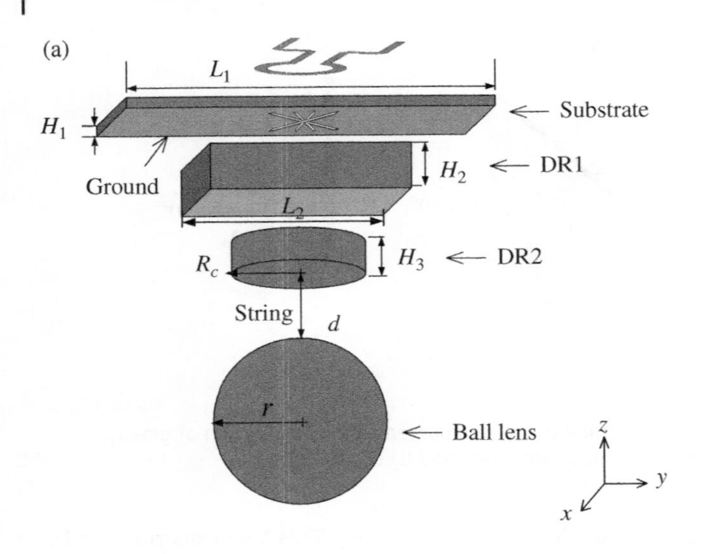

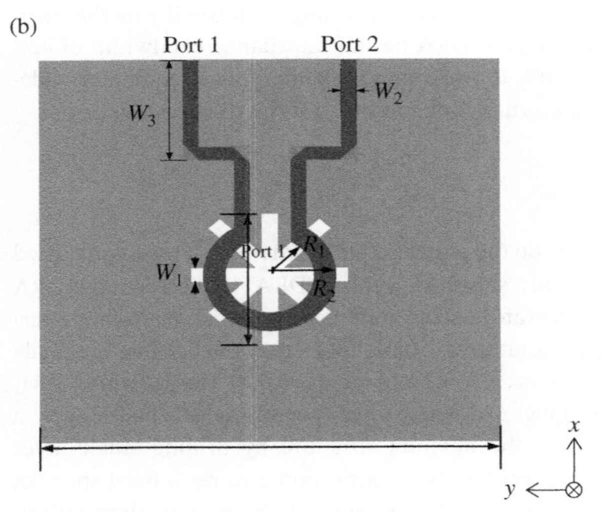

Figure 3.36 Configuration of the proposed DRA. (a) 3D view; (b) layout of the double-side printed substrate.

Table 3.2 Dimensions of the proposed DRA.

Parameter	L_1	H_1	L_2	H_2	H_3	R_c	W_2
Value (mm)	70	1.0	50	10	6.3	15	1.4
Parameter	r	L_3	R_1	R_2	W_1	W_3	d
Value (mm)	20	19.1	5.8	8	1.0	15	20

3.4.2 Working Principle

Figure 3.37 shows the working principle of the ball lens. The ball lens is a common element for enhancing antenna gain [30] and suitable for pre-processing materials and sensor applications [31]. Ball lens usually employs uniform material, which is positioned under the radiating source to create parallel propagation rays.

In a ball lens or aspheric lens, effective focal length (*EFL*) and back focal length (*BFL*) are the dominated parameters that determine the ball lens performance. As the distance between the source and the lens center, *EFL* is the quantity used in lens calculation for magnification. A general formula for the *EFL* is given by [30].

$$\frac{1}{EFL} = (n-1)\left(\frac{1}{r_1} - \frac{1}{r_2}\right) + \frac{(n-1)^2}{n}\frac{t_c}{r_1 r_2} \tag{3.5}$$

where r_1 is the radius of curvature of surface 1, r_2 is the radius of curvature of surface 2, t_c is the center thickness, and n is the index of refraction. For a ball lens,

$$r_2 = -r_1 = r \text{ and } t_c = 2r \tag{3.6}$$

where sign convention for the curvature radius has been used.

Substituting (3.6) into (3.5) leads to a simplified formula for the *EFL*,

$$EFL = \frac{nr}{2(n-1)} \tag{3.7}$$

BFL, which is the distance from the lens surface to the source along the optical axis, is calculated as,

$$BFL = \frac{r(2-n)}{2(n-1)} \tag{3.8}$$

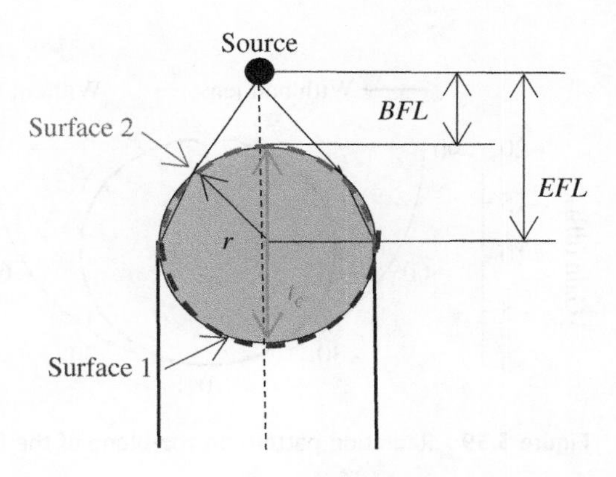

Figure 3.37 The propagation rays of the electromagnetic through the ball lens.

From (3.7) and (3.8), it can be seen that the *EFL* and *BFL* are proportional to the radius of ball lenses (r). In our proposed DRA, the use of alumina material gives the refractive index of $n = 1.4$. By using the ball lens of $r = 20$ mm, we have $BFL = d = 15$ mm from (3.8). In order to maintain good flexibility of the ball lens rotation and include the fringe field effect of DR radiation, the *BFL* is optimized as 20 mm for the final design.

Figure 3.38 compares the *E*-field distribution of the DRA design with and without the ball lens. It can be observed that the radiating power from the DRA is coupled to the ball lens and separated into two portions. As a result, in Figure 3.39, the 2D radiation pattern of DRA with the ball lens has around 1.2 dB gain enhancement compared with the DRA without the ball lens.

Figure 3.40 shows the *E*-field distributions of the proposed DRA with $\theta = 50°$ and $\theta = 90°$, where θ is the rotating angle of the ball lens. It is found that the

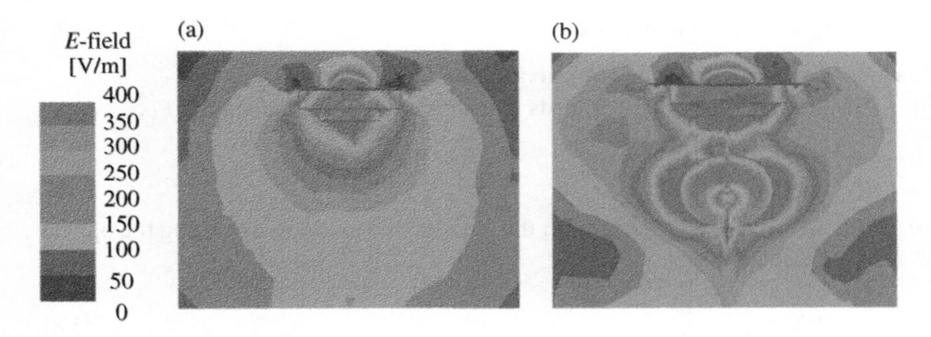

Figure 3.38 *E*-field distributions of DRA on *yoz* plane at 2.4 GHz. (a) Without ball lens; (b) with ball lens.

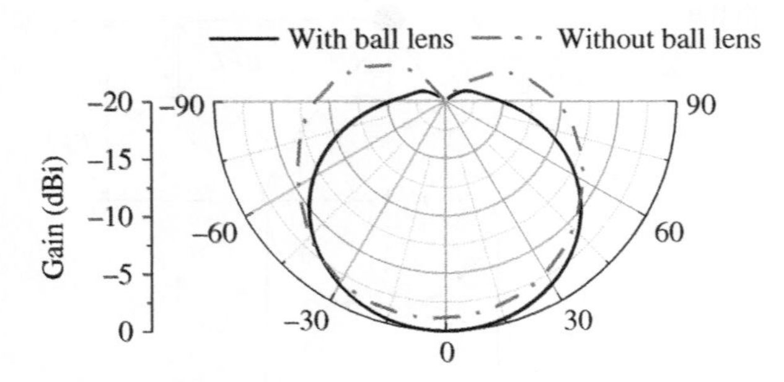

Figure 3.39 Radiation pattern on *yoz*-plane of the DRA with and without ball lens.

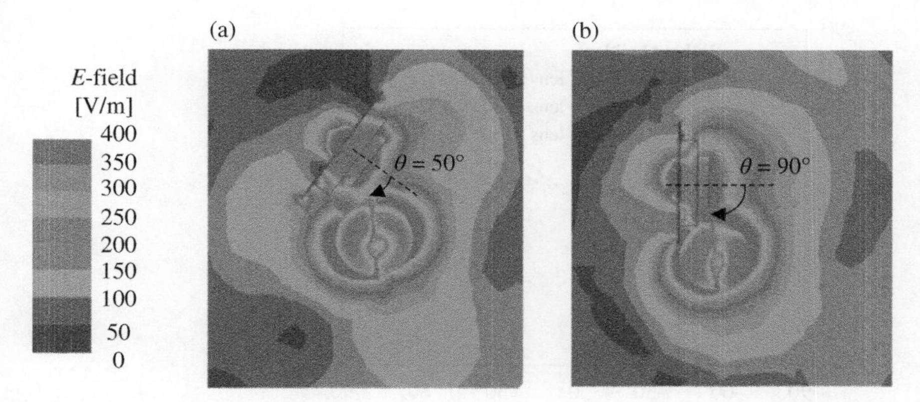

Figure 3.40 *E*-field distributions of DRA on *yoz* plane at 2.4 GHz. (a) θ = 50°; (b) θ = 90°.

antenna radiating field is changed by θ due to the coupling effects of the ball lens. In practical scenario, the antenna pattern adjustment can be obtained from the gravity action of ball lens, as was indicated in Figure 3.35.

Figure 3.41 presents the radiation pattern with different θ on *yoz* plane. It was found that the rotation angle (θ) of 50° and 90° gives a beam direction of 20° and 50°, respectively. This indicates the ball lens rotation on $\theta = \pm 90°$ results in a continuous beam scanning over $\pm 40°$ at 2.4 GHz. Because of the symmetry of the antenna structure, the scanning beams in the *xoz* and *yoz* planes are symmetrical.

Figure 3.42 demonstrated the main beam direction of the radiation patterns at different frequencies. Small variation can be observed from different frequencies,

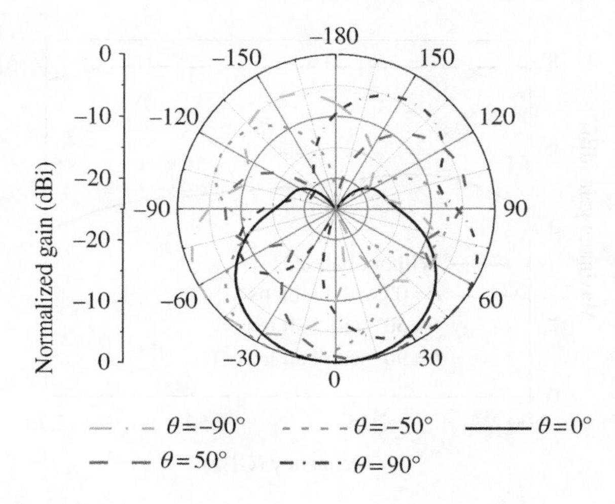

Figure 3.41 Beam-steering patterns on *yoz* plane at 2.4 GHz with different θ.

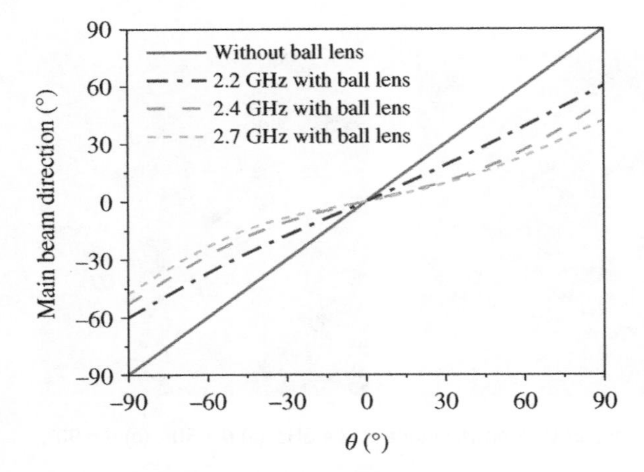

Figure 3.42 Main beam direction versus θ at different frequencies.

which is caused by the different refractive index. Even so, the beam steering of the proposed DRA is obvious compared with the case without the ball lens in red line. It is found that the maximum beam direction is adjusted from 90° to 50° for 2.4 GHz, to 60° for 2.2 GHz, and to 48° for 2.7 GHz.

Figure 3.43 presents the DRA gain versus frequency for different DRA prototypes and rotating angles. It can be seen that DRA reaches its maximum gain at 2.2 GHz with 7.9 dBic. During the rotation over ±90°, the antenna realized gain is in the range of 5.6–7.9 dBic or 2.3 dB variation. This is expected because the broadside coupling is strongest compared to the other tilted directions.

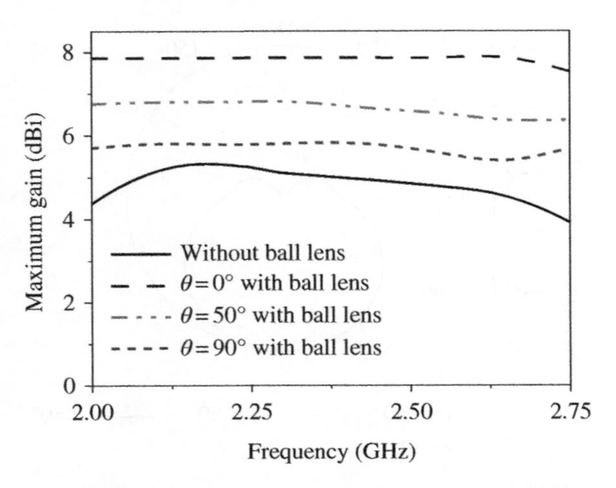

Figure 3.43 Maximum gain versus frequency for different DRA prototypes and θ.

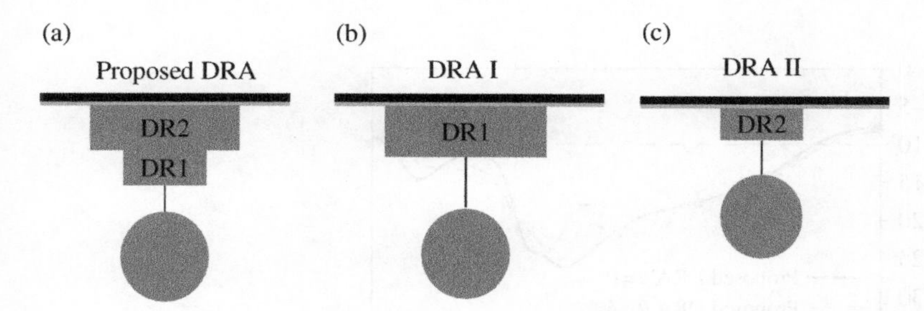

Figure 3.44 DRA configurations for comparison. (a) Proposed DRA; (b) DRA I; (c) DRA II.

Nevertheless, it is found that the proposed DRA has a stable gain performance with less than 0.5 dB variation over a wide frequency band from 2.0 to 2.75 GHz.

Figure 3.44 shows three antenna configurations with different stacked layers. Compared with the proposed DRA, DRA I uses DR1 only and removes DR2 component, whereas DRA II uses DR2 only and removes DR1 component. In DRA I and II, the distance (d) between the DR and the ball lens is identical to that of the proposed DRA.

Figure 3.45 gives the $|S_{11}|$ and the axial ratio comparison of three DRA prototypes with $\theta = 0°$ and the proposed DRA with different θ. It was noticed that the DRA I and DRA II have narrower impedance and axial ratio bandwidths compared with the proposed DRA. For the proposed DRA, its wideband and CP performance have little effects by its rotating angle θ. During the rotation, the $|S_{11}|$ shows a consistent wideband coverage from 1.95 to 3.20 GHz or impedance bandwidth of 48.5%. The axial ratio under 3 dB covers from 1.94 to 3.11 GHz or axial bandwidth of 46.3%.

The comparison on their steering performance is obvious but the figures are not given in here for brevity. Compared with the proposed DRA, the gain value of DRA I is reduced around 1.5 dB for all rotating angles. For DRA II, both scanning angle and gain value are reduced due to the poor coupling between the DRA base and the ball lens. The observation indicates that proper co-design of the stacked DRA can not only increase the antenna impedance and axial ratio bandwidths but also increase the scan angle and the antenna gain of the steering beams.

3.4.3 Fabrication and Measurement

In this section, the proposed DRA was fabricated and measured for validation. Figure 3.46a,b shows the photographs of the proposed DRA with the volume of $0.8 \times 0.8 \times 0.012\ \lambda_0^3$. The PCB is made of 1.0 mm thick FR4 epoxy ($\varepsilon_r = 4.4$ and $\tan\delta = 0.02$). The DRA was fabricated using low-cost, commercial composite alumina with $\varepsilon_r = 9.2$ and $\tan\delta = 0.008$.

A string, which passes through the drilled hole of the DRA, enables the structure reconfigurable by the action of gravity. Therefore, in Figure 3.46c,d, the DRA has

(a)

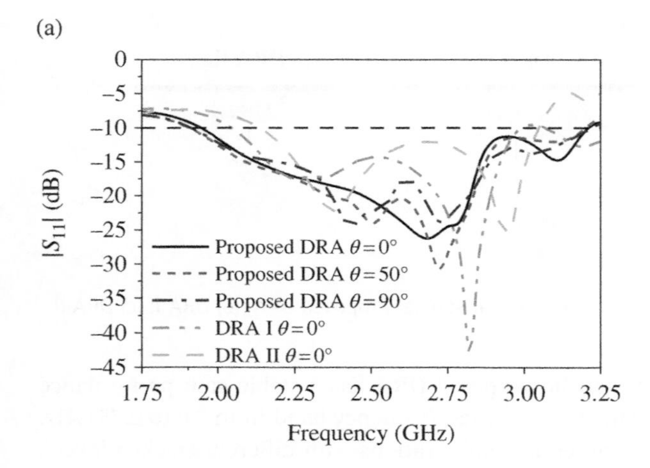

(b)

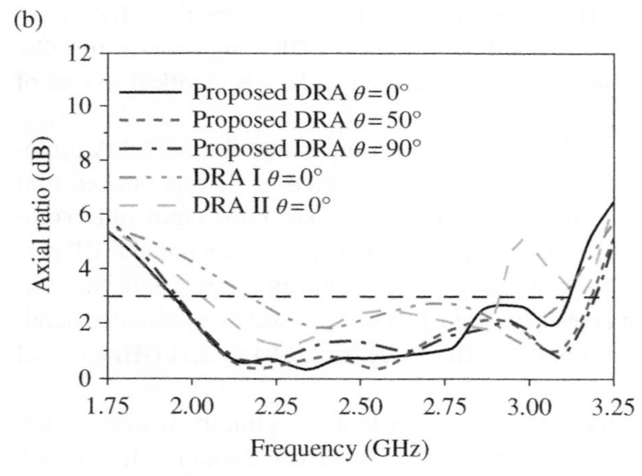

Figure 3.45 Simulated (a) S parameters of DRAs with different θ, (b) axial ratios of DRAs with different θ.

$\theta = 0°$ when mounted on the celling, whereas has $\theta = 90°$ when mounted on the side wall. The S-parameters were measured with an Agilent N5230C portable network analyzer-L (PNA-L), and radiating performance was measured in Satimo StarLab system. The proposed DRA exhibits left-hand (LH) or right-hand (RH) CP by exciting Port 1 or Port 2. The isolation between two ports is higher than 10 dB. For demonstration, Port 1 is measured whereas Port 2 is connected to the 50 Ω terminal for all measurement.

Figure 3.47 shows the measured and simulated $|S_{11}|$ at $\theta = 0°$, 50°, and 90°. The measured impedance bandwidth covers from 2.09 to 3.25 GHz (43.4%) for $|S_{11}| < -10$ dB. The measured $|S_{11}|$ bandwidth has good agreement to the

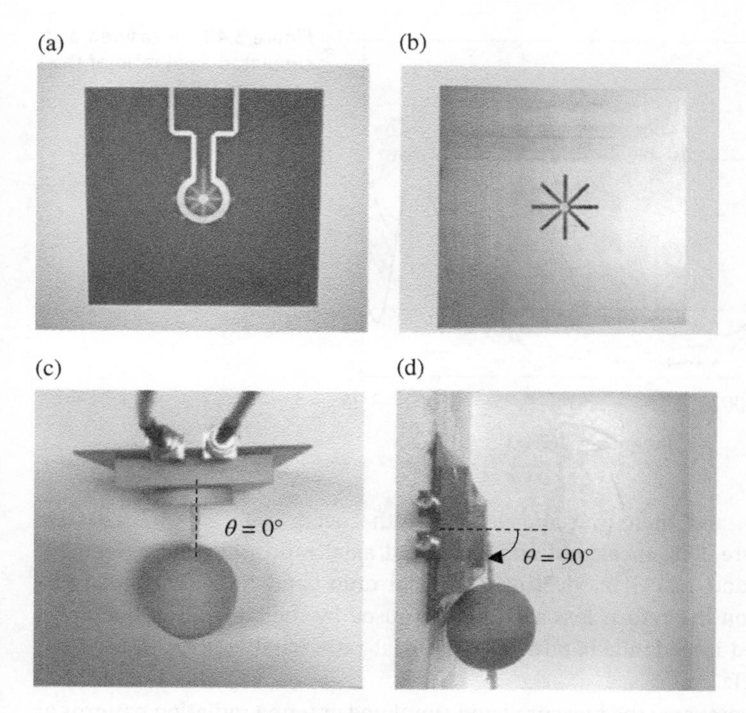

Figure 3.46 Photographs of the proposed DRA. (a) Top view of the substrate, (b) back view of the substrate, (c) DRA with $\theta = 0°$, and (d) DRA with $\theta = 90°$.

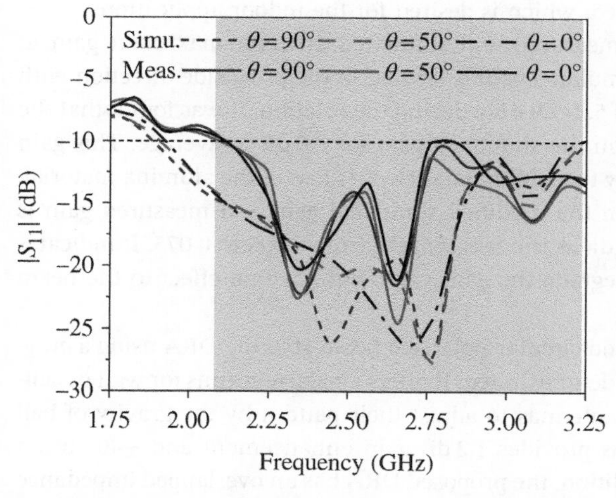

Figure 3.47 Measured and simulated $|S_{11}|$ of the proposed DRA with different θ.

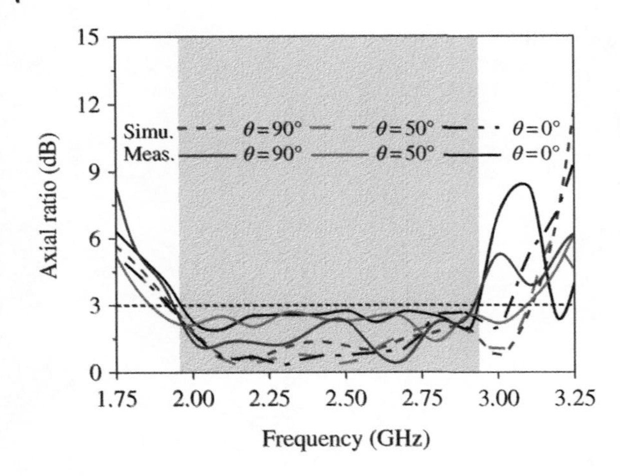

Figure 3.48 Measured and simulated axial ratio of the proposed DRA with different θ.

simulated one, and their overlapped bandwidth covers from 2.09 to 3.20 GHz (42%). In Figure 3.48, measured and simulated axial ratio has good agreement between 1.96 and 2.91 GHz, yielding 39% axial ratio bandwidth. The small frequency shifts on the return loss are mainly caused by the fabrication tolerance. The overlapped impedance bandwidth and axial ratio bandwidth is 32.8% from 2.09 to 2.91 GHz.

Figure 3.49 compares the measured and simulated antenna radiation patterns at $\theta = 0°$, 50°, and 90°, where good agreements are observed. It proves that the antenna can achieve continuous beam scanning in 2D planes with scan range over ±40° and provides pattern adjustment for different scenarios. Small cross-polarization level is observed, which is desired for the indoor applications.

Figure 3.50 shows the measured and simulated antenna maximum gain at 2.4 GHz. The maximum simulated gain is located in the broadside direction with 7.9 dBic and in the range of 5.2–7.9 dBic during the rotation. It was found that the measured gain is lower than the simulated gain for 3.0 dB in average. This gain degrade is mainly caused by the additional dielectric loss of the alumina material. A good agreement between the modified simulated gain and measured gain is obtained by modifying the dielectric loss tangent from 0.008 to 0.075. It indicates the dielectric loss would degrade the gain value but have no effect to the beam direction.

In this section, A wideband circular-polarized beam-steering DRA using a gravitational ball lens has been demonstrated. It offers a passive means for wall or ceiling-mounted base station antennas to adjust their pattern by the gravity of ball lenses. The use of ball lens provides 1.2 dB gain enhancement and ±40° beam steering of the DRA. In addition, the proposed DRA has an overlapped impedance and axial ratio bandwidths of 32.8% from 2.09 to 2.91 GHz during the rotation.

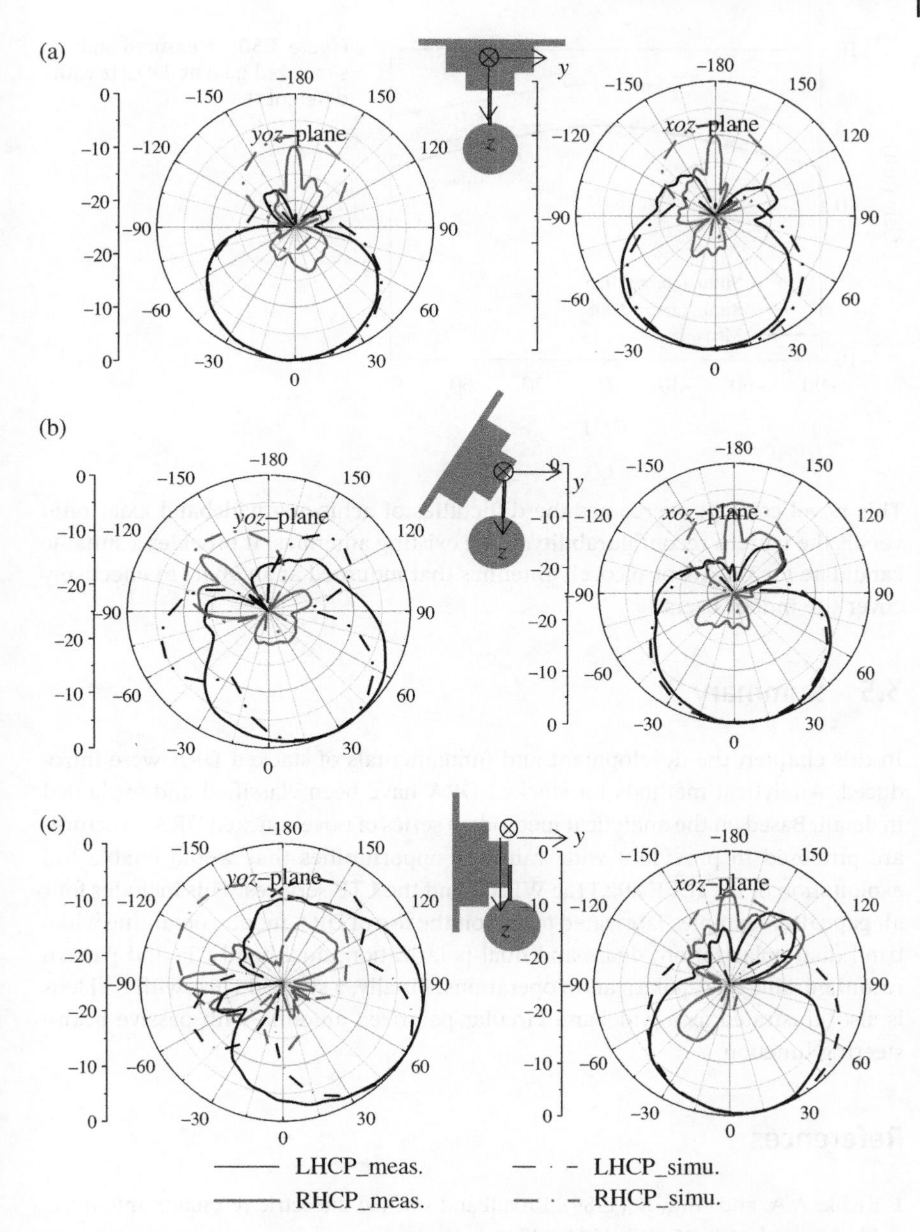

Figure 3.49 Measured and simulated normalized radiation pattern of the proposed DRA at 2.4 GHz. (a) $\theta = 0°$, (b) $\theta = 50°$, and (c) $\theta = 90°$.

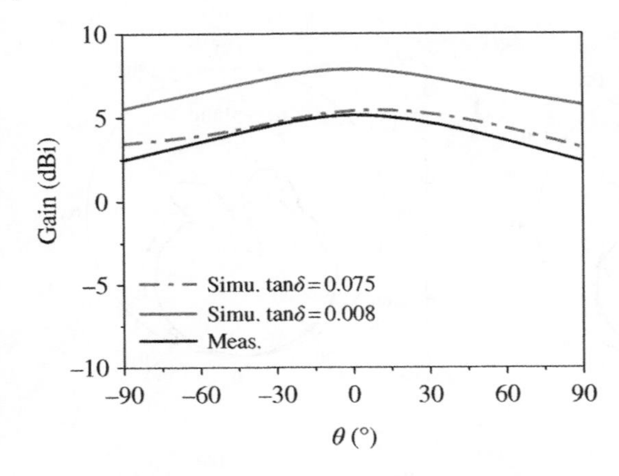

Figure 3.50 Measured and simulated gain at 2.4 GHz with different θ.

This novel concept overcomes the difficulties of achieving wideband axial ratio versus the pattern reconfigurability of the existing antennas. It provides a suitable candidate for WLAN or picocell antennas that mounted at an angle to effectively cover the indoor users.

3.5 Summary

In this chapter, the development and fundamentals of stacked DRA were introduced. Analytical methods for stacked DRA have been classified and explained in detail. Based on the analytical methods, a series of novel stacked DRA structures are proposed to provide a wide range of opportunities that would enable full exploitation of the IEEE 802.11ac WLAN and the LTE services. This includes four air gap DRA examples fabricated to exploit the use of DRA modes, obtaining wideband dual-polarization, dual-band dual-polarization, dual-band CP, and pattern reconfigurable dual-polarization operations. Finally, a stacked DRA with ball lens is also illustrated as a wideband circular polarized antenna with passive beam-steering function.

References

1 Kishk, A.A. and Ahn, B. (1989). Broadband stacked dielectric resonator antennas. *Electronics Letters* 25 (18): 1232–1233.

2 Kishk, A.A., Zhang, X., Glisson, A.W., and Kajfez, D. (2003). Numerical analysis of stacked dielectric resonator antennas excited by a coaxial probe for wideband applications. *IEEE Transactions on Antennas and Propagation* 51 (8): 1996–2006.

3 Xia, Z.X., Leung, K.W., and Lu, K. (2019). 3-D-printed wideband multi-ring dielectric resonator antenna. *IEEE Antennas and Wireless Propagation Letters* 18 (10): 2110–2114.

4 Gupta, S., Kshirsagar, P., and Mukherjee, B. (2019). A low-profile multilayer cylindrical segment fractal dielectric resonator antenna: usage for wideband applications. *IEEE Antennas and Propagation Magazine* 61 (4): 55–63.

5 Yang, W.W., Sun, W.J., Tang, H., and Chen, J.X. (2019). Design of a circularly polarized dielectric resonator antenna with wide bandwidth and low axial ratio values. *IEEE Transactions on Antennas and Propagation* 67 (3): 1963–1968.

6 Xia, Z.X. and Leung, K.W. (2022). 3-D-printed wideband circularly polarized dielectric resonator antenna with two printing materials. *IEEE Transactions on Antennas and Propagation* 70 (7): 5971–5976.

7 Sun, S., Jiao, Y., and Weng, Z. (2020). Wide-beam dielectric resonator antenna with attached higher-permittivity dielectric slabs. *IEEE Antennas and Wireless Propagation Letters* 19 (3): 462–466.

8 Fakhte, S., Oraizi, H., Matekovits, L., and Dassano, G. (2017). Cylindrical anisotropic dielectric resonator antenna with improved gain. *IEEE Transactions on Antennas and Propagation* 65 (3): 1404–1409.

9 Fakhte, S., Oraizi, H., and Matekovits, L. (2017). High gain rectangular dielectric resonator antenna using uniaxial material at fundamental mode. *IEEE Transactions on Antennas and Propagation* 65 (1): 342–347.

10 Denidni, T.A., Weng, Z., and Niroo-Jazi, M. (2010). Z-shaped dielectric resonator antenna for ultrawideband applications. *IEEE Transactions on Antennas and Propagation* 58 (12): 4059–4062.

11 Song, C., Bennett, E.L., Xiao, J. et al. (2020). Passive beam-steering gravitational liquid antennas. *IEEE Transactions on Antennas and Propagation* 68 (4): 3207–3212.

12 Lu, K., Leung, K.W., and Pan, Y.M. (2011). Theory and experiment of the hollow rectangular dielectric resonator antenna. *IEEE Antennas and Wireless Propagation Letters* 10: 631–634.

13 Sehrawat, N., Kanaujia, B.K., and Agarwal, A. (2018). Calculation of the resonant frequency of a rectangular dielectric resonator antenna using perturbation theory. *Journal of Computational Electronics* 18 (1): 211–221.

14 Sun, W.J., Yang, W.W., Chu, P., and Chen, J.X. (2019). Design of a wideband circularly polarized stacked dielectric resonator antenna. *IEEE Transactions on Antennas and Propagation* 67 (1): 591–595.

15 Fakhte, S., Oraizi, H., and Matekovits, L. (2017). Gain improvement of rectangular dielectric resonator antenna by engraving grooves on its side walls. *IEEE Antennas and Wireless Propagation Letters* 16: 2167–2170.

16 Boyuan, M., Pan, J., Yang, D., and Guo, Y.X. (2022). Investigation on homogenization of flat and conformal stacked dielectric resonator antennas. *IEEE Transactions on Antennas and Propagation* 70 (2): 1482–1487.

17 Yang, X., Liu, Y., and Gong, S.X. (2018). Design of a wideband omnidirectional antenna with characteristic mode analysis. *IEEE Antennas and Wireless Propagation Letters* 17 (6): 993–997.

18 Luo, Y., Chen, Z.N., and Ma, K. (2019). Enhanced bandwidth and directivity of a dual-mode compressed high-order mode stub-loaded dipole using characteristic mode analysis. *IEEE Transactions on Antennas and Propagation* 67 (3): 1922–1925.

19 Liu, S., Yang, D., Chen, Y. et al. (2020). Broadband dual circularly polarized dielectric resonator antenna for ambient electromagnetic energy harvesting. *IEEE Transactions on Antennas and Propagation* 68 (6): 4961–4966.

20 Chen, Z. (2016). *Design and Implementation of Compact Semi-Smart Indoor and Outdoor Base Station Antennas for Mobile Communications*. Beijing: Beijing University of Posts and Telecommunications Press.

21 Pan, Y.M., Leung, K.W., and Lu, K. (2012). Omnidirectional linearly and circularly polarized rectangular dielectric resonator antennas. *IEEE Transactions on Antennas and Propagation* 60 (2): 751–759.

22 Toh, B.Y., Cahill, R., and Fusco, V.F. (2003). Understanding and measuring circular polarization. *IEEE Transactions on Education* 46 (3): 313–318.

23 Zou, L., Abbott, D., and Fumeaux, C. (2012). Omnidirectional cylindrical dielectric resonator antenna with dual polarization. *IEEE Antennas and Wireless Propagation Letters* 11: 515–518.

24 Khalily, M., Rahim, M.K.A., and Kishk, A.A. (2011). Bandwidth enhancement and radiation characteristics improvement of rectangular dielectric resonator antenna. *IEEE Antennas and Wireless Propagation Letters* 10: 393–395.

25 Almpanis, G., Fumeaux, C., and Vahldieck, R. (2010). Dual-mode bridge-shaped dielectric resonator antennas. *IEEE Antennas and Wireless Propagation Letters* 9: 103–106.

26 Li, W.W. and Leung, K. W. (2013). Compact circularly polarized hollow dielectric resonator antenna with embedded vertical feed network. *2013 IEEE Antennas and Propagation Society International Symposium (APSURSI)*, Orlando, FL, USA, 1234–1235.

27 Fang, S.F., Leung, K.W., Lim, E.H. et al. (2010). Compact differential rectangular dielectric resonator antenna. *IEEE Antennas and Wireless Propagation Letters* 9: 662–665.

28 Chair, R., Kishk, A.A., and Lee, K.F. (2006). Hook- and 3-D J-shaped probe excited dielectric resonator antenna for dual polarisation applications. *IET Microwaves, Antennas and Propagation* 153 (3): 277–281.

29 Lim, E.H., Leung, K.W., and Fang, X.S. (2011). The compact circularly-polarized hollow rectangular dielectric resonator antenna with an underlaid quadrature coupler. *IEEE Transactions on Antennas and Propagation* 59 (1): 288–293.

30 Nguyen, N.T., Boriskin, A.V., Le Coq, L., and Sauleau, R. (2016). Improvement of the scanning performance of the extended hemispherical integrated lens antenna using a double lens focusing system. *IEEE Transactions on Antennas and Propagation* 64 (8): 3698–3702.

31 Ahn, B., Jo, H.-W., Yoo, J.-S. et al. (2019). Pattern reconfigurable high gain spherical dielectric resonator antenna operating on higher order mode. *IEEE Antennas and Wireless Propagation Letters* 18 (1): 128–132.

4

Dielectric Resonator Antenna Array

CHAPTER MENU

4.1 Overview

Dielectric resonator antenna (DRA) has drawn increasing attention in the past decade for its superior advantages of compact structure, flexible feeding methods, and high radiation efficiency [1, 2]. Most importantly diverse mode families and their higher modes can be excited in a single DRA structure [3–5]. It provides the antenna array with additional degrees of freedom (DoFs). Thus, using DRA in antenna arrays can obtain stable gain over wider angle and lower side lobe levels for the process of beam scanning. For the process of shaped beam synthesis, the intelligent optimization algorithm synthesizes the target beam via the amplitude and phase control. The use of DRA has the potential to effectively reduce the system costs and improve the beam control abilities.

Dielectric Resonator Antennas: Materials, Designs and Applications, First Edition.
Zhijiao Chen, Jing-Ya Deng, and Haiwen Liu.

Published 2024 by John Wiley & Sons, Inc.

The rest of the chapter is organized as follows. Section 4.2 describes the design procedure of pattern diverse DRA. It introduces operation of pattern diverse antenna and gives antenna design examples. In Section 4.3, pattern diverse DRA arrays are presented to effectively widen the scanning angle of an antenna array. Section 4.4 introduces the methodologies of shaped beam synthesis on a DRA array. A suitable shaped beam synthesis algorithm is chosen and shows the advantage of pattern diverse DRA array. Finally, the remarks are summarized in Section 4.5.

4.2 Pattern Diverse DRA

Pattern diverse antennas add new DoF in antenna array design and offer great potential for improving the scanning capability of the array. Pattern diverse antennas are classified here as multi-port excitation and single-port excitation. Both of these two types have been numerically studied in the past decades for improving antenna coverage or/and array scanning angle.

The method of using multi-port excitation is to design separate feeding networks for different ports in one antenna and try to reduce the mutual coupling between two ports [4, 6]. Conventional antennas with two-dimensional structure require high isolation of multi-port by designing complex feeding networks [7, 8]. It leads to large antenna size and high loss. In [9], decoupling radial slots are used to enhance port isolation (>20 dB), with very small effects on the excited resonant modes. DRA has a three-dimensional structure with a rich variety of modes, which can stimulate a variety of orthogonal modes to achieve high isolation of multi-port [10, 11]. DRA realizes the miniaturized design of multi-port pattern diverse antenna. At the moment, the maximum number of diversity ports is 4 as proposed in [12], where a four-port cylindrical pattern-diversity and polarization-diversity DRA is presented for the first time.

The method of using single-port excitation could be switching the feeding networks or reconfiguring the antenna structures. The feeding networks could be switched mechanically or electrically. Electrical tuning has a much shorter response time. However, the electrical switches are required, which increases the size and loss of the antenna. Mechanical tuning does not need additional components that cause size increase. However, manual adjustment is required, which takes more time and workforce than electrical tuning. The adjustment parts can be divided into two categories: feeding network and antenna structure.

A compact electrically tunable feeding network with four coupling slots is designed in [13]. This method increases the switching speed, but the losses caused by electrical switching reduce the antenna efficiency. Mechanical switches (SPDT,

SP3T) are employed on the feeding networks for diverse patterns of DRA in [14]. But insertion loss caused by different states of the switch is different, which results in different peak gain values. By using electric switches, the reconfigurable antenna is proposed in [15] by enabling and disabling the parasitic elements. But the large size of the parasitic element limits the number of states. In [16], reconfigurable DRA is designed by mechanical adjustment of the feeding cable, which may cause disconnection between the DRA and the cable in practice use. Also, the gap between the patch and the DRA might slow down the tunability.

The reconfigurable antenna could employ liquid-filled resonators, such as liquid metal [17], water [18], and dielectric liquid (e.g. ethyl acetate [19, 20], perfluorodecalin [21]). They achieve diverse patterns by changing the liquid distribution in the resonator. The weakness of this method is the slow tuning speed. To control the flowing liquid, an expensive servo system is required. In addition, the metal liquid is significantly affected by the temperature, which makes the liquid antenna unsuitable for the outdoor environment with large temperature difference.

The pattern diverse DRA proposed in Section 4.2.1 shows the advantages of low-profile, passive phase-controlled, and wide-angle beam scanning. The proposed DRA has a compact structure with 0.38λ side length and is excited with high-isolated fundamental modes of TE and TM families in the same frequency. By controlling the phase between these two modes, DRA can reconfigure its pattern by generating E-plane $\pm 66°$ tilted beams. The proposed DRA is extended to antenna array, and the array can be scan in a wide angle due to the pattern diverse characteristics of the proposed antenna. The flexible and convenient control also makes this antenna a suitable candidate for shaped beam synthesis applications.

4.2.1 Antenna Configuration

The 3D configuration of the proposed pattern reconfigurable DRA is shown in Figure 4.1. It consists of a dielectric resonator (DR) structure, a double-side printed substrate, and two Sub-MiniatureA (SMA) ports. The DR structure is constructed by a rectangular DR and four cubic DR supports, forming an air gap between the rectangular DR and the substrate. Rogers RT6002 material is used as the substrate. It has a thickness of h_3, relative permittivity $\varepsilon_r = 2.94$, and tan $\delta = 0.0012$. The topologies of the printed metal over the substrate are colored in yellow and shown in Figure 4.1b,c. On the top surface of the substrate, an annular patch with outer diameter of d_1 and inner diameter of d_2 is printed. This annular patch is fed by an offset via-hole probe that connected to Port 2 underneath the substrate. On the bottom side of the substrate, Ports 1 and 2 are mounted on the ground plane. Holes with diameter d_2 are printed to isolate the probes from the ground plane. Port 1 is placed in the center and connected to a probe with the height of h_2. This probe passes through the center hole of annular patch and inserted into rectangular DR.

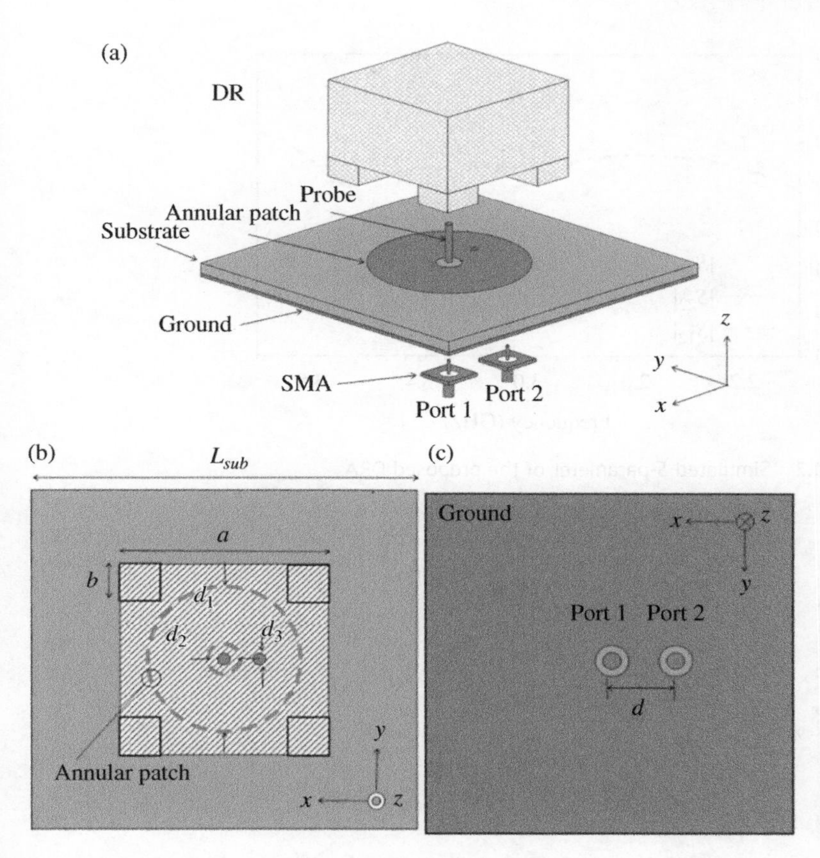

Figure 4.1 Configuration of the proposed DRA. (a) 3D view, (b) top view, and (c) bottom view.

Figure 4.2 shows the S-parameters of the proposed DRA. Port 1 has a wide impedance bandwidth of 2.50–3.68 GHz for $|S_{11}| < 10$ dB, whereas Port 2 has a bandwidth of 2.94–3.03 GHz. Two ports have their strongest resonance at 3.0 and 3.1 GHz, respectively. High isolation over 25 dB can be observed over the whole operating band. At 3.0 GHz, the isolation between two ports is as high as 27.9 dB. In Figure 4.3, the simulated radiation patterns of two ports are given for their E-field (*xoz* plane) and H-field (*xoy* plane) at 3.0 GHz. It was found that Port 1 has "∞" shape pattern on the E-plane and "O" shape on H-plane. In contrast, Port 2 has a nearly "O" shape on the E-plane and "∞" shape pattern on the H-plane. The "O" shape pattern of Port 2 is not as perfect as Port 1 due to the reflection from the ground plane.

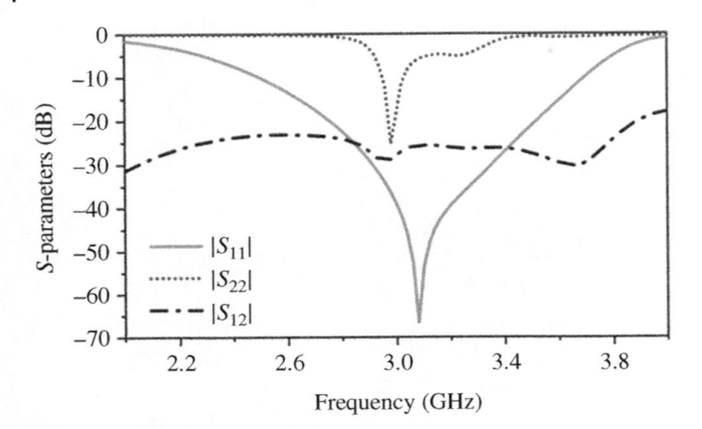

Figure 4.2 Simulated *S*-parameter of the proposed DRA.

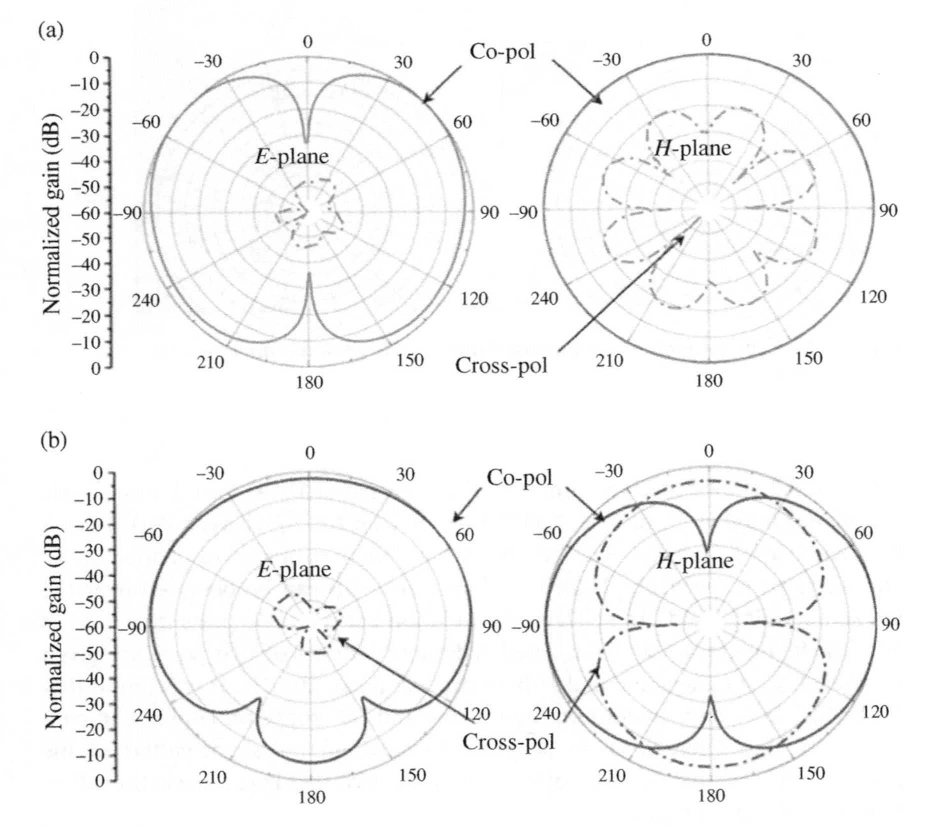

Figure 4.3 Simulated *E*-plane and *H*-plane radiation pattern of air-gap DRA at 3.0 GHz. (a) Port 1. (b) Port 2.

It is well known that the *E*- and *H*-plane radiation patterns of an electric dipole have "∞" shape and "O" shape, respectively, whereas patterns on these two planes are interchanged for a magnetic dipole. The complementary radiation patterns of two ports verify that Port 1 is a *z*-directional electric dipole, whereas Port 2 is a *y*-directional magnetic dipole. The electric dipole and magnetic dipole are orthogonal to each other, which explains the high isolation between two ports, as shown in Figure 4.2. The maximum realized gain of Ports 1 and 2 are 2.81 and 2.66 dB, respectively. The radiation efficiencies of both ports are higher than 95%.

Figure 4.4 shows the equivalent schematic diagram of the DRA, which shows two orthogonal modes. DRAs can be modeled as a *z*-direction electric dipole and a *y*-direction magnetic dipole. An electric dipole is represented as *J* and

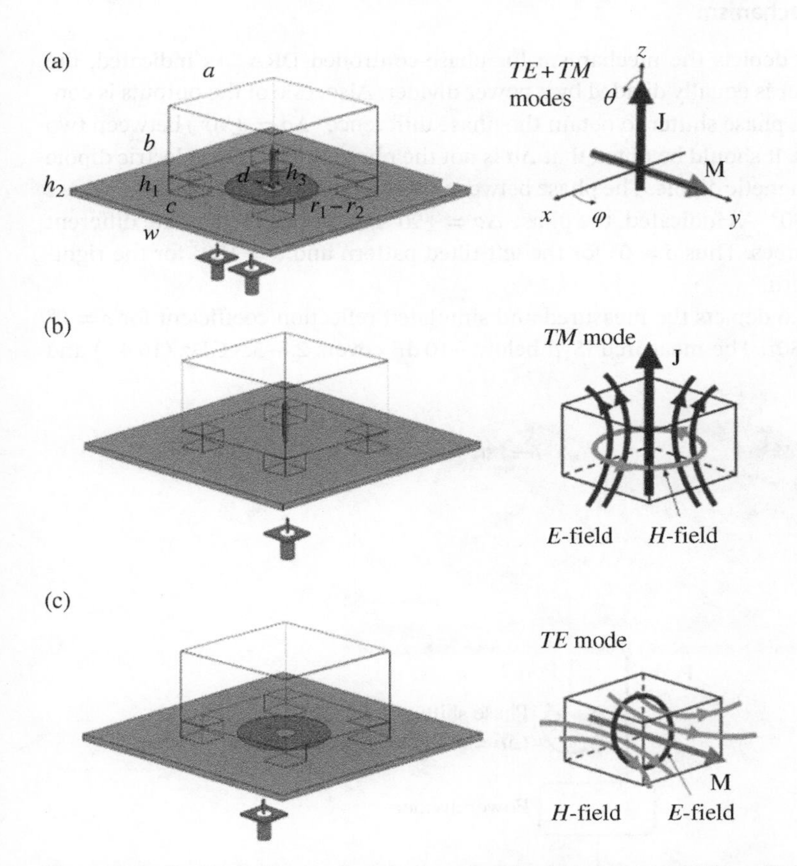

Figure 4.4 Configurations and equivalent models of the proposed DRA, which is the combination of a probe-fed DRA and a patch-fed DRA. (a) Proposed DRA; (b) probe-fed DRA; (c) patch-fed DRA.

magnetic dipole is represented as *M*. They are excited by independent ports, so the operation of two dipoles can be studied separately. By superimposing the far-fields of the electric dipoles and the magnetic dipoles, the total far-field of the proposed DRA can be calculated.

Due to the excitation of fundamental modes, DRA has a compact structure with the side length of 0.38λ and the patch diameter of 0.16λ. Electric dipole and magnetic dipole are excited independently by different ports. Their working frequency can be adjusted independently with very small effects to the other port. Moreover, due to the high isolation between two ports, this DRA can be regarded as two antenna elements that integrated into one DRA structure.

4.2.2 Mechanism

Figure 4.5 depicts the mechanism for phase-controlled DRA. As indicated, the input power is equally divided by a power divider. Also, one of the outputs is connected to a phase shifter to obtain the phase difference ($\Delta a = \pm 90°$) between two DRA ports. It should be noted that Δa is not the phase between the electric dipole and the magnetic dipole. The phase between two dipoles is represented as δ, where $\Delta a = \delta - 90°$. As indicated, the phase $\Delta a = \pm 90°$ is assigned to generate different pattern shapes. Thus $\delta = 0°$ for the left-tilted pattern and $\delta = 180°$ for the right-tilted pattern.

Figure 4.6 depicts the measured and simulated reflection coefficient for $\delta = 0°$ and $\delta = 180°$. The measured $|S_{11}|$ below -10 dB covers 2.8–3.3 GHz (16.4%) and

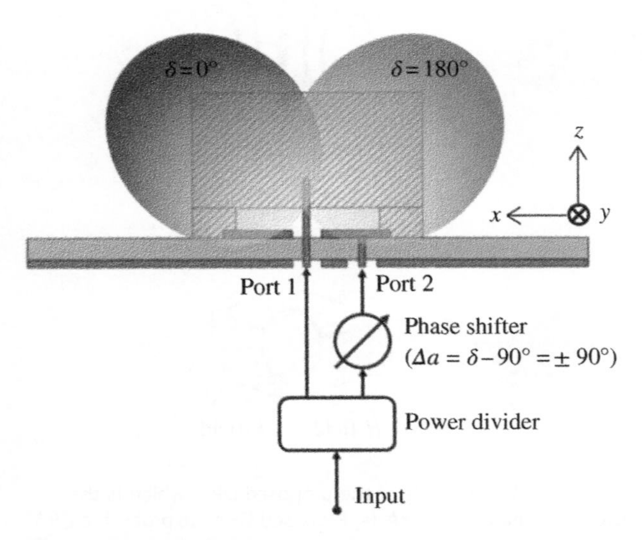

Figure 4.5 Mechanism for phase-controlled pattern-reconfigurable DRA.

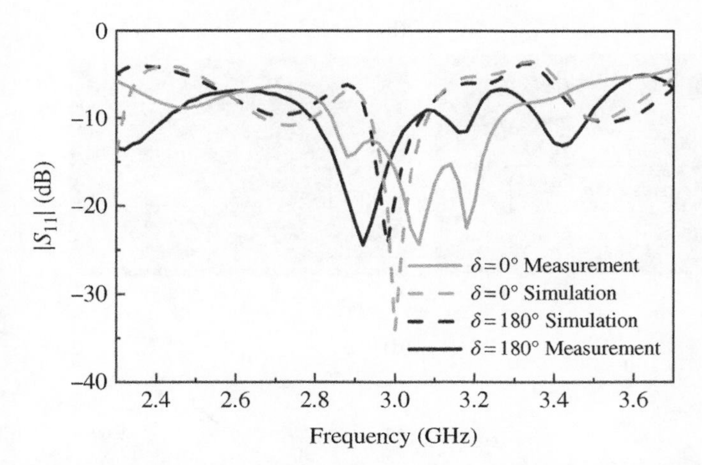

Figure 4.6 Measured and simulated reflection coefficient of the phase-controlled pattern-reconfigurable DRA.

2.8–3.1 GHz (10.2%) for two cases, whereas the simulated $|S_{11}|$ values are 2.9–3.1 and 2.9–3.0. Wider impedance bandwidth and extra resonances outside the working band are caused by interference from the feeding networks. In Figure 4.7a,b, the E-field distributions within the DRA are tilted to different sides for two states ($\delta = 0°$ and $\delta = 180°$). Normalized radiation patterns are plotted in Figure 4.7c,d and their maximum gains are pointed to $\pm 66°$, respectively. The maximum gains are 5.39 and 4.79 dB with a high front-to-back ratio of ~25 dB. The co-polarized fields of both planes are stronger than their cross-polarization by more than 30 dB in the main ($\pm x$) direction. By including the loss from the power divider and phase shifter, the measured efficiencies are 82.6% and 76.1% for two states of $\delta = 0°$ and $\delta = 180°$, respectively.

The total far-field of the proposed DRA is calculated by superimposing the far-fields of a z-directed electric dipole and a y-directed magnetic dipole. The total field in horizontal plane $E_{T\theta}$ and vertical plane $E_{T\varphi}$ is presented as [16],

$$E_{T\theta} = \frac{k}{4\pi r} e^{jw[t-(r/c)]} \left(j\eta I_e l_e \sin\theta - e^{j\delta} j I_m l_m \cos\varphi \right) \tag{4.1}$$

$$E_{T\varphi} = \frac{k}{4\pi r} e^{jw[t-(r/c)]} e^{j\delta} j I_m l_m \cos\theta \sin\varphi \tag{4.2}$$

where k is the wave number and δ is the phase difference of the two currents. l_e and l_m are the length of the electric dipole and magnetic dipole, and I_e and I_m are the current, respectively. By considering $\eta I_e l_e = I_m l_m$, $E_{T\theta}$ and $E_{T\varphi}$ can be simplified as,

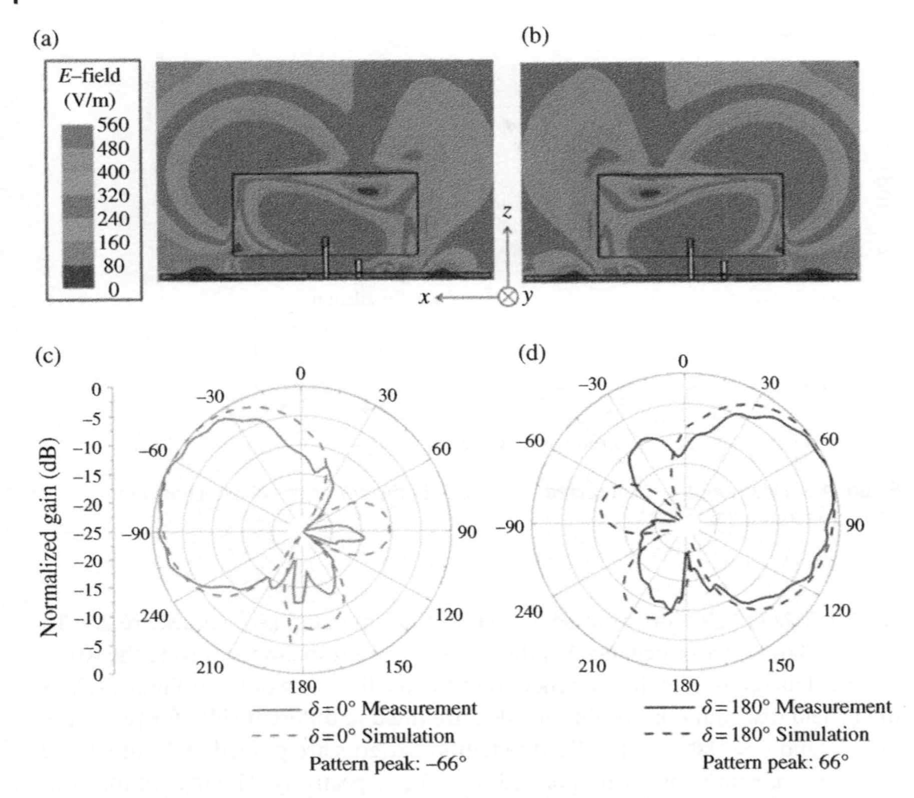

Figure 4.7 *E*-filed distributions and normalized *E*-plane radiation pattern of the phase-controlled pattern reconfigurable DRA at 3.0 GHz. (a) The *E*-field distribution for $\delta = 0°$. (b) The *E*-field distribution for $\delta = 180°$. (c) *E*-plane radiation pattern for $\delta = 0°$. (d) *E*-plane radiation pattern for $\delta = 180°$.

$$E_{T\theta} = \frac{k}{4\pi r} e^{jw[t-(r/c)]} Il(j\sin\theta - j\cos\delta\cos\varphi + 6\sin\delta\cos\varphi) \tag{4.3}$$

$$E_{T\varphi} = \frac{k}{4\pi r} e^{jw[t-(r/c)]} Il(j\cos\delta\cos\theta\sin\varphi - \sin\delta\cos\theta\sin\varphi) \tag{4.4}$$

And the total field E_T is given by,

$$\begin{aligned} E_T &= \sqrt{|E_{T\theta}|^2 + |E_{T\varphi}|^2} \\ &= \frac{k}{4\pi r} e^{jw[t-(r/c)]} Il\sqrt{\sin^2\theta + \cos^2\varphi + \cos^2\theta\sin^2\varphi - 2\cos\delta\sin\theta\cos\varphi} \end{aligned} \tag{4.5}$$

For $\delta = 0°$, the total fields E_T, $E_{T\theta}$, and $E_{T\varphi}$ can be simplified into,

$$E_{T\theta} = \frac{k}{4\pi r} e^{jw[t-(r/c)]} Il(j\sin\theta - j\cos\varphi) \tag{4.6}$$

$$E_{T\varphi} = \frac{k}{4\pi r} e^{jw[t-(r/c)]} Il(j\cos\theta\sin\varphi) \tag{4.7}$$

Therefore,

$$E_T \propto \sqrt{\sin^2\theta + \cos^2\varphi + \cos^2\theta\sin^2\varphi - 2\sin\theta\cos\varphi} \tag{4.8}$$

For $\delta = 180°$, likewise, the total fields E_T, $E_{T\theta}$, and $E_{T\varphi}$ can be simplified to,

$$E_{T\theta} = \frac{k}{4\pi r} e^{jw[t-(r/c)]} Il(j\sin\theta + j\cos\varphi) \tag{4.9}$$

$$E_{T\varphi} = \frac{k}{4\pi r} e^{jw[t-(r/c)]} Il(-j\cos\theta\sin\varphi) \tag{4.10}$$

Therefore,

$$E_T \propto \sqrt{\sin^2\theta + \cos^2\varphi + \cos^2\theta\sin^2\varphi + 2\sin\theta\cos\varphi} \tag{4.11}$$

According to formulas, the calculated maximum and minimum E_T is located along the $\pm x$ direction. For $\delta = 0°$, E_T reaches its maximum at $\theta_{peak} = -90°$ and $\varphi_{peak} = 180°$ and minimum at $\theta_{null} = 90°$ and $\varphi_{null} = 180°$. This results in a lateral radiation pattern as was found in [16].

In this design, due to the ground plane effect, θ_{peak} and θ_{null} shift from the calculated values $\theta_{peak} = -90°$, $\theta_{null} = -90°$ to $\theta_{peak} = -66°$, $\theta_{null} = -120°$, as was observed in Figure 4.7c. Actually, for the wide-angle beam scanning over $\pm 90°$, element with $\theta_{peak} = -66°$ is more suitable than that of $\theta_{peak} = -90°$ to minimize gain fluctuation over scan range over $\pm 90°$.

Compared with the design in [16], the proposed DRA reduced the DRA profile from $0.42\lambda \times 0.42\lambda$ to $0.38\lambda \times 0.38\lambda$. Moreover, the pattern methodology is modified from the p–i–n diode control to phase shifter control, making it straightforward to be integrated to the phase-only base station system.

4.3 DRA Array with Wide-Angle Beam Scanning

Antenna arrays are the key technology in 5G communication systems. Especially, phased array antenna can quickly scan and adjust the beam direction according to the surrounding environment. The scanning angle of the conventional phased array is limited to ±60°, which is constrained by the conditions listed below. First, the radiation pattern of the antenna elements is directional and has limited beam-width. Secondly, mutual coupling between antenna elements increases when the antenna array increases its scanning angle. Most importantly, the element spacing should be less than a specified value. For beam scanning is half a wavelength spacing to avoid grating lobe; for the broadside scanned array is a wavelength spacing; for the ordinary end-fire scanned array is half a wavelength spacing; and according to the array factor in theory, the beam scanning has 3 dB drops at about 60°. As a result, conventional phased arrays cannot avoid the limited scanning angle and the blind angle in some specific directions.

In order to increase the scanning angle of the array, a widebeam antenna element can be used. However, the gain of a widebeam antenna is relatively low and the array needs to increase the number of elements to achieve the required gain. An alternative solution for improving scanning angle is using pattern diverse antennas as antenna elements. It adds a new DoF in antenna array, offering great potential for improving the array scanning capability. In [22–25], the use of pattern diverse antenna as array element enables a wide 3 dB beam coverage as high as ±70°. The pattern diverse techniques include switching between dual modes of microstrip antenna [22], controlling parasitic slots beside element [23], tuning phase between dual interleaved subarrays [24], and loading metamaterial structure [25].

In this section, we investigate the potential of pattern diverse DRA for wide-angle beam scanning. The fundamentals of the uniform linear antenna array are reviewed first. Then a four-element pattern diverse DRA array is proposed and discussed on its beam-scanning performance.

4.3.1 Fundamentals on Antenna Array

Figure 4.8 shows an N element antenna array in linear format. Assuming that the mutual coupling between the antenna elements is ignored, the far-field radiation of the nth antenna element can be expressed as follows:

$$\vec{E} = A\vec{I_n}\frac{e^{-jkR_n}}{4\pi R_n}F_e(\theta, \varphi) \tag{4.12}$$

where A is the scaling factor related to the form of the antenna element, $\vec{I}_n = I_n e^{j\alpha_n}$ can be regarded as the excitation current of the unit antenna placed in the center of the small segment, I_n and α_n represent the magnitude and phase

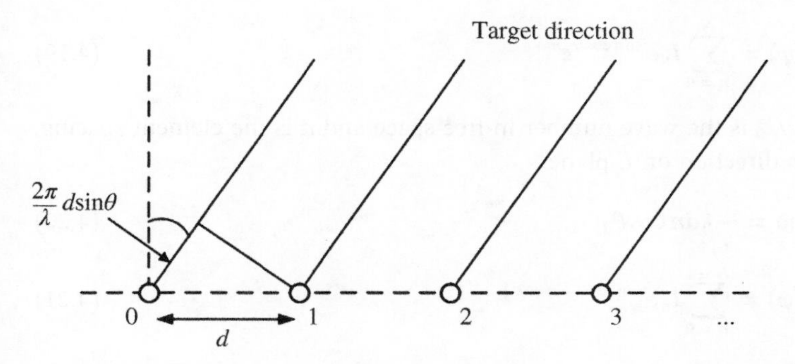

Figure 4.8 Geometry of an *N*-element linear array.

of the element antenna excitation, z_n is the coordinate position of the antenna unit, $F_e(\theta, \varphi)$ is the element radiation pattern,

$$\frac{1}{R_n} \approx \frac{1}{r_n} \tag{4.13}$$

$$R_n \approx r - z_n \cos\theta \tag{4.14}$$

$$\vec{E} = \sum_{n=0}^{N-1} \vec{E} = A\frac{e^{-jkr}}{4\pi r} F_e(\theta, \varphi) \sum_{n=0}^{N-1} \vec{I_n} e^{jkz_n \cos\theta} \tag{4.15}$$

From this, the array factor of the antenna can be obtained as,

$$F(\theta, \varphi) = F_e(\theta, \varphi) \sum_{n=0}^{N-1} \vec{I_n} e^{jkz_n \cos\theta} \tag{4.16}$$

$$AF(\theta, \varphi) = \sum_{n=0}^{N-1} \vec{I_n} e^{jkz_n \cos\theta} = \sum_{n=0}^{N-1} I_n e^{j(kz_n \cos\theta + a_n)} \tag{4.17}$$

$$F(\theta, \varphi) = F_e(\theta, \varphi) AF(\theta, \varphi) \tag{4.18}$$

The output of an array antenna equals to the element radiation pattern multiplied by the array factor. The element radiation pattern depends only on the radiation characteristics of the antenna elements. The array factor is determined by the array geometry and excitation (amplitude and phase). Since this process doesn't account for the mutual coupling effects, the element and array factors are independent of each other.

For the linear array $z_n = nd$, each element is connected to a phase shifter, and the phase of each unit is 0, a, $2a$, $3a$ ... $(N-1)\,a$. The phase difference is represented as $a_n = na$. The array factor AF is expressed as,

$$AF(\theta, \varphi) = \sum_{n=0}^{N} I_n e^{jkdn\cos\theta} e^{jna} \tag{4.19}$$

where $k = 2\pi/\lambda_0$ is the wave number in free space and d is the element spacing, θ is the beam direction on E-plane.

$$a_n = na = -kdn\cos\theta_0 \tag{4.20}$$

$$AF(\theta, \varphi) = \sum_{n=0}^{N-1} I_n e^{j\pi n(\cos\theta - \cos\theta_0)} \tag{4.21}$$

In a uniformly excited linear array, let $I_0 = I_1 = I_2 = \ldots I_N$, and $\psi = kd\cos\theta + a = kd(\cos\theta - \cos\theta_0)$. Introducing these conditions into (4.9) is simplified to,

$$AF = I_0 \sum_{n=0}^{N-1} e^{jn\psi} = I_0\left(1 + e^{j\psi} + \cdots + e^{j(N-1)\psi}\right) = I_0 e^{j(N-1)\psi} \frac{\sin(N\psi/2)}{\sin(\psi/2)} \tag{4.22}$$

The phase factor can be omitted for the centered array. The final array factor becomes,

$$AF = I_0 \frac{\sin(N\psi/2)}{\sin(\psi/2)} \tag{4.23}$$

From equation (4.22), the array factor of the antenna is a periodic function with a period of 2π. A grating lobe will occur at intervals of 2π. To suppress the grating lobe, it is necessary to ensure that,

$$|\psi|_{max} = kd|\cos\theta - \cos\theta_m| < 2\pi \tag{4.24}$$

$$d < \frac{\lambda}{1 + |\cos\theta_m|} \tag{4.25}$$

where θ_m is the maximum scanning angle. As the ratio of array element spacing to the wavelength (d/λ) increases, the antenna array will produce grating lobes, resulting in decreased radiation efficiency [26]. Table 4.1 gives the conditions for antenna arrays that can avoid the grating lobe.

For example, a compact, low-profile, broadside radiating two-element Huygens dipole array with 0.3λ spacing is proposed in [28]. In [30], an 8 by 8 end-fire array is designed with 0.45λ spacing. In [32], a wide-angle beam-scanning antenna array is proposed to scan from $-70°$ to $+70°$ in the operating band, and the element spacing is around 0.44λ. It can be seen that small element spacing offers potentials for wide-angle beam scanning.

Table 4.1 The conditions for antenna array to avoid the synthesis of grating lobes.

Array types	θ_m	Element spacing	Example
Broadside array	$\pm\pi/2$	$d<\lambda$	[27, 28]
End-fire array	0 or π	$d<\lambda/2$	[29, 30]
Beam-scanning array	$\lvert cos\theta_m \rvert <1$	$d<\lambda/2$	[31, 32]

For uniform linear arrays, the radiation pattern is distorted when the scanning angle reaches ±60°. There are several ways to increase the array scanning angle. One is to optimize the array factor, such as using non-uniform excitation. When the excitation amplitude decreases from the middle to the sides, the sidelobe reduces and achieves wide scan angle. Another way to optimize the array factor is to use a random antenna array. This non-periodic arrangement eliminates grating lobes and reduces the sidelobe. However, these two approaches increase the system complexity.

Another way to widen the scanning angle is to optimize the radiation pattern of the antenna array elements. Using antenna elements with diverse patterns can increase the DoF of the array element factor in (4.18). Then scanning angle of the array elements increases to compensate for the steep drop in array factor that occurs at ±60°. The method improves the scanning capability of the array without increasing the system complexity.

Based on this methodology, a phase-controlled pattern diverse antenna is designed in [24], making it simple to be integrated into the passive base station system. However, the use of the phase-controlled subarray as antenna element yields a large profile of $0.81\lambda_c$ and limits its scanning angle to ±75°. To achieve the wide-angle beam scanning, the pattern reconfigurable element should have a low profile ($<0.4\lambda_c$) to have an element spacing of $0.5\lambda_c$ [26]. For a well-designed base station, the pattern reconfigurable antenna element is desired to be controlled by the passive system without the use of additional components.

Using pattern diverse DRA as array element can obtain wider scanning angle, less system loss in low costs. In [33], a low-cost wide-angle beam-scanning phased array is proposed based on P-I-N controlled pattern reconfigurable DRA. The antenna array achieves two-dimensional beam scanning with high gain and low sidelobe levels.

4.3.2 Pattern Diverse Antenna Array

By using the compact phase-controlled pattern diversity DRA designed in Section 4.2, a four-element DRA array is implemented for beam scanning on *E*-plane. Figure 4.9a shows the prototypes of the four-element antenna array.

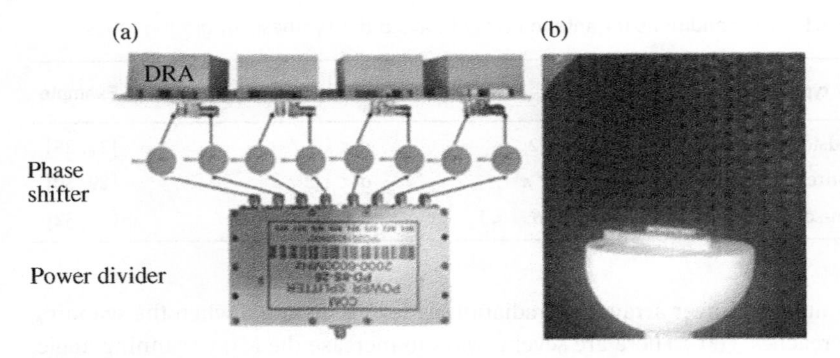

Figure 4.9 Photographs of the proposed DRA array. (a) Schematic of passive array system. (b) Measurement setup.

The distance between the elements in the x-direction is set to be 0.5λ or 50 mm. The total size of the passive phased array is 70 mm × 223 mm × 23.5 mm.

For demonstration, the DRA array is connected to seven mechanical phase shifters (ARRA 5418-180S) and 1 : 8 microstrip power divider (PD-8S-26) to verify its beam scan operation. Figure 4.9b shows the DRA measurement setup in an anechoic chamber. DRA array is mounted on semisphere support that is made of forms ($\varepsilon_r = 1$). The proposed DRA array is measured with the phase shifter and power divider as a whole structure to verify its beam scanning in real-life scenarios.

Figure 4.10 shows the schematics of DRA array and the control mechanism. A 1 : 8 power divider is connected to seven phase shifters before they feed the DRA ports. The power divider, phase shifters, and coaxial cables are fixed in the hollow of support to avoid their interference during the measurement. To explore its wide-angle beam scan ability, a simplified analysis is presented by using the left-tilted patterns ($\Delta a = 90°$) for all DRAs to demonstrate negative half E-plane scan performance ($0° < \theta < 90°$).

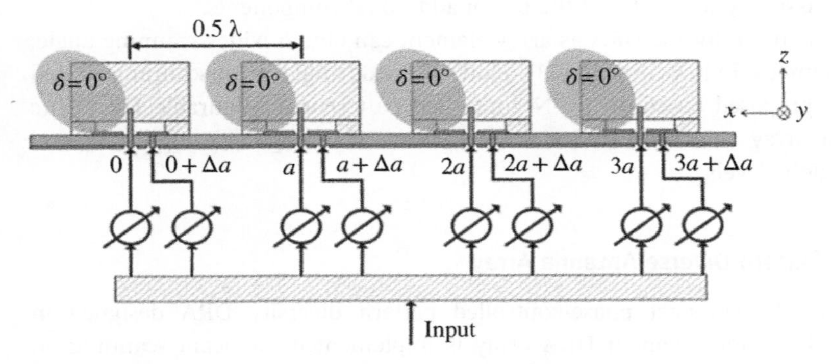

Figure 4.10 Schematics of DRA array and its control mechanism.

For a four-element linear array with $0.5\lambda_0$ spacing, the number of element N in (4.18) is 3 and d is $0.5\lambda_0$. As passive controlled system is employed, the amplitude I_n is uniformed as 1. When we assign the phase difference $a_n = na = -kdn\cos\theta_0$, the array factor AF is simplified into,

$$AF(\theta, \varphi) = \sum_{n=0}^{3} e^{j\pi n(\cos\theta - \cos\theta_0)} \tag{4.26}$$

It was found that $AF(\theta, \varphi)$ reaches its maximum when the beam direction is assigned to,

$$\theta_0 = \arccos\left(-\frac{a}{\pi}\right) \tag{4.27}$$

It can be seen that for the beam scanning over $-90° < \theta_0 < 90°$, $AF(\theta)$ has a consistent value. The grating lobe is avoided due to the small element spacing of $d = 0.5\lambda_0$.

The field pattern synthesis is calculated as,

$$E_{Total}(\theta) = E_{Tn}(\theta) \cdot AF(\theta) \tag{4.28}$$

where $E_{Tn}(\theta)$ is the field pattern of antenna element. As $AF(\theta)$ is a uniform value at its scan angle, the $E_{Total}(\theta)$ has a linear relationship with $E_{Tn}(\theta)$. Here, $E_{Tn}(\theta)$ is assigned with left-tilted DRA with maximum gain at $-66°$ to verify its scan ability over $-90° < \theta < 0°$. The scan ability of the other side ($0° < \theta_0 < 90°$) can be verified by using the DRA pattern with maximum gain at $66°$, which is omitted for brevity.

Figure 4.11 shows the relationship between the scan angle θ and $E_{Total}(\theta)$, which is calculated from (4.28) in MATLAB. It was found that $E_{Total}(\theta)$ has its maximum gain at ~ $-66°$, which is expected as $E_{Total}(\theta)$ has a linear relationship with $E_{Tn}(\theta)$.

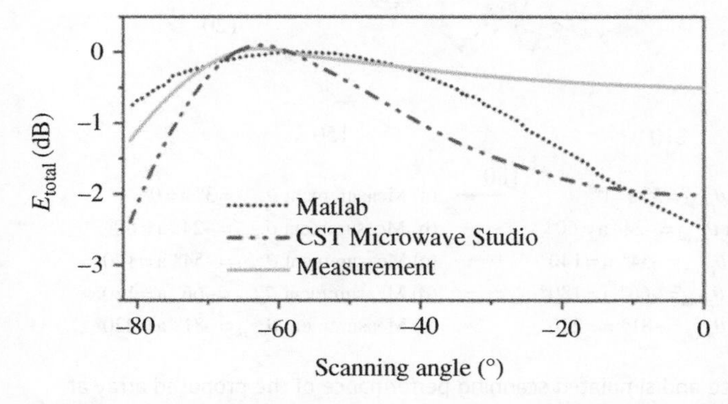

Figure 4.11 Relationship between scan angle θ and $E_{Total}(\theta)$ in MATLAB, CST Microwave Studio, and measurement.

Moreover, the calculated maximum gain fluctuation is <2.5 dB over the scan angle of $-90° < \theta < 0°$.

Figure 4.12 plots the measured and simulated radiation patterns of scanning beams with $a = 0°$, 60°, 140°, 180°, and 220°. Figure 4.13 shows that the proposed DRA array has $S_{11} < -10$ dB impedance matching over 2.95–3.05 GHz (3.3%) for all steering conditions. It verifies its scanning ability from −3° to −81° in the *E*-plane, with maximum gain located on 66° for both measured and simulated results. The 3 dB beamwidth covers +9° to −105° for negative half *E*-plane scan and would cover ±105° when both reconfigurable states of DRA are enabled. For all plots, the simulated patterns are normalized to a maximum gain of 10.8 dB at 66°, whereas the measured patterns are normalized to a maximum gain of 7.22 dB at 66°. In general, the measured gain is lower than simulated one by

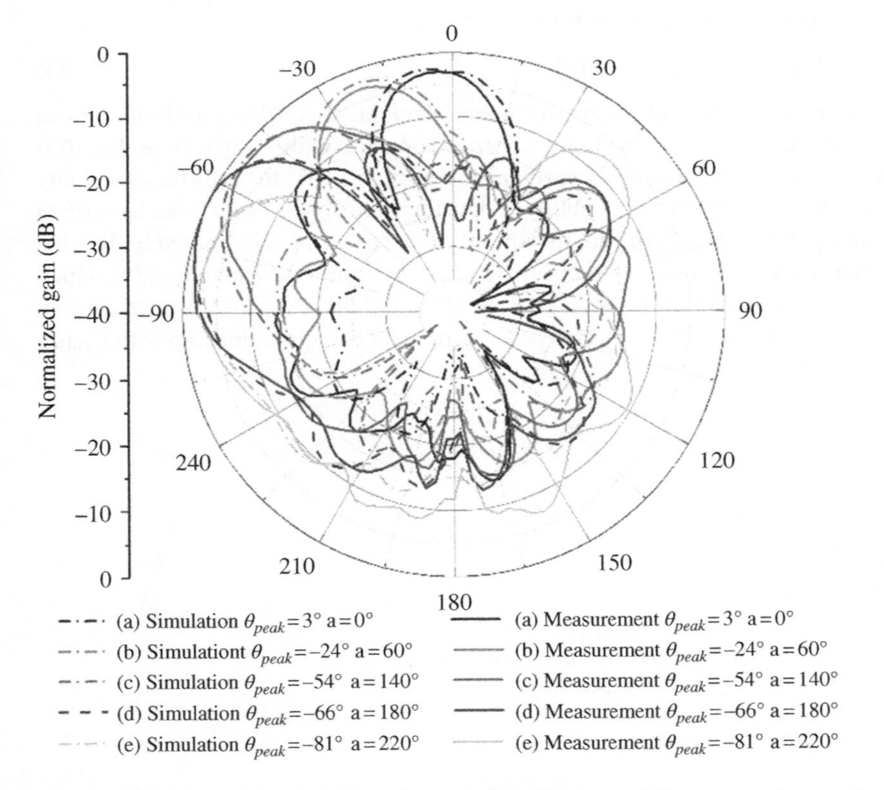

Figure 4.12 Measured and simulated scanning performance of the proposed array at 3.0 GHz.

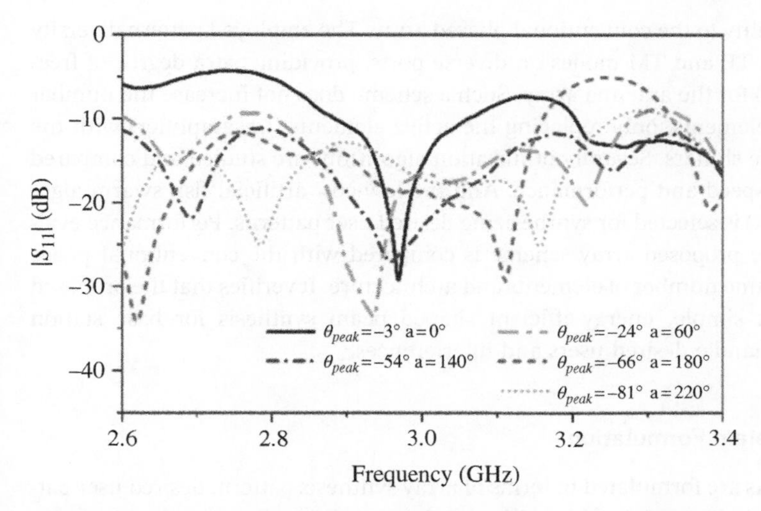

Figure 4.13 Measured reflection coefficient for all steering conditions.

around 3.5 dB due to the loss from dielectric material, power divider, and phase shifters.

In Figure 4.12, good agreement is observed between the measured and simulated normalized patterns. The measured gain fluctuation over 0° to −81° is 1.25 dB, whereas the simulated gain fluctuation is 1.90 dB. Both the measured and simulated results show their maximum gain on −66°. The sidelobe level is −10.7 dB in simulation and −11.7 dB in measurement. The sidelobe level reaches its worst at its maximum scan angle on 81°, which has −8.8 dB in simulation and − 6.8 dB in measurement. The sidelobe level could be optimized by enabling the amplitude control, which is not discussed for the passive system. The front back ratio (FBR) for all steering angles is <−12.0 dB in the simulation. The higher measured FBR in some specific angles (i.e. −3° and −81°) is observed mainly due to the measurement setup. These results demonstrate the potential of using phase-controlled pattern diverse antenna to achieve gain variation <1.25 dB over ±105°. This solves the limited scan range problem of the passive array in the base station antenna.

4.4 DRA Array for Shaped Beam Synthesis

The pattern diversity DRA array proposed above shows the scanning ability with ±105°. In this section, the same DRA array model with four elements is explored with its shaped beam synthesis ability. The pattern diversity DRA can be utilized for shaped beam synthesis of a passive phase-only antenna array, and compare its synthesis ability to the conventional phased array. The employed pattern diversity DRA excites TE and TM modes on diverse ports, providing extra degree of freedoms (DoFs) for the antenna array. Such a scheme does not increase the number of antenna elements, only replacing the active elements (i.e. amplifier) with the passive phase shifters. Several optimization algorithms are studied and compared in terms of speed and performance. And the low-cost artificial fish swarm algorithm (AFSA) is selected for synthesizing desired user patterns. Performance evaluation of the proposed array scheme is compared with the conventional phase array with same number of elements and architecture. It verifies that the proposed scheme is a simple, energy-efficient shaped beam synthesis for base station antenna to handle desired users and interferences.

4.4.1 Problem Formulation

The problems are formulated in terms of array synthesis pattern, desired user pattern, and optimization algorithm. The synthesis abilities of the proposed passive array in six patterns with different positions of peaks and nulls are evaluated and compared with that of the conventional phase array.

Figure 4.14 shows two configurations of four-element linear arrays with 0.5λ spacing. System 1 is a traditional antenna array with four phase shifters and four amplifiers to control the phase and amplitude of the four array elements. In comparison, the antenna array in System 2 employs four pattern diverse DRA as element, which connect eight phase shifters for phase-only control. In fact, the proposed pattern diverse DRA integrates electric and magnetic dipoles on a single antenna element, thus having two ports for each element. System 2 is proposed to avoid the use of amplifier since the amplifier is more expensive than the phase

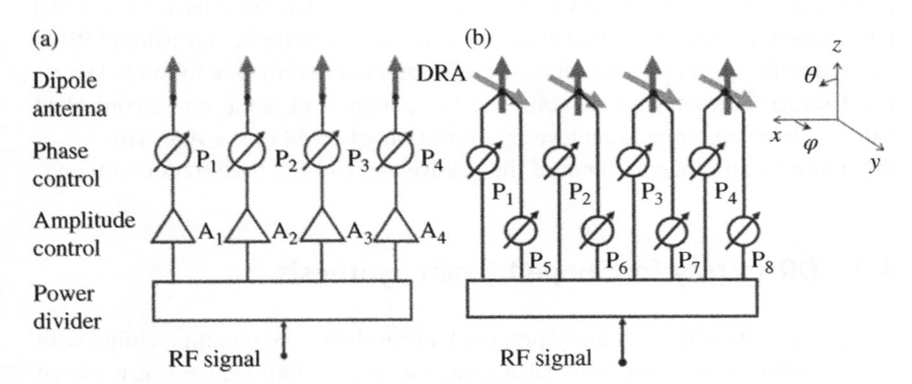

Figure 4.14 Configurations of four-element linear array. (a) Conventional phased array. (b) Proposed passive antenna array.

shifter, and the amplitude control will seriously affect the equivalent isotropically radiated power (EIRP) of the antenna system.

For a conventional phased array with $N = 4$ elements located at positions ($r_n = x_n$, y_n, z_n) like System 1, the array factor is expressed as,

$$AF(\theta, \varphi) = \sum_{n=1}^{N} w_n e^{jkr_n \vec{a}(\theta,\varphi)} \tag{4.29}$$

where $\vec{a}(\theta, \varphi) = \sin\theta\cos\varphi\vec{x} + \sin\theta\sin\varphi\vec{y} + \cos\theta\vec{z}$. Its DoFs are provided by its excitation $w_n = A_n e^{jP_n}$, which consists of amplifiers (A_{1-4}) and phase shifters (P_{1-4}). The far-filed radiation is calculated as,

$$\vec{E}_{Total}(\theta, \varphi) = E(\theta, \varphi) \times AF(\theta, \varphi) = E(\theta, \varphi) \sum_{n=1}^{N} A_n e^{j[knd\cos\theta + P_n]} \tag{4.30}$$

where $k = 2\pi/\lambda$ and $d = 0.5\lambda$. $E(\theta, \varphi)$ is the far-field radiation pattern of the antenna element, which is unified as 1 for the z-direction dipole antenna.

In Figure 4.14b, the proposed passive antenna array utilized dual-port pattern diverse DRA as antenna element, where dual-modes of TE and TM families are excited. The analytical model of the DRA has been presented in Section 4.1. These modes can be represented as electric dipoles in the z-direction and magnetic dipoles in the y-direction. Therefore, diverse radiation patterns of omnidirectional pattern and directional pattern are generated by alternative ports. Due to the orthogonal excited modes, high isolation is observed between two ports without any additional decoupling structures.

The DoFs of the proposed passive array are provided by eight phase shifters, representing as P_{1-8}. By decomposing $E(\theta, \varphi)$ into vertical and horizontal polarization components, far-field radiation pattern is represented as,

$$\vec{E}_{Total}(\theta, \varphi) = \left[\vec{\theta} E_\theta(\theta, \varphi) + \vec{\varphi} E_\varphi(\theta, \varphi)\right] \sum_{n=1}^{N} A_n e^{jkr_n \vec{a}(\theta,\varphi)} \tag{4.31}$$

where

$$\vec{\theta} = \cos\theta\cos\varphi\vec{x} + \cos\theta\sin\varphi\vec{y} - \sin\theta\vec{z} \tag{4.32}$$

$$\vec{\varphi} = -\sin\varphi\vec{x} + \cos\varphi\vec{y} \tag{4.33}$$

$E_\theta(\theta, \varphi)$ and $E_\varphi(\theta, \varphi)$ are the vertical and horizontal polarization components of the far-field pattern.

The realization of beam synthesis needs to establish a simplified desired pattern first. Figure 4.15 presents an example of desired user pattern with ±90° coverage, which is synthesized by the proposed passive antenna array. The desired user

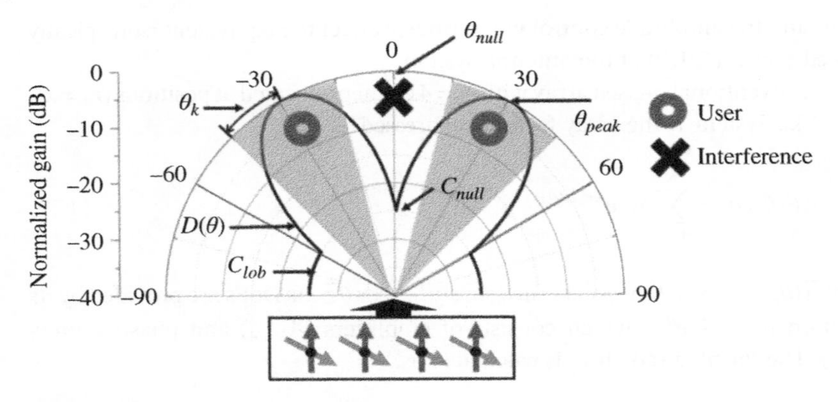

Figure 4.15 Target user orientation.

pattern is divided into 12 portions at an interval of 15°. In the figure, O and X are denoted as the users at θ_{peak} and direction of interference at θ_{null}, respectively. The beam in black represents the desired pattern, and the blue shadow represents the range that needs to be covered.

In this schematic, the synthetic beam covers the user on θ_{peak}, forming a shadow area with the beamwidth of $2\theta_k$. The main beam in shadowed portion can be formulated as a rectangular shape. However, the synthesis of a rectangular pattern is impossible for a small-scaled antenna array. Therefore, the main beam is formulated as a sin(θ) function in Figure 4.15 to be better approached by the antenna synthesis pattern with defined beam width. Then, the desired pattern is defined as,

$$D(\theta) = C_{null}\sin(A\theta + B) - C_{null} \tag{4.34}$$

where $\theta \in [\theta_{peak} - 2\theta_k, \theta_{peak} + 2\theta_k]$

A is the beamwidth, B is the location of beam peak. They are defined as,

$$A = \frac{\pi}{2\theta_k}, B = \frac{\pi}{2}\left(1 - \frac{\theta_{peak}}{2\theta_k}\right) \tag{4.35}$$

The portions outside the shadows are desired to be as low as possible. The value on the null depth is defined as C_{null}, and the value of sidelobe is defined as C_{lob}. $F(\theta, x)$ is the gap between the synthesis pattern and the desired user pattern and is defined as,

$$F(\theta,x) = \begin{cases} q_1 \mid E(\theta,x) - D(\theta) \mid \theta \in \left[\theta_{peak} - 2\theta_k, \theta_{peak} + 2\theta_k\right] \\ q_2[E(\theta,x) - C_{null}] \quad \theta = \theta_{null} \\ q_3[E(\theta,x) - C_{lob}] \text{ else} \end{cases} \tag{4.36}$$

where $E(\theta, x)$ is the synthesis pattern calculated from phase x and $D(\theta)$ is the desired user beam defined in (4.32). The weights are defined as q_1, q_2, and q_3, which are specified according to different scenarios.

Shaped beam synthesis can be regarded as a real-time process of minimizing the distance between the array synthesized pattern and the desired pattern. The array DoFs are optimized through a step-by-step process, until the synthesized pattern reaches the desired user beam pattern. The gap between the synthetic pattern and the desired pattern is represented as Mean Square Error (MSE).

For instant, the MSE for the proposed passive array is formulated as,

$$f(x) = \frac{1}{COV} \sum_{\theta=0}^{COV} | F(\theta, x) | \tag{4.37}$$

where $COV = 180$ for $\pm 90°$ coverage. Finally, an optimization model can be built in relation to an objective,

$$min.f(x) \tag{4.38}$$
$$st\, 0 < x_n < 2\pi(n = 1, 2, \cdots 8)$$

Here, $f(x)$ is minimized to reduce the gap between the synthesized pattern and the desired pattern until reach the thresholds, then output the value of x.

4.4.2 Algorithm Selection

It is a non-convex, multi-dimensional, multi-objective problem to minimize the gap between the synthesized pattern and the desired pattern $F_0(\theta, x)$. The function is optimized by changing the amplitude and phase of each array element. The objective function is established as,

$$F(\theta, x) = \sum_{i=0}^{M} \left[|\overline{S}(\theta_i)| - |F_0(\theta_i)| \right]^2 \tag{4.39}$$

By assuming $x = (\alpha_0, \alpha_1, \cdots, \alpha_{N-1})$, the optimization methods is presented as,

$$F(\theta, x^*) = min\ F\left(\alpha_0^*, \alpha_1^*, \cdots, \alpha_{N-1}^*\right) \tag{4.40}$$

A set of solutions x^* is used to make objective function $F(\theta, x^*) \rightarrow 0$. In multi-objective beam synthesis, there are more than one objective functions. Therefore, the following formula is used to construct the fitness function.

$$Fitness = w_1 \times fitness_1 + w_2 \times fitness_2 + \cdots + w_i \times fitness_i \tag{4.41}$$

Let $fitness_i$ be the sub-objective function. The amplitude and phase of each array element in the array are optimized by using the optimization algorithm. The objective function sets the weight w_i of each sub-objective function to better approach to the optimization target on the pattern shaping, achieving the multi-objective beam synthesis. For example, if the target is a cosecant square pattern, the fitness function is,

$$fitness = \sum_{n=1}^{n=360} \sqrt{(Gain_{opt_i} - Gain_{obj_i})^2} \quad (4.42)$$

In the process of sidelobe minimization, the fitness function is defined as,

$$fitness = \frac{1}{|SLL_{max}|} \quad (4.43)$$

There are two types of optimization algorithms. One type is the heuristic algorithm, which is based on the rules of the problem itself to achieve feasible solution. This kind of algorithm is faster but usually falls into local optimization because it is a problem-dependent process. Another is the meta-heuristic algorithm. This type of algorithm is inspired by biological evolution, physics, chemistry, and other processes to obtain a search strategy for the solution space. It is a problem-independent process and achieves global optimization. Table 4.2 compares the features of two types of algorithms.

Table 4.3 shows the advantages and disadvantages of three specific algorithms. Quasi-newton method, Genetic Algorithm (GA), and Artificial Fish Swarm Algorithm (AFSA).

AFSA is a new bionic swarm intelligence optimization algorithm proposed in 2002 [39]. The algorithm starts with the analysis of fish activities, which adopts a bottom-up method. The bottom-up method treats each data point as a single

Table 4.2 Comparison of heuristic algorithm with metal-heuristic algorithm.

Algorithms	Characteristic	Features	Common algorithms
Heuristic algorithm	Local optimization	1) Problem specific 2) Narrow application 3) Simple algorithm	Local search algorithm, newton algorithm, quasi-newton method [34]
Meta-heuristic algorithm	Global optimization	1) Problem independent 2) Wide application 3) High-level aggregation algorithm	Genetic algorithm [35, 36], particle swarm optimization algorithm [37], invasive weed algorithm [38], artificial fish swarm algorithm [39]

Table 4.3 Comparison of the features of three algorithms.

Algorithms	Advantages	Disadvantages
Quasi-newton algorithm	1) Fast convergence 2) Use first derivative	1) Local algorithm 2) Computational complexity
Genetic algorithm	1) Global search 2) The selection of individuals is random 3) Strong scalability	1) Programming is more complex 2) Determined by experience 3) Initial population dependence is strong
Artificial fish swarm algorithm	1) Strong versatility 2) Fast convergence 3) Global optimal value	1) Encoding of network weights 2) Local search imbalance 3) Specific problems do not guarantee solution quality

cluster at the beginning and then gradually merges (or coalesces) clusters until all clusters are merged into a single cluster. The modules in traditional top-down methods are not well connected, thus causing redundancies. In contrast, the bottom-up methods provide information hiding and reusability, which is more efficient. As a bottom-up method, AFSA provides a balance between the speed and accuracy of the optimization, which is suitable for a small element number array configuration.

In the AFSA, it is assumed that there are N artificial fishes, $X = (X_1, X_2, \ldots, X_n)$ is the artificial fishes, $Y_i = F(X_i)$ is a fitness function, $F(X)$ is the objective function, $d_{i,j} = \|X_i - X_j\|$ is the distance between artificial fishes. Step is the moving step, and δ is the crowding factor. The algorithm includes the following behaviors: preying, swarming, and following. All fishes use these three behaviors to find their food.

1) Preying

When an artificial fish is looking for food, it perceives the food in the environment to determine the movement and then determine its steps,

$$X_{i/next} = \begin{cases} X_i + Rand \cdot Step \cdot \dfrac{X_j - X_i}{\|X_c - X_i\|} Y_i < Y_j \\ X_i + Rand \cdot Step \, Y_i > Y_j \end{cases} \tag{4.44}$$

Let X_i be the artificial fish i current state and select a state X_j randomly in its visual distance. Y is its food, and Rand is a random number between 0 and 1. If $Y_i < Y_j$, the fish moves one step in the direction of X_j to the next position $X_{i/next}$. Otherwise, select a state X_j randomly again and judge whether it satisfies the forward condition. If it cannot satisfy after several tries, it moves randomly in the visible distance.

2) Swarming

The artificial fish assemble in groups naturally in the moving process and avoid too much crowding. Its mathematical expression is,

$$X_{i/next} = \begin{cases} X_i + Rand \cdot Step \cdot \dfrac{X_c - X_i}{\|X_c - X_i\|} \dfrac{Y_c}{n_f} > \delta Y_j \\ Foraging\ behavior\ \dfrac{Y_c}{n_f} < \delta Y_j \end{cases} \tag{4.45}$$

Assumming n_f is the number of partners in the current neighborhood ($d_{i,j}$ < Visual) of the artificial fish and X_c is the state of the center position. If $Y_c/n_f < \delta Y_j$, it means that the companion center has more food (higher fitness function value) and is not very crowded. The fishes are then steps to the companion center. Otherwise, it executes the preying behavior.

3) Following

The following behavior is a directional behavior. When a fish or several fishes find food, the neighborhood partners will trail and reach the food quickly. For each fish, it was necessary to move to a place with a higher concentration of food while avoiding overcrowding. Its mathematical expression is,

$$X_{i/next} = \begin{cases} X_i + Rand \cdot Step \cdot \dfrac{X_{max} - X_i}{\|X_{max} - X_i\|} \dfrac{Y_{max}}{n_f} > \delta Y_j \\ Foraging\ behavior\ \dfrac{Y_{max}}{n_f} < \delta Y_j \end{cases} \tag{4.46}$$

Assume X_{max} is the maximum state of food concentration explored in the neighborhood ($d_{i,j}$ < Visual). If $Y_{max}/n_f < \delta Y_j$, it means that the companion X_{max} state has higher food concentration (higher fitness function value) and the surroundings are not very crowded, it steps to the companion X_{max}. Otherwise, it executes the preying behavior.

The AFSA should provide a bulletin board recording the optimal position and fitness of the artificial fish during each iteration. Each fish updates and compares its status before making a new iterative action. If the current fish is in better position, the value on the bulletin will be replaced. Finally, after a certain iteration, the optimal state and fitness on the output announcement are used as the final solution of the optimization task.

Figure 4.16 shows the flowchart of the AFSA. In the process of preying, fish will constantly switch between these behaviors according to the perception of environmental information. AFSA has a strong ability to avoid local minima by analyzing the random behaviors of fish such as preying, swarming, and following. These behaviors speed up the convergence of the fish school during optimization. For

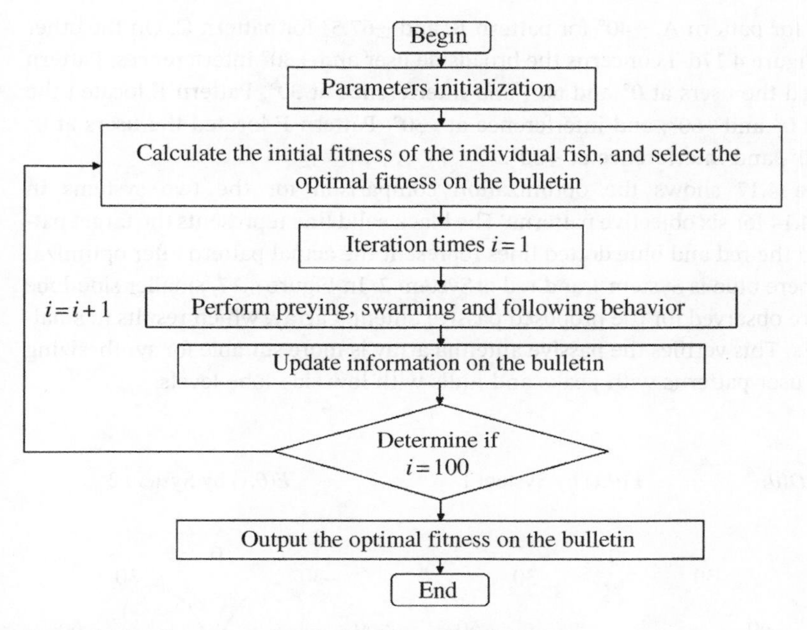

Figure 4.16 Flowchart of artificial fish swarm algorithm.

these reasons, AFSA is more suitable for solving high-dimensional optimization problems such as shaped beam synthesis.

In AFSA example in Figure 4.16, the program first initializes random values to generate 100 artificial fish. Then, fishes are coming towards the food via three steps of behavior: preying, swarming, and following. Preying behavior enforces artificial fish to reach the extreme value quickly, and it jumps out of the local extremum by providing artificial fish with the opportunity to move randomly. The swarming behavior drives a small number of artificial fish to trap in the local extremum and to tend toward the global extremum, thus escaping from the local extremum. This behavior speeds up the artificial fish to approach to the global extremum. After 100 iterations, all fish stop moving and output the optimal fitness on the bulletin.

4.4.3 Array Pattern Verifications

The AFSA is applied to verify the effectiveness of the proposed pattern reconfigurable DRA array. A set of experiments is performed for patterns with different positions of users and interferences. Figure 4.17a–f concerns the patterns with broadside interference and symmetrical desired users, where the users are located at ±30° for pattern A, ±40° for pattern B, and ±67.5° for pattern C. On the other hand, Figure 4.17d–f concerns the broadside user and ±30° interferences. Pattern D located the users at 0° and 60°, and interference at 30°. Pattern E located the users at 0° and –60°, and interference at –30°. Pattern F located the users at 0° and ±60°, and interference at ±30°.

Figure 4.17 shows the optimization comparison for the two systems in Figure 4.14 for six objective patterns. The black solid line represents the target pattern and the red and blue dotted lines represent the actual pattern after optimization, where blue is System 1 and red is System 2. In Figure 4.17, smaller side-lobe levels are observed for the proposed passive antenna array, which results in smaller MSEs. This verifies the passive antenna array is more suitable for synthesizing desired user patterns with peaks and nulls with low side-lobe levels.

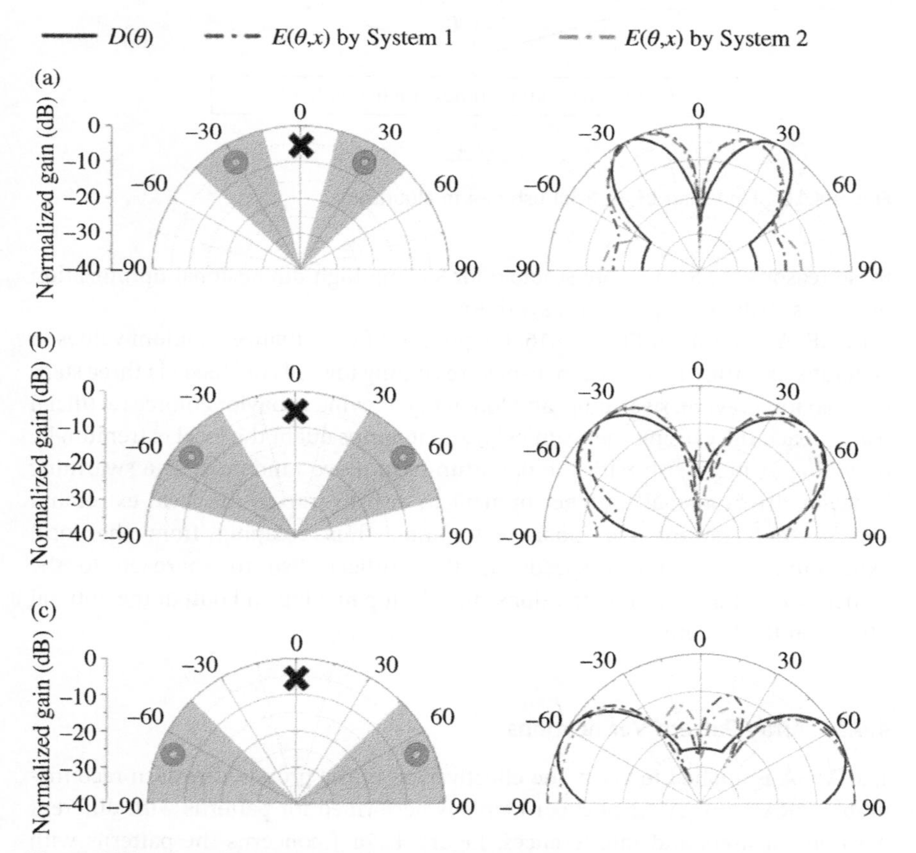

Figure 4.17 Shaped beam synthesis on patterns. (a) Pattern A. (b) Pattern B. (c) Pattern C. (d) Pattern D. (e) Pattern E. (f) Pattern F.

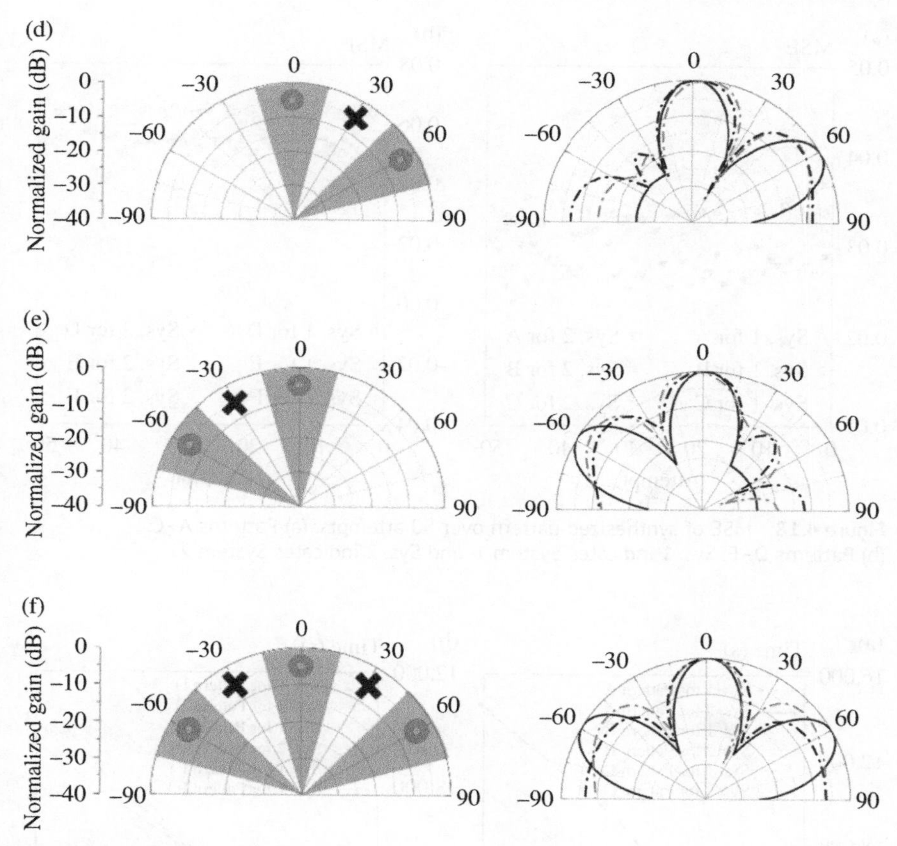

Figure 4.17 (Continued)

Figure 4.18 shows the MSE of the synthesized pattern for over 50 attempts. It was found that different MSE values are obtained for different patterns and different attempts. On average, the proposed passive antenna array has smaller MSE than that of the conventional phased array antenna. It indicates pattern synthesized by the proposed passive antenna array has better approach to the desired user pattern than that of the conventional phased array.

Figure 4.19 shows the time consumption of these two systems. It was found that System 2 requires less time and less memory for beam synthesis. System 2 has advantages in beam optimization and user change.

The reflection coefficients for all patterns of the antenna array are shown in Figure 4.20. This verifies its feasibility of implementing passive-shaped beam synthesis in six patterns. Figure 4.21 shows the 3D far-field pattern, measured, and

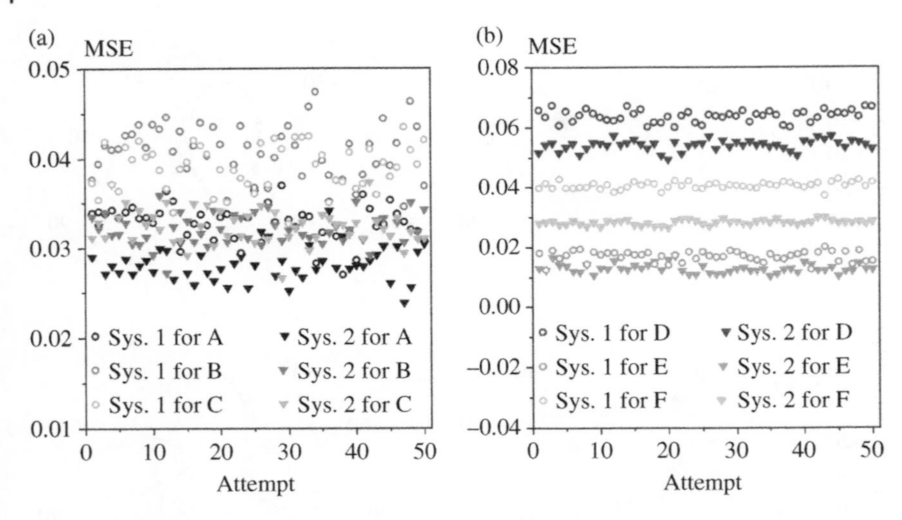

Figure 4.18 MSE of synthesized pattern over 50 attempts. (a) Patterns A–C. (b) Patterns D–F. Sys. 1 indicates System 1, and Sys. 2 indicates System 2.

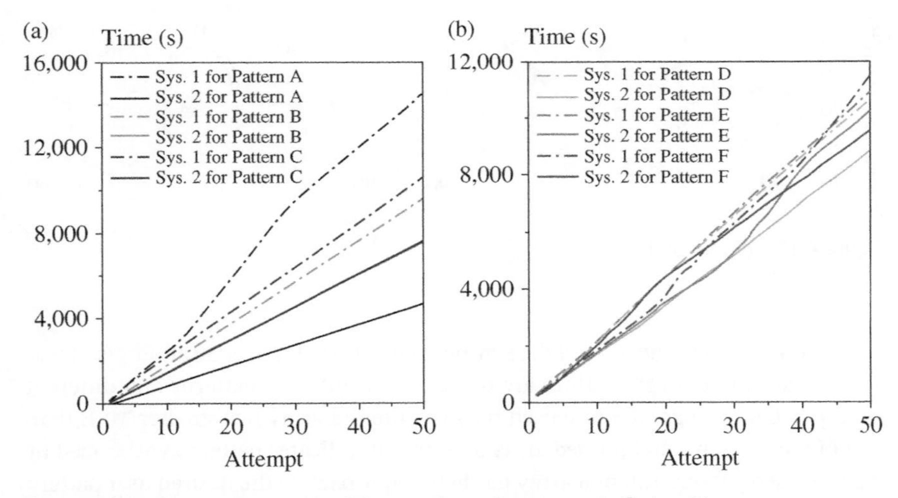

Figure 4.19 Time consumed by 50 operations of patterns A–F algorithm. (a) Patterns A–C. (b) Patterns D–F. Sys. 1 indicates System 1, and Sys. 2 indicates System 2.

simulated *E*-plane pattern of the proposed passive pattern diversity DRA array. Good agreements are observed between the simulated and measured results. The users can be successfully covered by the synthesized pattern in all cases, with less than 3 dB difference from the desired beam.

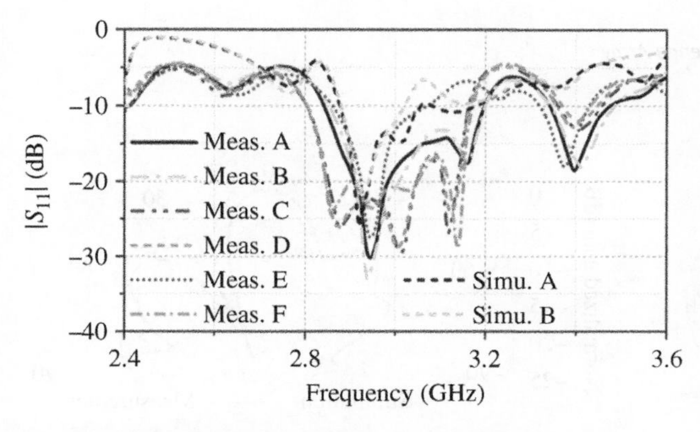

Figure 4.20 Reflection coefficient for all patterns.

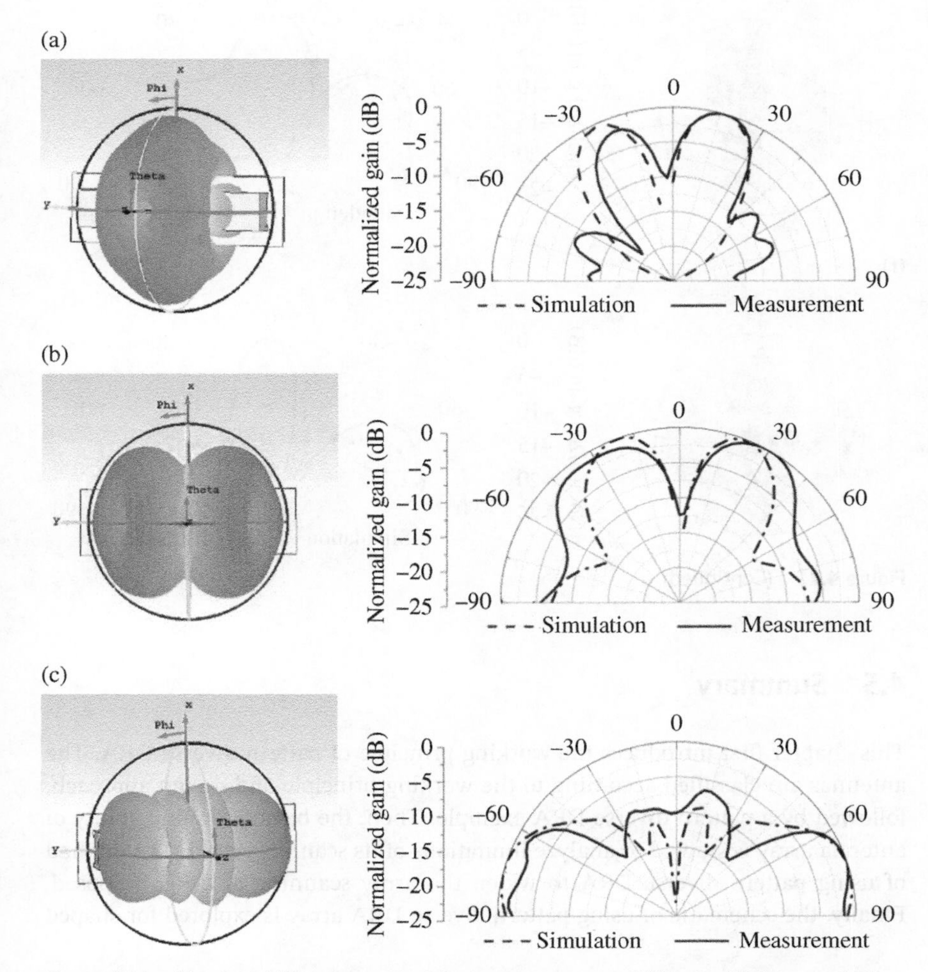

Figure 4.21 Simulated 3D patterns and *E*-plane patterns at 3 GHz. (a–f) Patterns A–F.

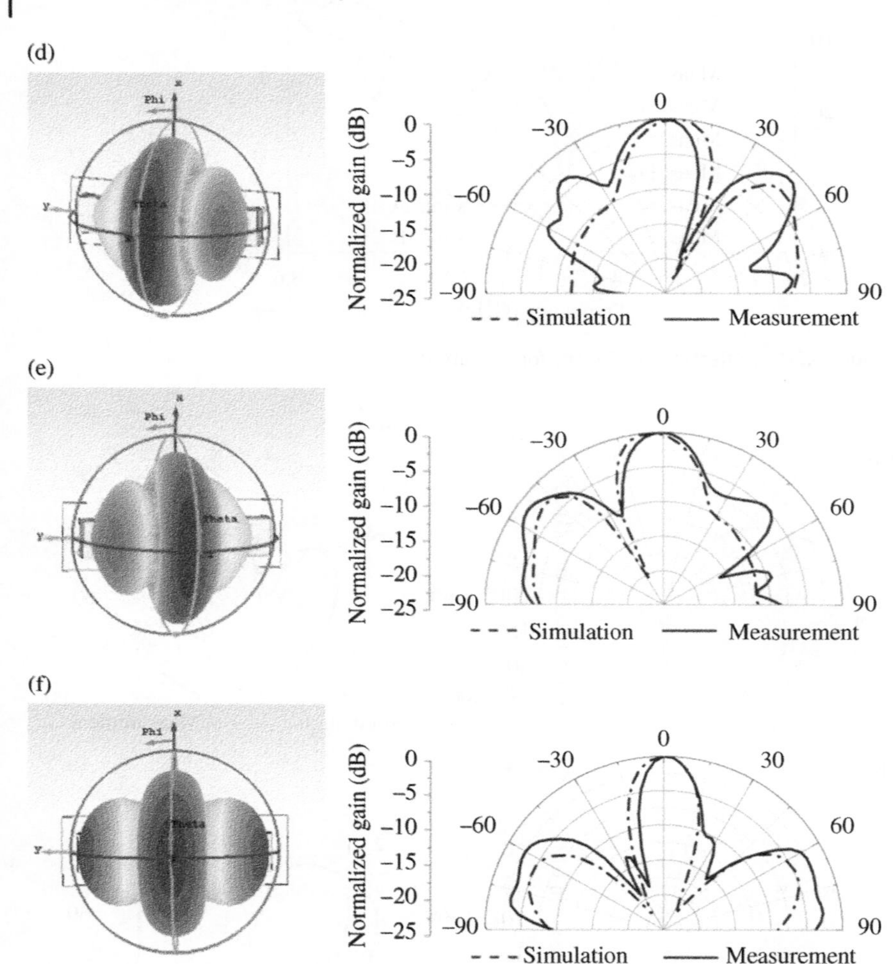

Figure 4.21 (Continued)

4.5 Summary

This chapter first introduces the working principle of pattern diverse DRA. The antennas are classified according to the working principle and design approach, followed by a pattern diverse DRA example. Then, the beam scanning ability of antenna array is studied to analyze limitations of its scanning angle. The method of using pattern diverse DRA to widen the array scanning angle is proposed. Finally, the schematic of using pattern diverse DRA array is explored for shaped beam synthesis. The comparation with conventional antenna array shows its superior synthesis performance with smaller MSE. Optimization algorithms are compared, and AFSA is selected for both schematics. The results validate the feasibility of using pattern diversity DRA as an energy-efficient means for the base station antenna array to handle users and interferences.

References

1 Petosa, A. (2007). *Dielectric Resonator Antenna Handbook*. Norwood, MA: Artech House.

2 Luk, K.M. and Leung, K.W. (2003). *Dielectric Resonator Antennas*. Baldock: Research Studies.

3 Chen, Z., Shoaib, I., Yao, Y. et al. (2016). Pattern-reconfigurable dual-polarized dielectric resonator antenna. *IEEE Antennas and Wireless Propagation Letters* 15: 1273–1276.

4 Yang, N., Leung, K.W., and Wu, N. (2019). Pattern-diversity cylindrical dielectric resonator antenna using fundamental modes of different mode families. *IEEE Transactions on Antennas and Propagation* 67 (11): 6778–6788.

5 Guha, D., Banerjee, A., Kumar, C., and Antar, Y.M.M. (2012). Higher order mode excitation for high-gain broadside radiation from cylindrical dielectric resonator antennas. *IEEE Transactions on Antennas and Propagation* 60 (1): 71–77.

6 Gu, L., Yang, W., Xue, Q., and Che, W. (2021). A dual-band steerable dual-beam metasurface antenna based on common feeding network. *IEEE Transactions on Antennas and Propagation* 69 (10): 6340–6350.

7 Wu, X. and Sengupta, K. (2016). On-chip THz spectroscope exploiting electromagnetic scattering with multi-port antenna. *IEEE Journal of Solid-State Circuits* 51 (12): 3049–3062.

8 Kim, D., Kim, J., and Nam, S. (2019). Beam steering of a multi-port chassis antenna using the least squares method and theory of characteristic modes. *IEEE Transactions on Antennas and Propagation* 67 (8): 5684–5688.

9 Wong, K., Tso, Z., and Li, W. (2022). Very-wide-band six-port single-patch antenna with six uncorrelated waves for MIMO access points. *IEEE Access* 10: 69555–69567.

10 Abdalrazik, A., El-Hameed, A.S.A., and Abdel-Rahman, A.B. (2017). A three-port MIMO dielectric resonator antenna using decoupled modes. *IEEE Antennas and Wireless Propagation Letters* 16: 3104–3107.

11 Fang, X.S., Leung, K.W., and Luk, K.M. (2014). Theory and experiment of three-port polarization-diversity cylindrical dielectric resonator antenna. *IEEE Transactions on Antennas and Propagation* 62 (10): 4945–4951.

12 Zhang, B., Ren, J., Sun, Y.-X. et al. (2022). Four-port cylindrical pattern- and polarization-diversity dielectric resonator antenna for MIMO application. *IEEE Transactions on Antennas and Propagation* 70 (8): 7136–7141.

13 Liu, X., Leung, K.W., Zhang, T. et al. (2021). An electrically controlled pattern- and polarization-reconfigurable cylindrical dielectric resonator antenna. *IEEE Antennas and Wireless Propagation Letters* 20 (12): 2309–2313.

14 Ahn, B.K., Jo, H., Yoo, J. et al. (2019). Pattern reconfigurable high gain spherical dielectric resonator antenna operating on higher-order mode. *IEEE Antennas and Wireless Propagation Letters* 18 (1): 128–132.

15 Liu, B., Qiu, J., Wang, C., and Li, G. (2017). Pattern-reconfigurable cylindrical dielectric resonator antenna based on parasitic elements. *IEEE Access* 5: 25584–25590.

16 Pan, Y.M., Leung, K.W., and Guo, L. (2017). Compact laterally radiating dielectric resonator antenna with small ground plane. *IEEE Transactions on Antennas and Propagation* 65 (8): 4305–4310.

17 Bai, X., Su, M., Liu, Y., and Wu, Y. (2018). Wideband pattern-reconfigurable cone antenna employing liquid-metal reflectors. *IEEE Antennas and Wireless Propagation Letters* 17 (5): 916–919.

18 Ren, Z., Qi, S., Wu, W., and Shen, Z. (2021). Pattern-reconfigurable water horn antenna. *IEEE Transactions on Antennas and Propagation* 69 (8): 5084–5089.

19 Chen, Z. and Wong, H. (2017). Wideband glass and liquid cylindrical dielectric resonator antenna for pattern reconfigurable design. *IEEE Transactions on Antennas and Propagation* 65 (5): 2157–2164.

20 Ren, J., Zhou, Z., Wei, Z.H. et al. (2020). Radiation pattern and polarization reconfigurable antenna using dielectric liquid. *IEEE Transactions on Antennas and Propagation* 68 (12): 8174–8179.

21 Song, C., Bennett, E.L., Xiao, J. et al. (2020). Passive beam-steering gravitational liquid antennas. *IEEE Transactions on Antennas and Propagation* 68 (4): 3207–3212.

22 Cheng, Y., Ding, X., Shao, W. et al. (2017). A novel wide-angle scanning phased array based on dual-mode pattern-reconfigurable elements. *IEEE Antennas and Wireless Propagation Letters* 16: 396–399.

23 Ding, X., Cheng, Y., Shao, W. et al. (2017). A wide-angle scanning planar phased array with pattern reconfigurable magnetic current element. *IEEE Transactions on Antennas and Propagation* 65 (3): 1434–1439.

24 Cheng, Y., Ding, X., Shao, W., and Wang, B. (2018). Dual-band wide-angle scanning phased array composed of SIW-cavity backed elements. *IEEE Transactions on Antennas and Propagation* 66 (5): 2678–2683.

25 Wang, Z., Dong, Y., Peng, Z., and Hong, W. (2022). Hybrid metasurface, dielectric resonator, low-cost, wide-angle beam-scanning antenna for 5G base station application. *IEEE Transactions on Antennas and Propagation* 70 (9): 7646–7658.

26 Balanis, C.A. (2005). *Antenna Theory Analysis and Design*, 3e. Hoboken, NJ: Wiley.

27 Omar, A.A., Choi, J., Kim, J. et al. (2022). A planar, polarization-switchable endfire and ±broadside millimeter-wave antenna array without lumped components. *IEEE Transactions on Antennas and Propagation* 70 (5): 3864–3869.

28 Tang, M.-C., Chen, X., Shi, T. et al. (2021). A compact, low-profile, broadside radiating two-element Huygens dipole array facilitated by a custom-designed decoupling element. *IEEE Transactions on Antennas and Propagation* 69 (8): 4546–4557.

29 Li, Y. and Luk, K.-M. (2016). A multibeam end-fire magnetoelectric dipole antenna array for millimeter-wave applications. *IEEE Transactions on Antennas and Propagation* 64 (7): 2894–2904.

30 Cetinoneri, B., Atesal, Y.A., and Rebeiz, G.M. (2011). An 8-8 Butler matrix in 0.13-CMOS for 5–6-GHz multibeam applications. *IEEE Transactions on Microwave Theory and Techniques* 59 (2): 295–301.

31 Li, Y., Xue, Q., Yung, E.K.-N., and Long, Y. (2007). A fixed-frequency beam-scanning microstrip leaky wave antenna array. *IEEE Antennas and Wireless Propagation Letters* 6: 616–618.

32 Yang, G., Li, J., Zhou, S.G., and Qi, Y. (2017). A wide-angle E-plane scanning linear array antenna with wide beam elements. *IEEE Antennas and Wireless Propagation Letters* 16: 2923–2926.

33 Wang, Z., Zhao, S., and Dong, Y. (2022). Miniaturized, vertically polarized, pattern reconfigurable dielectric resonator antenna and its phased array for wide-angle beam steering. *IEEE Transactions on Antennas and Propagation* 70 (10): 9233–9246.

34 Balanis, C.A. (2005). *Antenna Theory*. Wiley.

35 Holland, J.H. (1992). *Adaptation in Natural and Artificial Systems: An Introductory Analysis with Applications to Biology, Control, and Artificial Intelligence*. MIT Press.

36 Deb, K., Pratap, A., Agarwal, S., and Meyarivan, T. (2002). A fast and elitist multiobjective genetic algorithm: NSGA-II. *IEEE Transactions on Evolutionary Computation* 6 (2): 182–197.

37 Kennedy, J. and Eberhart, R. (1995). Particle swarm optimization. *IEEE International Conference on Neural Networks Proceedings*, vol. 4, pp. 1942–1948.

38 Mehrabian, A.R. and Lucas, C. (2006). A novel numerical optimization algorithm inspired from weed colonization. *Ecological Informatics* 1 (4): 355–366.

39 Ge, H., Sun, L., Chen, X., and Liang, Y. (2016). An efficient artificial fish swarm model with estimation of distribution for flexible job shop scheduling. *International Journal of Computational Intelligence Systems* 9 (5): 917–931.

5

MIMO DRA with Improved Isolation

CHAPTER MENU

5.1 Overview

With the rapid development of global mobile market such as intelligent terminals, Internet of Things (IoT), and the Internet of Vehicles (IoV), the demands for mobile data transmission are increasing [1]. 5th-Generation (5G) has become a hot research topic to provide users with higher speed and high-quality communication services. In 5G, multiple-input multiple-output (MIMO) technology is the main research trend due to the advantages of high system capacity and spectrum utilization. The MIMO antenna can also achieve high antenna gain and flexible beam shaping by integrating 3D information.

With requirements of improved isolation, efficiency, directivity, and gain performances [2]. Diversity technology such as space diversity, pattern diversity, and polarization diversity [3] are widely used in MIMO system. Multi-element antennas in MIMO systems are also required to be more compact to be integrated into smart devices. However, strong mutual coupling generated between the antenna

Dielectric Resonator Antennas: Materials, Designs and Applications, First Edition.
Zhijiao Chen, Jing-Ya Deng, and Haiwen Liu.

Published 2024 by John Wiley & Sons, Inc.

elements would deteriorate the MIMO system performance. Hence, isolation enhancement becomes one of the main challenges in MIMO antenna designs [4].

The MIMO antennas have been implemented by microstrip antennas [5], patch antennas [6], and dielectric resonator antennas (DRAs) [7]. DRAs have several benefits for MIMO antenna due to their small size, lightweight, high radiation efficiency, small conductive loss, and ease of excitation and fabrication. In addition, DRA simplifies the antenna array by adding metamaterials, metal plates, and metal holes with small structural change. DRA can be integrated with broadband neutral line to offset the original coupling current between antenna elements, thus reducing the mutual coupling. DRA can also be excited with multiple orthogonal modes such as transverse electric (TE)/transverse magnetic (TM) modes to generate decoupled antenna ports with high isolation.

The rest of the chapter is organized as follows. In Section 5.2, decoupling methods of MIMO antenna have been classified by coupling models. Then the MIMO DRA examples, along with the analysis on their decoupling operations, are introduced. Section 5.3 introduces a MIMO DRA example with detailed design process, operation, and verifications. The proposed antenna is a preferable candidate for the future 5G millimeter wave (mmW) applications. Finally, remarks are concluded in Section 5.4.

5.2 MIMO DRA Research Trends and Classifications

MIMO antennas are usually implemented by multi-port antenna and reconfigurable antenna. For multi-port antennas, different modes are excited at different ports, and the MIMO antenna can be realized by means of pattern or polarization diversity. For reconfigurable antennas, MIMO antenna can be realized by changing the antenna structure, such as pattern/polarization reconfigurable antenna, multi-sector selection antenna, and switch parasitic antenna.

In MIMO systems, multiple antenna elements with high isolation are essential to improve the rate and link reliability. MIMO antennas have been reported with different methods, and we introduce three research trends here. The first trend is to design compact antenna array by reducing element coupling for transmission efficiency improvement. The second trend is to co-design the field and circuit by reducing the coupling between the elements. The third trend achieves the MIMO antenna self-decoupling by analyzing the multi-mode/multi-port of antenna, which has the advantages of simple structure, high efficiency, and low loss.

Here, we divide the decoupling methods into three categories with direct decoupling, neutral structure decoupling method, and self-decoupling method. Their schematic diagrams are compared in Figure 5.1.

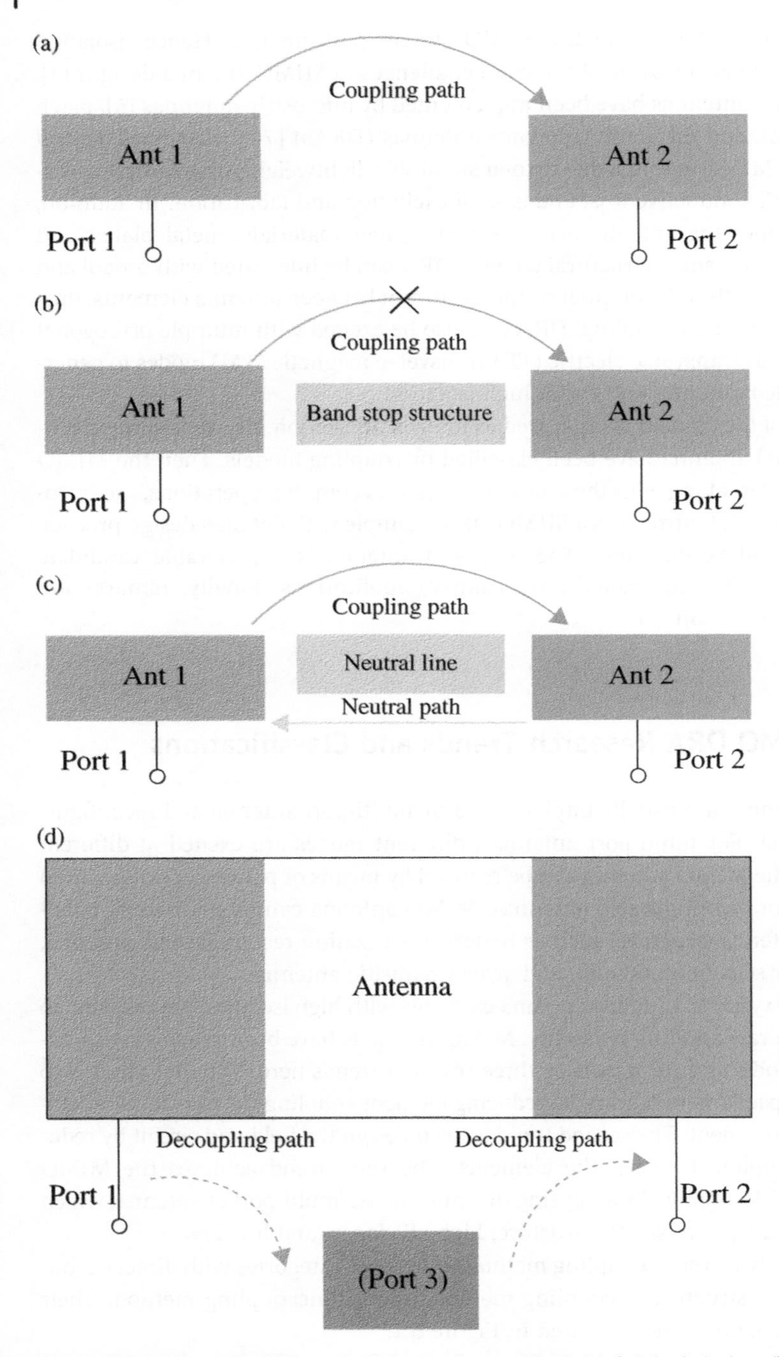

Figure 5.1 Schematic diagrams of decoupling method. (a) Original design. (b) Direct decoupling method. (c) Neutral line decoupling method. (d) Self-decoupling method.

The direct decoupling method reduces the current or electric fields between antenna elements by using metamaterials [8, 9], electromagnetic bandgap (EBG) structures [10–12], or defected ground planes [13, 14]. The neutral structure decoupling method introduces an extra coupling path to counteract the original coupling by using parasitic scattering [15], distributed decoupling networks [16], lumped decoupling networks [17, 18], as well as neutral lines [19, 20]. Self-decoupling method uses multiple modes of resonators instead of decoupling structure to achieve high-level isolation [21, 22]. DRA is a competitive candidate for MIMO antenna design for these three methods, which are detailed with examples in the following sections.

5.2.1 Compact Antenna Array Design with Direct Decoupling

Uniform planar antenna array in MIMO design usually requires the element spacing to be half wavelength. It is possible to realize less than half wavelength spacing with the utilization of metamaterials and novel manufacturing technology. The compact antenna array can include as many antenna elements as possible in a limited space, so as to achieve higher antenna design flexibility and higher information transmission efficiency.

Compared with the traditional antenna array, the compact antenna array has a significant improvement in array directivity, spatial degree of freedom, and near-field communication capacity [23]. At present, some scholars have created the mutual coupling model through electromagnetic full-wave simulation and measurement. A mathematical model for the mutual coupling has been established to accurately adjust the excitation current of antenna array for higher array gain than traditional array. It is expected that the coupling relationship between antenna elements can be precisely exploited to further improve the performance of MIMO communication systems.

Direct decoupling method reduces the element coupling by controlling 3D electromagnetic field of DRA with additional decoupling structure. It provides a simple way to prohibit the couplings from flowing between the elements at antenna's resonant frequency. For DRA structure, the isolation level can be further increased without increasing the array volume, i.e. changing the DRA structure by adding segmented slots or vias.

Figure 5.2 illustrates a MIMO DRA example using direct decoupling method. Vias are designed inside DRA element to change the electric field distributions and thus weaken the electric field generated by mutual coupling. In this way, the isolation between MIMO DRA elements can be greatly enhanced. At the same time, the decoupling vias have a slight effect on the far-field pattern of the excited antenna. The decoupling vias barely affect antenna performances, and the losses caused by vias are negligible. Since the vias are placed inside the DRA element, no

(a)

Proposed antenna H_p

DRA-1

DRA-2

Via

(b)

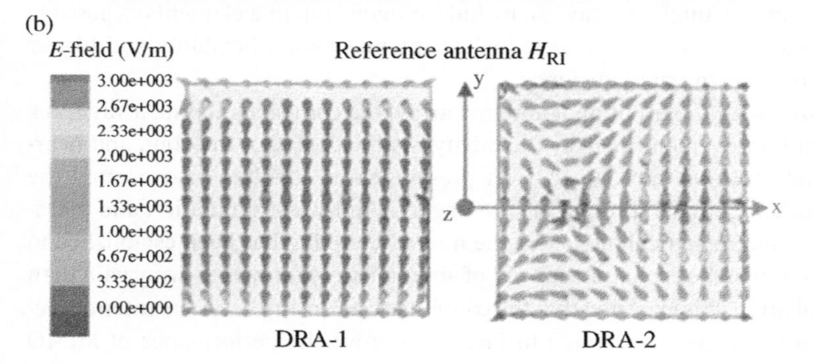

(c)

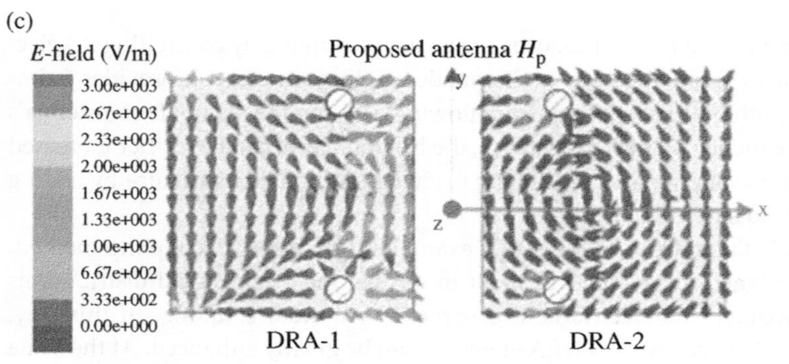

Figure 5.2 Example of direct decoupling. (a) Configurations of MIMO antennas. (b) *E*-field inside the DRAs (without vias). (c) *E*-field inside the DRAs (with via). *Source:* Pan et al. [7]/IEEE.

additional space is required, ensuring that the MIMO antenna system is simple and compact.

5.2.2 Co-design on DRA Field and Circuit

In large-scaled MIMO antenna design, strong coupling between complex feeding circuits and antenna elements would affect the MIMO system performance. In order to reduce the coupling between the antenna element, the co-design of antenna field domain and circuit domain can effectively reduce the coupling between antenna elements. The existing electromagnetic theory and engineer experience have difficulty in finding an optimum solution quickly for reducing coupling. Therefore, the high-precision full-wave electromagnetic simulation model (fine model) is frequently invoked to evaluate the antenna performance. However, the time consumed is long, especially for multi-target antenna optimization. The key idea of model-assisted antenna optimization is to use mathematical or physical models to simulate the electromagnetic response behavior of radio frequency (RF)/microwave devices in low cost. This technology transforms the high-cost antenna optimization problem into a low-cost optimization model with higher efficiency. The physics-based model is constructed from a coarse model that should have the same physical meaning as the fine model, so as to ensure high reliability. At present, the mainstream methods include space mapping [24], response prediction [25], and multi-fidelity local surrogate model [26]. A fine model is specified with the help of high frequency simulator structure (HFSS), whereas the coarse model is established by equivalent transmission line model. After a coarse model is built, the key parameters are optimized by using space mapping method. The established model should be highly consistent with the simulation results of the fine model of the array.

For antenna field and circuit co-design, the relationship between the electromagnetic field and the equivalent circuit is established based on Maxwell equations. The electromagnetic characteristics can be presented by a series of equivalent circuits to realize the overall analysis of the radiating element [23]. This co-design method can be established by equivalent circuit models, analytical models, partial element equivalent circuit (PEEC), etc. The equivalent circuit method is used to establish a transmission line model, using the impedance to represent the effective electromagnetism of the antenna. The impedance matrix network is built to simulate and optimize the matching of the antenna design. However, this method has limited application and only works for some certain structures. In contrast, analytical models provide a good insight into searching of initial values in the optimization processes with different structure analysis. Starting from the electromagnetic equations, PEEC algorithm equates each subregion of the antenna to a subcircuit composed of inductors, capacitors, and resistors, so as to realize the co-design of the field and circuit analysis. All these methods are suitable for the field and circuit co-design of the antenna and also provide an effective means for MIMO antenna decoupling.

Figure 5.3a presents a center-fed substrate integrated waveguide (SIW) slot-coupled rectangular DRA array as an example. $[S]_{single}$ and $[S]_{coupling}$ represent

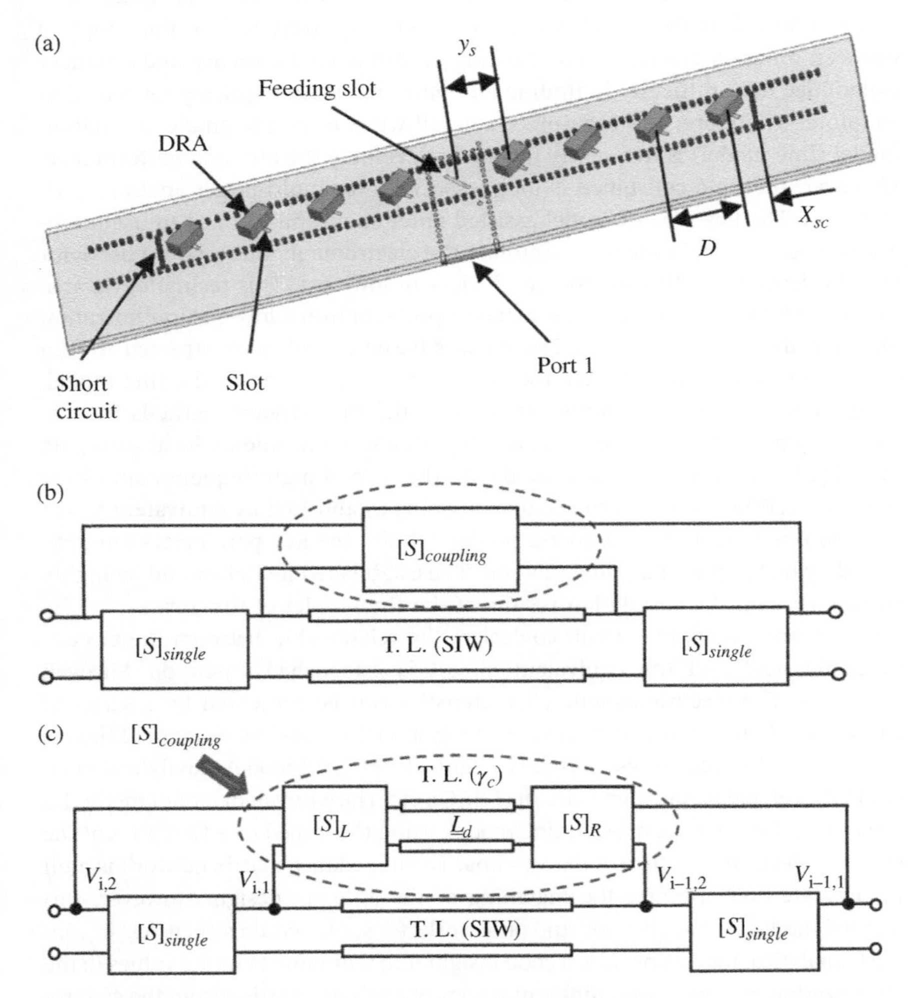

Figure 5.3 A center-fed SIW slot coupled rectangular DRA array. *Source:* Abdallah et al. [27]/IEEE. (a) Configuration. (b) Equivalent model of transmission line. *Source:* Abdallah et al. [28]/IEEE. (c) Equivalent model of lossy transmission line. *Source:* Abdallah et al. [28]/IEEE.

element S-parameters and coupling S-parameters, respectively. SIW is equivalent to a transmission line, and $[S]_{single}$ can be directly obtained by electromagnetic model simulation (Figure 5.3b). To obtain $[S]_{coupling}$, it is further split into $[S]_L$, $[S]_R$, and a section of lossy transmission line with propagation constant γ_c, as shown in Figure 5.3c. $[S]_L$ and $[S]_R$ are calculated from the S-parameters obtained by simulating two adjacent cells under two elements spacing D_{ref} and D_1.

Then $[S]_L$ and $[S]_R$ are solved by the following steps:

Denote $[S]_L$ and $[S]_R$ by S'_{11}, S'_{12}, S'_{21}, and S'_{22} in combination:

$$[S]_L = \begin{bmatrix} S'_{22} & S'_{21} \\ S'_{12} & S'_{11} \end{bmatrix} \quad [S]_R = \begin{bmatrix} S'_{22} & S'_{21} \\ S'_{12} & S'_{11} \end{bmatrix} \tag{5.1}$$

The S-parameters obtained by simulation under D_{ref} and D_1 spacing are denoted by $[S]_A$ and $[S]_B$, respectively:

$$[S]_A = \begin{bmatrix} S_{11A} & S_{12A} \\ S_{21A} & S_{22A} \end{bmatrix} \quad [S]_B = \begin{bmatrix} S_{11B} & S_{12B} \\ S_{21B} & S_{22B} \end{bmatrix} \tag{5.2}$$

Denote $[S]_A$ and $[S]_B$ by the combination of S'_{11}, S'_{12}, S'_{21}, and S'_{22}:

$$\begin{bmatrix} S_{11} & S_{12} \\ S_{21} & S_{22} \end{bmatrix} = \begin{bmatrix} S'_{11} + k \times S'_{12} \times S'_{21} \times S''_{11} & k \times S'_{12} \times S''_{12} \\ k \times S'_{21} \times S''_{21} & S'_{22} + k \times S''_{12} \times S''_{21} \times S'_{22} \end{bmatrix} \tag{5.3}$$

$$S_{11A} = S'_{11} + \frac{S'_{12} \times S'_{21} \times S'_{22}}{1 - S'_{22} \times S'_{22}} \quad S_{21A} = \frac{S'_{21} \times S'_{12}}{1 - S'_{22} \times S'_{22}} \tag{5.4}$$

$$S_{11B} = S'_{11} + \frac{S'_{12} \times S'_{21} \times e^{-2\gamma L_d} \times S'_{22}}{1 - S'_{22} \times e^{-2\gamma L_d} \times S'_{22}} \quad S_{21B} = \frac{S'_{21} \times S'_{12} \times e^{-\gamma L_d}}{1 - S'_{22} \times e^{-\gamma L_d \times S'_{22}}}, \tag{5.5}$$

k is equal to:

$$k = \frac{1}{1 - S'_{22} \times S''_{11}} \tag{5.6}$$

L_d is equal to:

$$L_d = D_1 - D_{ref} \tag{5.7}$$

By solving S'_{11}, S'_{12}, S'_{21}, and S'_{22}, $[S]_L$ and $[S]_R$ are obtained.

Figure 5.4a compares the results between the equivalent circuit model and the EM simulation. The model with and without the mutual coupling is also included for comparison. Figure 5.4b compares the results between the

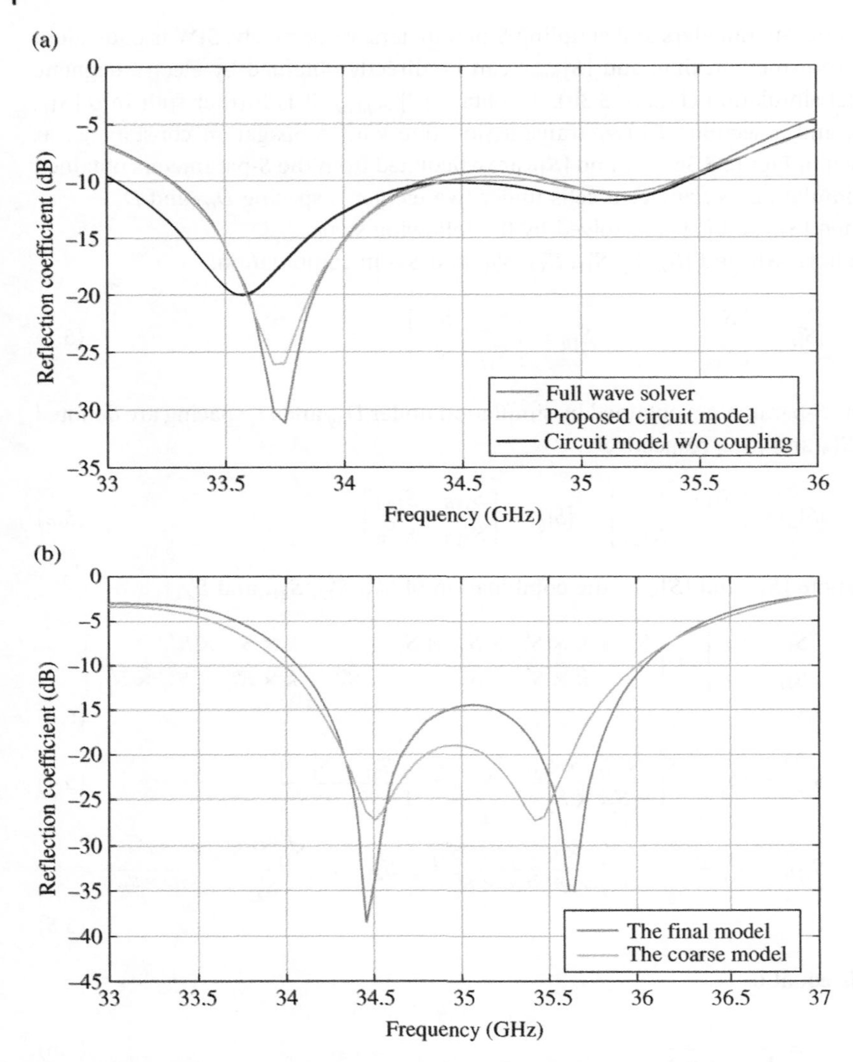

Figure 5.4 Comparison of circuit model and electromagnetic simulation results. *Source:* [27]/IEEE. (a) Without space mapping method. (b) After mapping method.

equivalent model and the EM simulation after two iterations optimization of the key parameters in the circuit by using the space mapping method.

The neutral structure decoupling method counteracts mutual coupling by introducing a neutralization path. Broadband neutral structures are designed to introduce an additional current path to cancel the coupling current on the ground

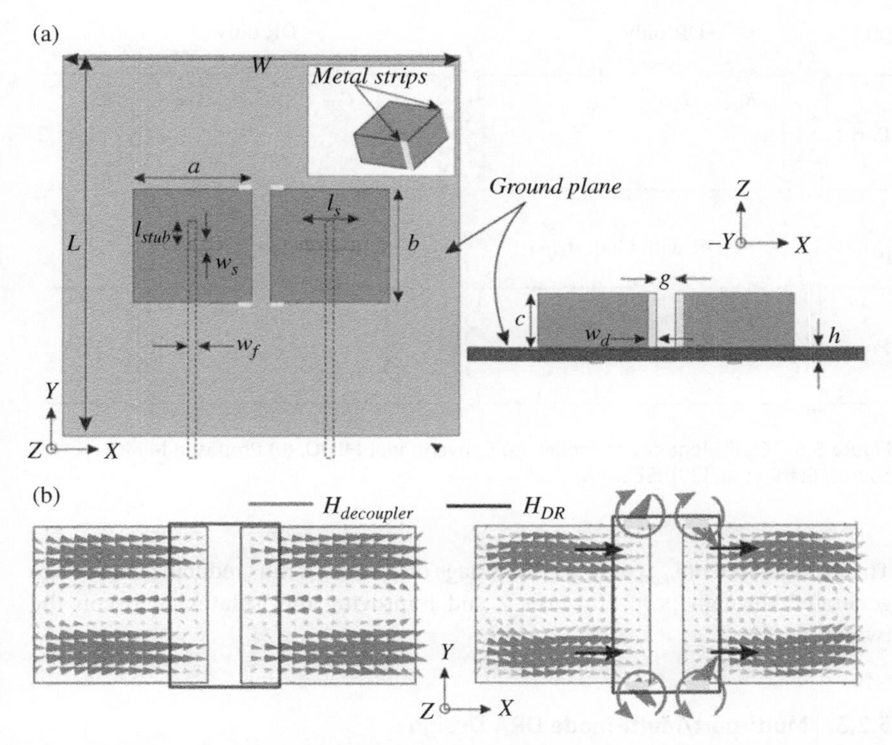

Figure 5.5 Example of neutral structure method. (a) Geometry of the proposed DRA. (b) Electromagnetic field compared to conventional DRA. *Source:* Elahi et al. [29]/IEEE.

plane. Neutralization structure occupies a small space, thus can effectively increase the isolation between DRA cells with simple and compact structure. At the same time, the neutralization structure has little effect on the radiation pattern of the antenna.

Figure 5.5 illustrates an example of MIMO DRA using neutral structure. To improve the isolation between the DRA elements, a pair of rectangular metal strips, which were short-circuited from the ground, was printed on the adjacent edges of the sidewalls of each DR element. The metal strip generates a vertical current, resulting in a magnetic field resistant to the coupling field between the DRA.

The equivalent circuit model of the structure is shown in Figure 5.6. A series RLC resonator (R_{DR}, L_{DR}, C_{DR}) is used to model the DRA. T_1 represents the combined coupling among the microstrip feedline, slot, and DR; T_2 characterizes the mutual coupling between the DRA elements; and L_{gnd} represents the total inductance of the two grounded strip on the lateral walls of the DR in proposed design.

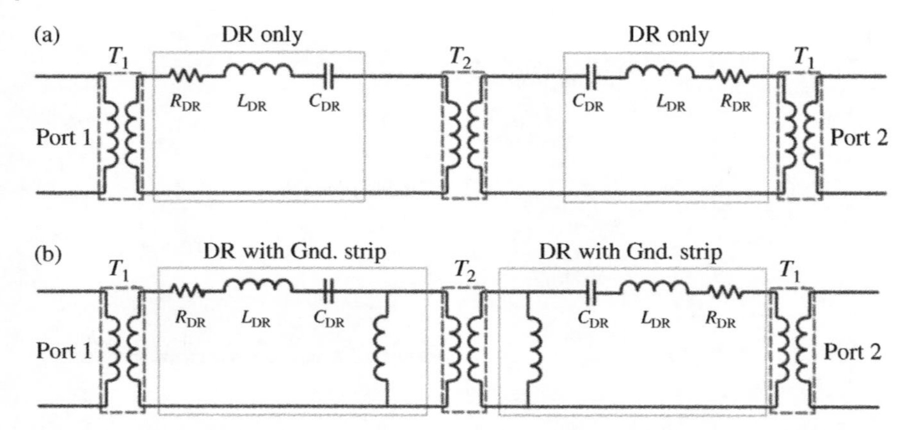

Figure 5.6 Equivalent circuit model: (a) Conventional MIMO. (b) Proposed MIMO. *Source:* Elahi et al. [29]/IEEE.

The introduction of L_{gnd} reduces the voltage of T_1 and T_2, thus reducing the voltage generated through port 1 to port 2 and improving the isolation between the two ports.

5.2.3 Multi-port/Multi-mode DRA Design

Multi-port antenna is a single structure fed by independent ports, featuring the same polarization but with omnidirectional/directional pattern in the same frequency band. The multi-mode and multi-port antenna can effectively reduce the coupling between multiple inputs and improve the MIMO system performance. It should be noted that multi-port DRA is one of the most popular solutions for MIMO system due to the diverse modes and high radiation efficiency at high frequency.

Figure 5.7 shows the orthogonal modes in cylindrical DRA [23] which is explained as follows.

1) $TM_{01\delta}$ mode has omnidirectional radiation with small propagation fading since its polarization direction is perpendicular to the ground. When the DRA is placed on the ground and surrounded by free space, $TE_{01\delta}$ mode cannot be excited due to the physical limitations of the boundary conditions. This mode is difficult to be used in antenna radiation scenarios, but it is widely used in metal cavity devices such as cavity dielectric resonator filters.
2) $TE_{011+\delta}$ mode is an omnidirectional radiation mode of the TE mode family, which is similar to $TM_{01\delta}$ mode but its electric and magnetic fields distribution

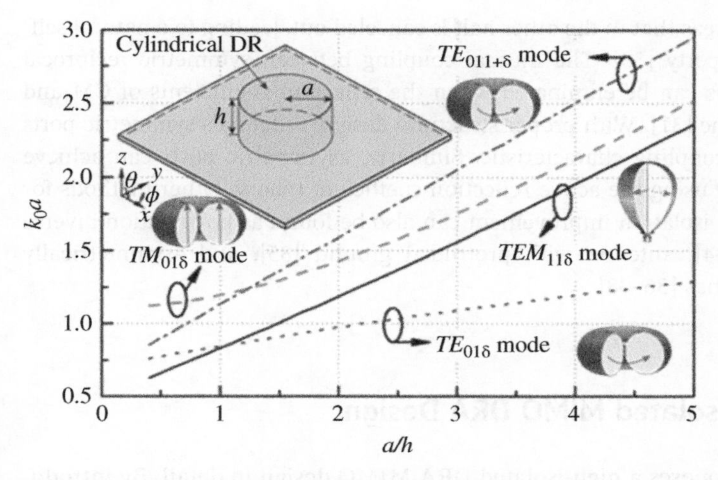

Figure 5.7 Orthogonal modes in a cylindrical DRA. *Source:* [23]/IEEE.

is opposite with that of $TM_{01\delta}$ mode. Thus they are a pair of orthogonal modes with different polarizations.

3) $HEM_{11\delta}$ mode is commonly used in the pattern diversity antenna as the directional radiation mode. It is also common to use the two orthogonal direction degenerate modes of $HEM_{11\delta}$ mode, such as $HEM^x_{11\delta}$ mode and $HEM^y_{11\delta}$ mode to obtain the polarization diversity antenna.
4) When the omnidirectional radiation is required, $TM_{01\delta}$ and $TE_{011+\delta}$ can be selected as a pair of orthogonal polarization with vertical and horizontal polarizations. In the pattern diversity scenario, since the vertically polarized $TM_{01\delta}$ mode is more commonly used, it is usually combined with the $HEM_{11\delta}$ mode to achieve pattern diversity.

The use of the multi-mode DRA to achieve isolation between antenna elements is a typical self-decoupling method for MIMO system. In recent years, the antenna self-decoupling technology has attracted attention. The self-decoupling method does not need to use additional decoupling structure, so it does not introduce additional insertion loss and does not increase the complexity of the original antenna system. It has the significant advantages of simple structure, low loss, and high efficiency.

The other approach to implement self-decoupling is based on differential mode (DM) and common mode (CM) that are excited by multiple ports for symmetric reciprocal dual-port antenna system. According to mode cancellation theory, the current mode fed through one port could be equivalent to the synthesis of the CM and DM currents. Hence, the current in half of the shared radiator is enhanced, whereas that in the other half is canceled out, leading to a natural self-decoupling property [30]. The mutual coupling between symmetric reciprocal antenna systems can be eliminated when the reflection coefficients of CM and DM are the same [31]. With proper structural design, antenna's symmetric ports present self-decoupling characteristic. Similarly, asymmetric ports can achieve self-isolation by using the active reflection coefficient theory. Other methods for MIMO antenna isolation improvement can also be found as polarization diverse antennas [32–34], antenna with protruded ground [35], and asymmetrically mirrored antennas [36–38].

5.3 High-Isolated MIMO DRA Design

This section proposes a high-isolated DRA MIMO design in detail. By introducing a metal strip printed on the upper surface of each DR, the strongest part of the coupling field moves away from the adjacent exciting slot of the DRA. The reflection coefficient of the proposed MIMO DRA with metal strips is less than −10 dB over 27.25–28.59 GHz, showing that the introduction of metal strips does not affect the impedance matching of DRA significantly. We also illustrate the design procedure and the mechanism of how metal strips enhance the isolation between two elements of the proposed MIMO DRA. Experimental results and discussions are carried out as a validation to the simulated results. Three methods proposed in the above sections can be found in this design.

The geometry and dimensions of the proposed MIMO DRA with enhanced isolation for 5G mm-wave application are shown in Figure 5.8 and Table 5.1, respectively. Two rectangular DRs with relative permittivity of 9.8 are mounted on the Rogers 5880 substrate with ε_r of 2.2, tan δ of 0.0009, and thickness of 0.254 mm. A microstrip-fed rectangular exciting slot is set underneath each DR for excitation purpose. A metal strip with length of L_p and width of W_p is printed on the upper surface of each DR to enhance the isolation between two antenna elements. The detailed design process of the proposed MIMO DRA is illustrated as follows.

Two identical DRs with the dimension $a \times b \times d$ are mounted on the metal ground plane with the dimension of 20 mm × 20 mm. The $TE^y{}_{311}$ mode is excited by a microstrip-fed rectangular slot. The resonant frequency of a rectangular DR for the $TE^y{}_{pqr}$ mode can be derived from the wavenumbers k_x, k_y, and k_z inside the DR, and the subscripts p, q, and r are numbers of the standing waves along the x-axis, y-axis, and z-axis, respectively. The wavenumbers can be calculated using the Marcatili's approximation method [39]. The dimension of the DR can be set to $a = 9.5$ mm, $b = 7.5$ mm, and $d = 2.54$ mm, which makes the DR resonate at

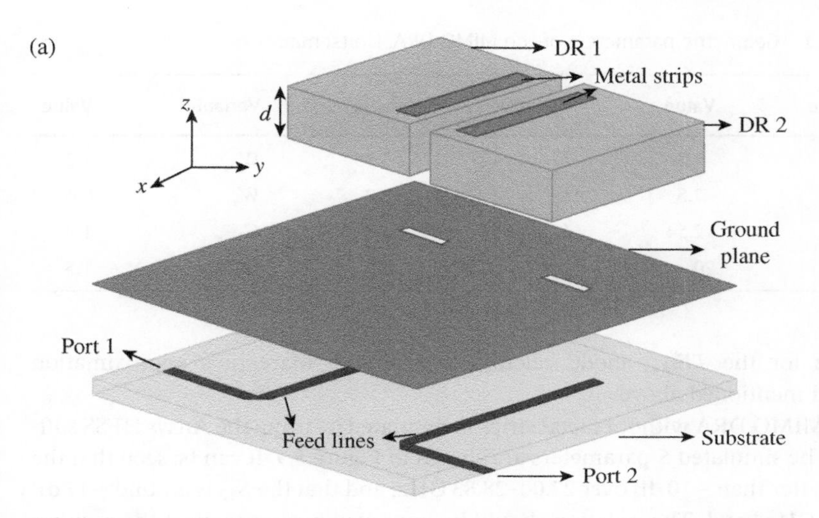

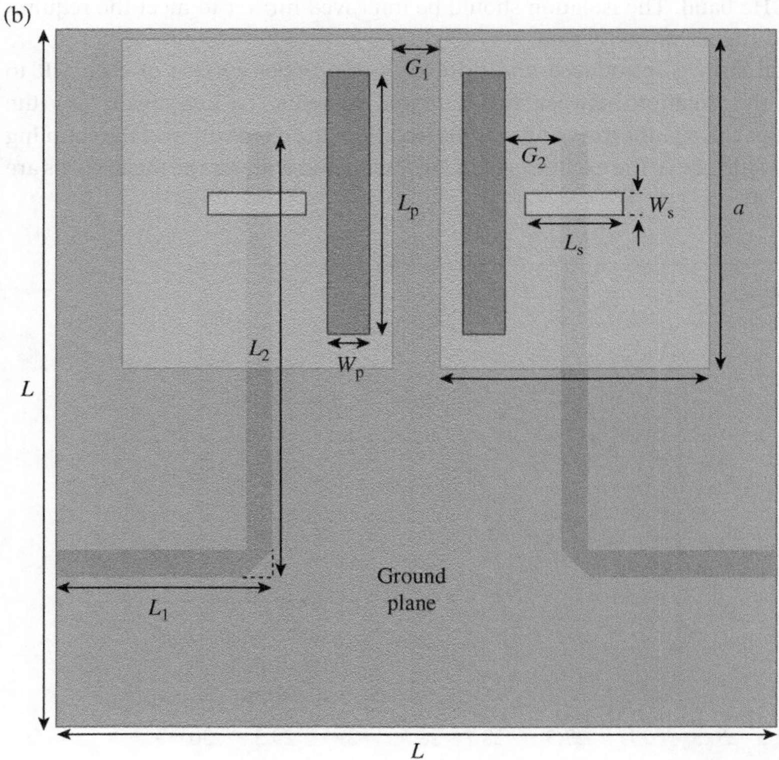

Figure 5.8 Geometry structure of the proposed MIMO DRA. (a) Exploded 3D view. (b) Top view.

Table 5.1 Geometric parameters of the MIMO DRA. Units: mm.

Variable	Value	Variable	Value	Variable	Value
a	9.5	L_1	5.95	W_p	1.2
b	7.5	L_2	12.2	W_s	0.6
d	2.54	L_s	2.7	G_1	1.3
L	20	L_p	7.5	G_2	1.5

28 GHz for the TE^y_{311} mode calculated using the Marcatili's approximation method mentioned above.

The MIMO DRA without metal strips is simulated by using the Ansys HFSS software. The simulated S-parameters are shown in Figure 5.9. It can be seen that the S_{11} is better than −10 dB over 27.00–28.83 GHz, and that the S_{12} is around −17 dB in a 28 GHz band. The isolation should be improved further to meet the requirement of a higher diversity gain.

A metal strip is introduced and printed on the upper surface of each DR to improve the isolation between two antenna elements. To investigate why the metal strips can significantly improve the isolation, the distribution of the coupling electrical field above the exciting slot of DR 2 with and without the metal strips are

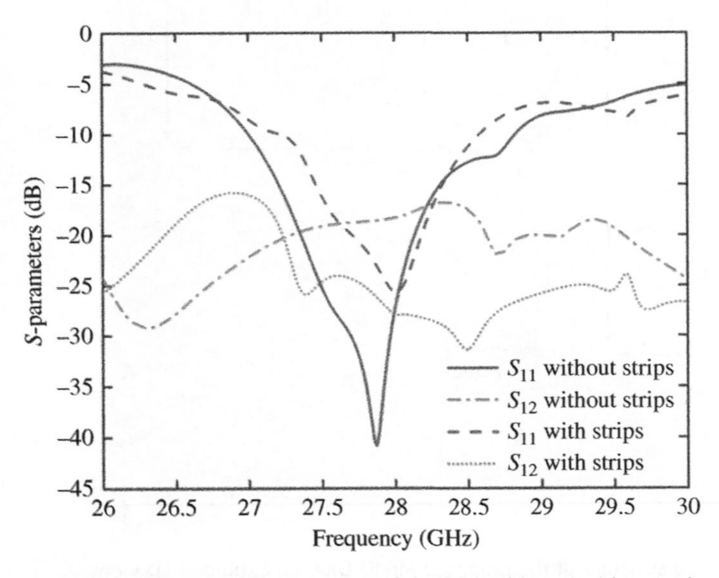

Figure 5.9 Simulated S-parameters of the DRA with and without strips.

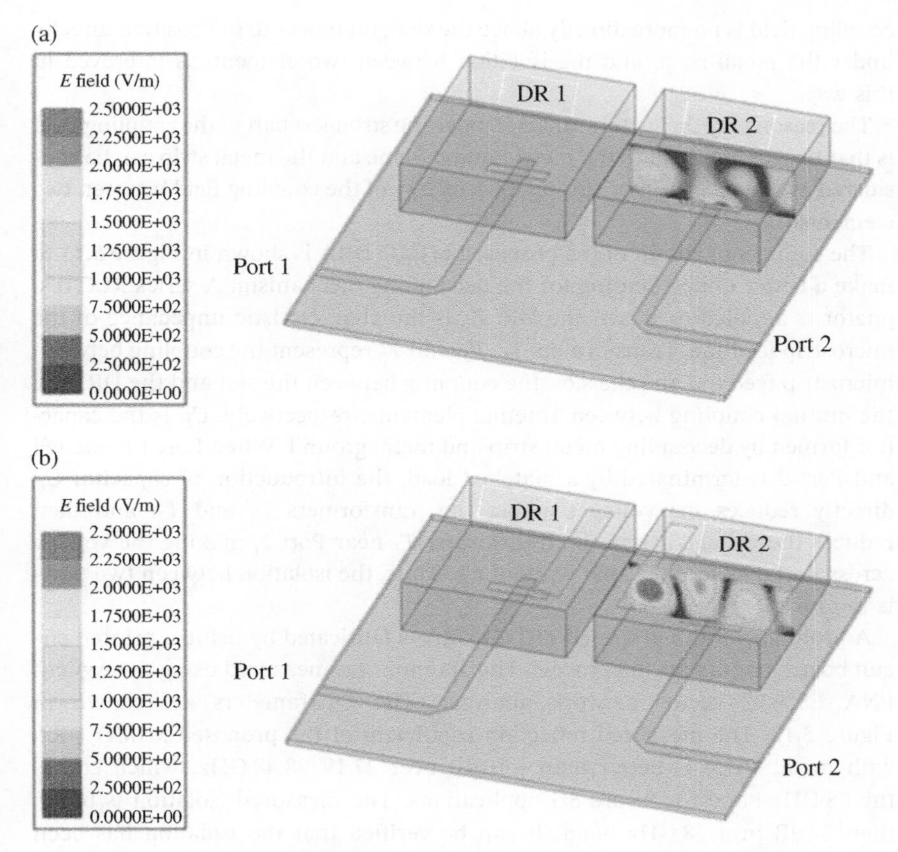

Figure 5.10 Amplitude distribution of the coupling field (only Port 1 is excited). (a) DRs are without metal strips. (b) DRs are with metal strips.

simulated and compared in Figure 5.10 when only Port 1 is fed. It can be observed in Figure 5.10a that the strongest part of the coupling field is directly above the microstrip-fed rectangular slot. Therefore, the energy of the coupling field can be transmitted to Port 2 via the slot resulting in strong coupling.

The *S*-parameters of antennas with and without metal strips are shown in Figure 5.9. It can be seen that the MIMO DRA with metal strips has a minimum improvement of 6 dB and a maximum improvement of 12 dB on the isolation over 27.5–28.35 GHz, and the metal strips do not observably influence the impedance matching.

Figure 5.10b shows the distribution of the coupling field in DR 2 with a metal strip printed on its upper surface. The figure shows that the strongest part of the coupling field is no more directly above the slot and moves to the position directly under the metal strip, and the isolation between two elements is improved in this way.

The reason why the metal strips can move the strongest part of the coupling field is that the combination of the metal ground plane and the metal strip can be considered as a plate capacitor storing most energy of the coupling field between two elements.

The equivalent circuit of the proposed MIMO DRA is shown in Figure 5.11 to make a better understanding for the decoupling mechanism. A series RLC resonator is adopted to model the DR. Z_0 is the characteristic impedance of the microstrip feedline. Transformers T_1, T_2, and T_3 represent the coupling between microstrip feedline and the slot, the coupling between the slot and the DR, and the mutual coupling between antenna elements, respectively. *Cp* is the capacitor formed by decoupling metal strip and metal ground. When Port 1 is excited and Port 2 is terminated by a matched load, the introduction of capacitor *Cp* directly reduces the voltages across the transformers T_3 and T_2, and then reduces the voltage across the transformer T_1 near Port 2, making the voltage across Port 2 lower than that without *Cp*. Thus, the isolation between two ports is improved.

A prototype of the proposed MIMO DRA is fabricated by using a printed circuit board manufacturing process. The antenna was measured using the Agilent PNA E8363C vector network analyzer. The *S*-parameters are shown in Figure 5.12. The measured reflection coefficient of the proposed MIMO DRA with metal strips is better than −10 dB over 27.19–28.48 GHz, which covers the 28 GHz band for future 5G applications. The measured isolation is better than 24 dB in a 28 GHz band. It can be verified that the isolation has been

(a)

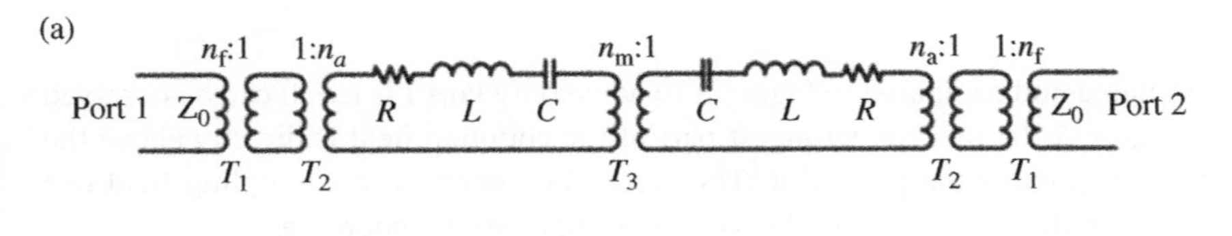

(b)

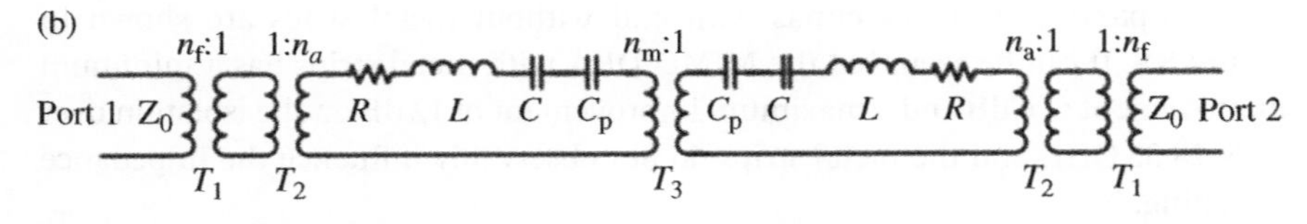

Figure 5.11 Equivalent circuit model of the proposed MIMO DRA. (a) DRs are without metal strips. (b) DRs are with metal strips.

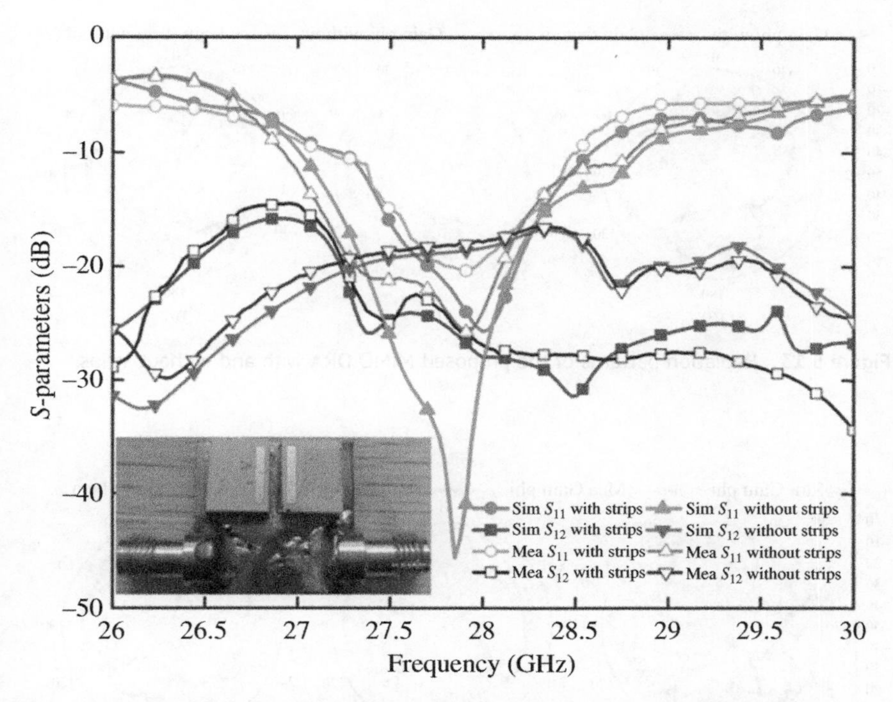

Figure 5.12 Simulated and measured S-parameters of the proposed MIMO DRA with and without metal strips.

improved by introducing the metal strips. The deviation of the scattering coefficients between simulation and measurement is mainly due to the connector soldering and practical dielectric constant of the medium varying with frequency.

Metal strips do not significantly affect the radiation pattern. This is due to the location of metal strips where the electric field of DR is weak. The radiation patterns with and without metal strips are shown in Figure 5.13, which shows there is just a slight influence of metal strips on radiation patterns.

The simulated and measured radiation patterns of MIMO DRA with metal strips at 28 GHz are shown in Figure 5.14. The realized gain of the proposed MIMO DRA is shown in Figure 5.15. The discrepancy between simulated and measured results is due to the connectors whose size is comparable to that of the DRA. The testing cable also affects the measured radiation pattern.

The envelope correlation coefficient (ECC) is a critical figure in the MIMO communication system. For a two-element MIMO antenna, the ECC can be calculated

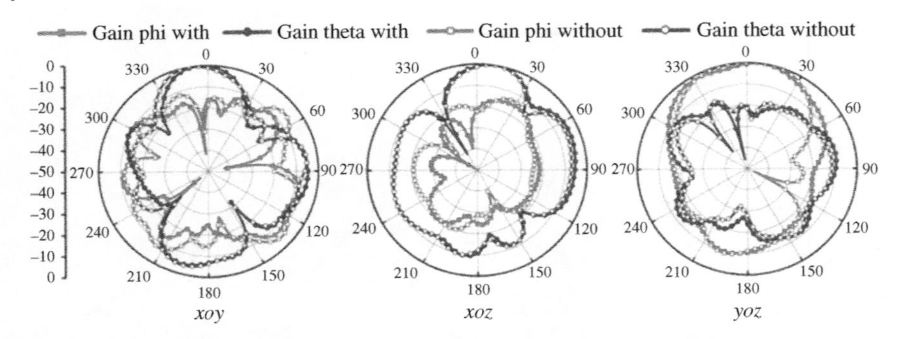

Figure 5.13 Radiation patterns of the proposed MIMO DRA with and without strips.

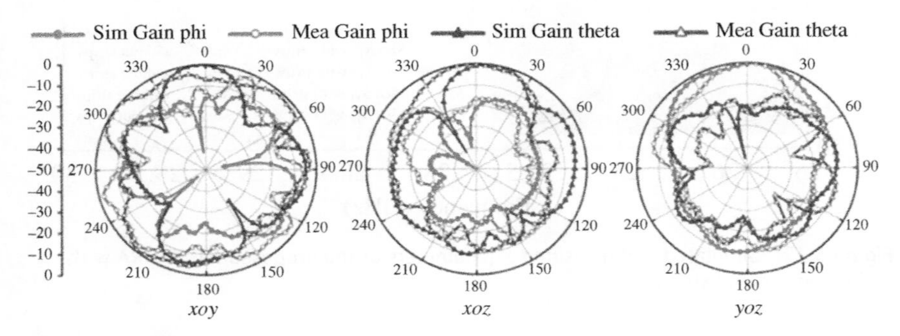

Figure 5.14 Measured and simulated radiation pattern of the proposed MIMO DRA.

according to radiation patterns. The ECC of the proposed MIMO DRA is lower than 0.013 in a 28 GHz band as shown in Figure 5.15, which contributes to a large channel capacity and diversity gain of the MIMO communication system.

The diversity gain is another critical parameter for evaluating the MIMO system performance. It can be obtained by [40]

$$DG = 10\sqrt{1 - ECC^2} \tag{5.8}$$

where 10 is the maximum apparent diversity gain at the 1% probability level for selection combining. Since ECC of the proposed MIMO DRA is less than 0.013, the diversity gain is more than 9.9 dB.

The channel capacity can be calculated by the following equation [41]:

$$C = \log_2\{|(I - S_R^H S_R)(I - S_T^H S_T)\gamma|/2\} - 1.18 \tag{5.9}$$

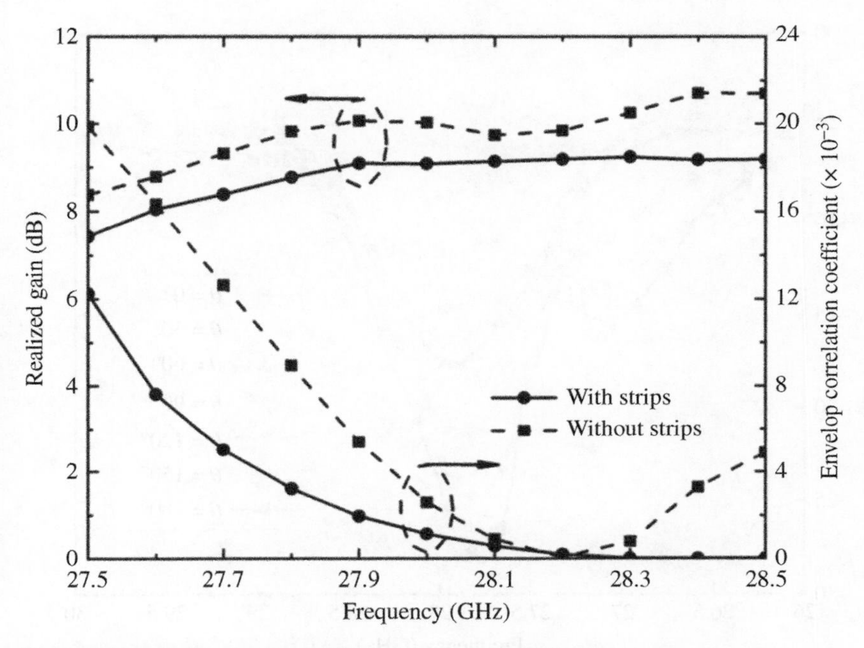

Figure 5.15 Realized gain and ECC of the proposed MIMO DRA.

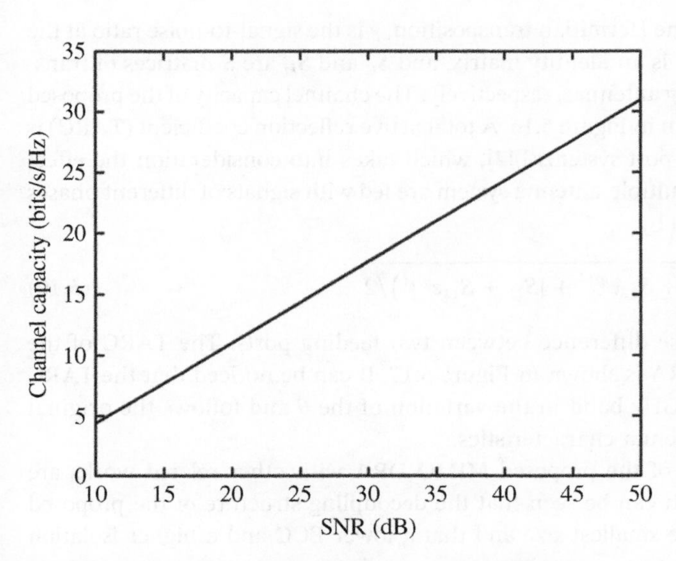

Figure 5.16 Channel capacity of the proposed MIMO DRA.

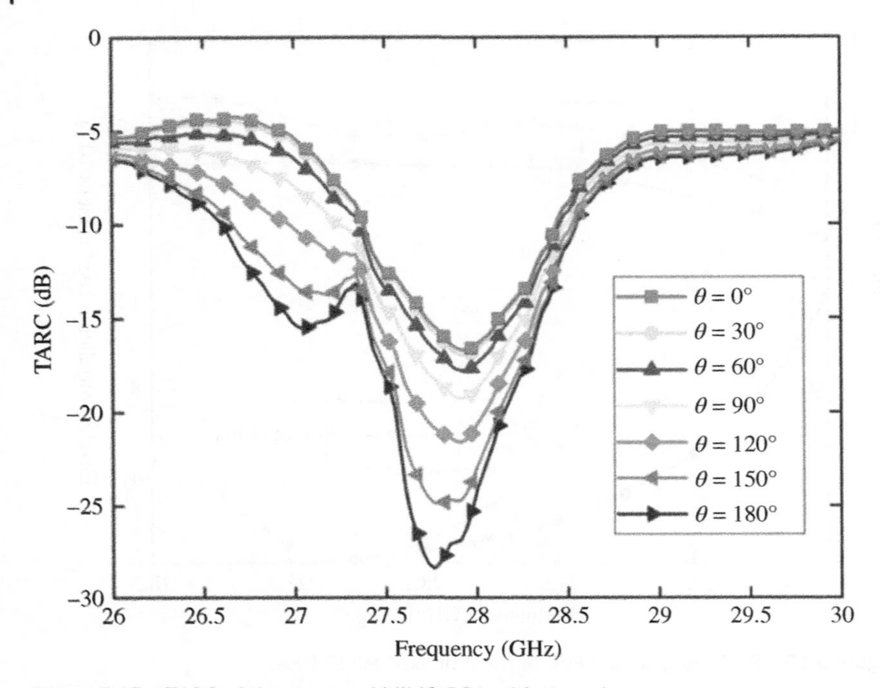

Figure 5.17 TARC of the proposed MIMO DRA with θ varying.

where $\{\cdot\}^H$ denotes the Hermitian transposition, γ is the signal-to-noise ratio at the receiving antenna, I is an identity matrix, and S_R and S_H are S-matrices of transmitting and receiving antennas, respectively. The channel capacity of the proposed MIMO DRA is shown in Figure 5.16. A total active reflection coefficient (TARC) is presented for multi-port systems [42], which takes into consideration the effect when ports of the multiple-antenna system are fed with signals of different phases and can be obtained by

$$\Gamma_a^t = \sqrt{\left(|S_{11} + S_{12}e^{j\theta}|^2 + |S_{21} + S_{22}e^{j\theta}|^2\right)/2} \tag{5.10}$$

where θ is the phase difference between two feeding ports. The TARC of the proposed MIMO DRA is shown in Figure 5.17. It can be noticed that the TARC always covers a 28 GHz band in the variation of the θ and follows the original behavior of the antenna characteristics.

The comparisons of the proposed MIMO DRA with other related works are listed in Table 5.2. It can be seen that the decoupling structure of the proposed MIMO DRA has the smallest size and that a lower ECC and a higher isolation are obtained.

Table 5.2 Comparison with other works.

References	Decoupling structure	Size of decoupling structure	Isolation (dB)	ECC
[43]	Orthogonal feed network	$1.125\lambda_0 \times 0.567\lambda_0 \times 0.221\lambda_0$	20	0.002
[44]	Metallic entity	$0.773\lambda_0 \times 0.387\lambda_0 \times 0.387\lambda_0$	23.3	0.167
[45]	Metasurface shield	$1.680\lambda_0 \times 0.240\lambda_0 \times 0.051\lambda_0$	30	—
[46]	FSS shield	$1.680\lambda_0 \times 0.420\lambda_0 \times 0.100\lambda_0$	30	5e-6
This work	Metal strips	$0.700\lambda_0 \times 0.112\lambda_0 \times 0.002\lambda_0$	24	0.013

5.4 Summary

This chapter overviews the requirements of MIMO antenna in the 5G environment, and then classifies the design methods of MIMO antenna based on their principles. This is followed by several DRA MIMO antenna examples operating with different isolation enhancement methods. The first method is to use the decoupling structure to suppress the coupling between antenna elements directly, the second method is to use the neutralization structure to cancel the coupling between antenna elements, and the third method is to use the multi-port/multi-mode with self-decoupling theory to achieve high isolation. It can be seen that DRA has great potential to implement MIMO antennas due to the compact size, high radiation efficiency, and versatility in shape and feeding mechanism. Especially, the 3D structure of the DRA offers additional degrees of freedom in exciting various modes in one antenna structure. Finally, a MIMO DRA with enhanced isolation and symmetrical pattern is proposed for future 5G mm Wave applications.

References

1 Wang, W., Zhao, Z., Fang, Z. et al. (2019). Compact broadband four-port MIMO antenna for 5G and IoT applications. *2019 IEEE Asia-Pacific Microwave Conference (APMC)*, Singapore. IEEE, pp. 1536–1538.

2 Zhao, L., Chen, A., Zhang, J. et al. (2017). A single radiator with four decoupled ports for four by four MIMO antennas and systems. *2017 IEEE International Symposium on Antennas and Propagation & USNC/URSI National Radio Science Meeting*, San Diego, CA, USA. IEEE, pp. 1659–1660.

3 Fangfang, F., Zehong, Y., Weihong, X. et al. (2007). Compact patch antenna adopted at mobile terminal in MIMO system. *2007 International Symposium on Microwave, Antenna, Propagation and EMC Technologies for Wireless Communications*, Hangzhou, China. IEEE, pp. 601–604.

4 Kang, L., Li, H., Wang, X., and Shi, X. (2015). Compact offset microstrip-fed MIMO antenna for band-notched UWB applications. *IEEE Antennas and Wireless Propagation Letters* 14: 1754–1757.

5 Huang, J., Li, X., and Li, Y. (2021). A millimeter-wave microstrip MIMO antenna for n260 frequency-band. *2021 International Conference on Microwave and Millimeter Wave Technology (ICMMT)*, Nanjing, China. IEEE, pp. 1–3.

6 Narbudowicz, A. and Ammann, M.J. (2018). Low-cost multimode patch antenna for dual MIMO and enhanced localization use. *IEEE Transactions on Antennas and Propagation* 66 (1): 405–408.

7 Pan, Y.M., Qin, X., Sun, Y.X., and Zheng, S.Y. (2019). A simple decoupling method for 5G millimeter-wave MIMO dielectric resonator antennas. *IEEE Transactions on Antennas and Propagation* 67 (4): 2224–2234.

8 Qamar, Z., Naeem, U., Khan, S.A. et al. (2016). Mutual coupling reduction for high-performance densely packed patch antenna arrays on finite substrate. *IEEE Transactions on Antennas and Propagation* 64 (5): 1653–1660.

9 Yang, X.M., Liu, X.G., Zhou, X.Y., and Cui, T.J. (2012). Reduction of mutual coupling between closely packed patch antennas using waveguided metamaterials. *IEEE Antennas and Wireless Propagation Letters* 11: 389–391.

10 Yang, X., Liu, Y., Xu, Y.-X., and Gong, S.-X. (2017). Isolation enhancement in patch antenna array with fractal UC-EBG structure and cross slot. *IEEE Antennas and Wireless Propagation Letters* 16: 2175–2178.

11 Expósito-Domínguez, G., Fernández-Gonzalez, J.-M., Padilla, P., and Sierra-Castañer, M. (2012). Mutual coupling reduction using EBG in steering antennas. *IEEE Antennas and Wireless Propagation Letters* 11: 1265–1268.

12 Farahani, H.S., Veysi, M., Kamyab, M., and Tadjalli, A. (2010). Mutual coupling reduction in patch antenna arrays using a UC-EBG superstrate. *IEEE Antennas and Wireless Propagation Letters* 9: 57–59.

13 Ouyang, J., Yang, F., and Wang, Z.M. (2011). Reducing mutual coupling of closely spaced microstrip MIMO antennas for WLAN application. *IEEE Antennas and Wireless Propagation Letters* 10: 310–313.

14 Wei, K., Li, J.-Y., Wang, L. et al. (2016). Mutual coupling reduction by novel fractal defected ground structure bandgap filter. *IEEE Transactions on Antennas and Propagation* 64 (10): 4328–4335.

15 Farsi, S., Aliakbarian, H., Schreurs, D. et al. (2012). Mutual coupling reduction between planar antennas by using a simple microstrip U-section. *IEEE Antennas and Wireless Propagation Letters* 11: 1501–1503.

16 Zhao, L., Yeung, L.K., and Wu, K.-L. (2014). A coupled resonator decoupling network for two-element compact antenna arrays in mobile terminals. *IEEE Transactions on Antennas and Propagation* 62 (5): 2767–2776.

17 Wu, C.-H., Chiu, C.-L., and Ma, T.-G. (2016). Very compact fully lumped decoupling network for a coupled two-element array. *IEEE Antennas and Wireless Propagation Letters* 15: 158–161.

18 Venkatasubramanian, S.N., Li, L., Lehtovuori, A. et al. (2017). Impact of using resistive elements for wideband isolation improvement. *IEEE Transactions on Antennas and Propagation* 65 (1): 52–62.

19 Wang, S. and Du, Z. (2015). Decoupled dual-antenna system using crossed neutralization lines for LTE/WWAN smartphone applications. *IEEE Antennas and Wireless Propagation Letters* 14: 523–526.

20 Zhang, S. and Pedersen, G.F. (2016). Mutual coupling reduction for UWB MIMO antennas with a wideband neutralization line. *IEEE Antennas and Wireless Propagation Letters* 15: 166–169.

21 Ren, A., Liu, Y., and Sim, C.-Y.-D. (2019). A compact building block with two shared-aperture antennas for eight-antenna MIMO array in metal-rimmed smartphone. *IEEE Transactions on Antennas and Propagation* 67 (10): 6430–6438.

22 Sun, L., Li, Y., Zhang, Z., and Feng, Z. (2020). Wideband 5G MIMO antenna with integrated orthogonal-mode dual-antenna pairs for metal-rimmed smartphones. *IEEE Transactions on Antennas and Propagation* 68 (4): 2494–2503.

23 N. Yang, K. W. Leung and N. Wu, "Pattern-diversity cylindrical dielectric resonator antenna using fundamental modes of different mode families," in *IEEE Transactions on Antennas and Propagation*, vol. 67, no. 11, pp. 6778–6788, 2019, https://doi.org/10.1109/TAP.2019.2922873.

24 Zhang, Z., Chen, H.C., and Cheng, Q.S. (2021). Surrogate-assisted quasi-newton enhanced global optimization of antennas based on a heuristic hypersphere sampling. *IEEE Transactions on Antennas and Propagation* 69 (5): 2993–2998.

25 Koziel, S., Cheng, Q.S., and Bandler, J.W. (2014). Fast EM modeling exploiting shape-preserving response prediction and space mapping. *IEEE Transactions on Microwave Theory and Techniques* 62 (3): 399–407.

26 Song, Y., Cheng, Q.S., and Koziel, S. (2019). Multi-fidelity local surrogate model for computationally efficient microwave component design optimization. *Sensors* 19 (13): 3023.

27 Abdallah, M.S., Wang, Y., Abdel-Wahab, W.M., and Safavi-Naeini, S. (2018). Design and optimization of SIW center-fed series rectangular dielectric resonator antenna array with 45° linear polarization. *IEEE Transactions on Antennas and Propagation* 66 (1): 23–31.

28 Abdallah, M., Wang, Y., Abdel-Wahab, W.M., and Safavi-Naeini, S. (2016). A tunable circuit model for the modeling of dielectric resonator antenna array. *IEEE Antennas and Wireless Propagation Letters* 15: 830–833.

29 Elahi, M., Altaf, A., Yousaf, J., and Majali, E.R.A. (2021). Isolation improvement in MIMO dielectric resonator antennas. *2021 IEEE International Symposium on Antennas and Propagation and USNC-URSI Radio Science Meeting (APS/URSI)*, Singapore, Singapore. IEEE, pp. 1151–1152.

30 Sun, L.B., Li, Y., Zhang, Z.J., and Wang, H.Y. (2021). Antenna decoupling by common and differential modes cancellation. *IEEE Transactions on Antennas and Propagation* 69 (2): 672–682.

31 Yang, B., Xu, Y., Tong, J. et al. (2022). Tri-port antenna with shared radiator and self-decoupling characteristic for 5G smartphone application. *IEEE Transactions on Antennas and Propagation* 70 (6): 4836–4841.

32 Li, M.Y., Ban, Y.L., Xu, Z.Q. et al. (2016). Eight-port orthogonally dual-polarized antenna array for 5G smartphone applications. *IEEE Transactions on Antennas and Propagation* 64 (9): 3820–3830.

33 Li, M.Y., Xu, Z.Q., Ban, Y.L. et al. (2017). Eight-port orthogonally dual-polarized MIMO antennas using loop structures for 5G smartphone. *IET Microwaves, Antennas & Propagation* 11: 1810–1816.

34 Li, M.Y., Ban, Y.L., Xu, Z.Q. et al. (2018). Tri-polarized 12-antenna MIMO array for future 5G smartphone applications. *IEEE Access* 6: 6160–6170.

35 Wang, Y. and Du, Z. (2014). A wideband printed dual-antenna with a protruded ground for mobile terminals. *Proceedings of International Symposium on IEEE Antennas and Propagation Society*, IEEE Antennas and Propagation Society, pp. 1133–1134.

36 Wong, K.L., Tsai, C.Y., and Lu, J.Y. (2017). Two asymmetrically mirrored gap-coupled loop antennas as a compact building block for eight antenna MIMO array in the future smartphone. *IEEE Transactions on Antennas and Propagation* 65 (4): 1765–1778.

37 Wong, K.L., Lin, B.W., and Li, W.Y. (2017). Dual-band dual inverted-F/loop antennas as a compact decoupled building block for forming eight 3.5/5.8-GHz MIMO antennas in the future smartphone. *Microwave and Optical Technology Letters* 59: 2715–2721.

38 Tsai, C.Y., Wong, K.L., and Li, W.Y. (2018). Experimental results of the multi-Gbps smartphone with 20 multi-input multi-output (MIMO) antennas in the 20 × 12 MIMO operation. *Microwave and Optical Technology Letters* 60: 2001–2010.

39 Mongia, R.K. (1992). Theoretical and experimental resonant frequencies of rectangular dielectric resonators. *IEE Proceedings H (Microwaves, Antennas and Propagation)*, Vol. 139, pp. 98–104. IET Digital Library.

40 Gao, Y., Chen, X., Ying, Z., and Parini, C. (2007). Design and performance investigation of a dual-element PIFA array at 2.5 GHz for MIMO terminal. *IEEE Transactions on Antennas and Propagation* 55 (12): 3433–3441.

41 Honma, N., Sato, H., Ogawa, K., and Tsunekawa, Y. (2015). Accuracy of MIMO channel capacity equation based only on S-parameters of MIMO antenna. *IEEE Antennas and Wireless Propagation Letters* 14: 1250–1253.

42 Manteghi, M. and Rahmat-Samii, Y. (2005). Multiport characteristics of a wideband cavity backed annular patch antenna for multipolarization operations. *IEEE Transactions on Antennas and Propagation* 53 (1): 466–474.

43 Abdalrazik, A., El-Hameed, A.S.A., and Abdel-Rahman, A.B. (2017). A threeport MIMO dielectric resonator antenna using decoupled modes. *IEEE Antennas and Wireless Propagation Letters* 16: 3104–3107.

44 Sharawi, M.S., Podilchak, S.K., Khan, M.U., and Antar, Y.M. (2017). Dualfrequency DRA-based MIMO antenna system for wireless access points. *IET Microwaves, Antennas & Propagation* 11 (8): 1174–1182.

45 Dadgarpour, A., Zarghooni, B., Virdee, B.S. et al. (2017). Mutual coupling reduction in dielectric resonator antennas using metasurface shield for 60 GHz MIMO systems. *IEEE Antennas and Wireless Propagation Letters* 16: 477–480.

46 Karimian, R., Kesavan, A., Nedil, M., and Denidni, T.A. (2017). Low-mutualcoupling 60 GHz MIMO antenna system with frequency selective surface wall. *IEEE Antennas and Wireless Propagation Letters* 16: 373–376.

6

3D Printed Dielectric-Based Antenna

CHAPTER MENU

6.1 Overview

3D printing was originally defined as "Rapid Prototyping Technology" by the American Society of Testing Materials (ASTM). It is an additive manufacturing (AM) process that stacks the materials layer by layer through a three-dimensional model, which is opposite from the traditional method of material reduction manufacturing. 3D printing has been defined by various names, such as layered assembling, added substance manufacture, added substance procedure, computerized fabricating, added substance form, free structure manufacture, and added

Dielectric Resonator Antennas: Materials, Designs and Applications, First Edition.
Zhijiao Chen, Jing-Ya Deng, and Haiwen Liu.

Published 2024 by John Wiley & Sons, Inc.

substance layered producing [1]. 3D printing technology has developed rapidly in the past decades due to the following advantages:

1) high flexibility to deal with complex structures that cannot be manufactured by traditional processes,
2) low cost by reducing labor costs on operation experience requirement,
3) high material utilization with reused unsintered or molded raw materials, and
4) short turnaround time due to the saving of the tool and mold preparation time.

3D printing has been applied for various high-precision projects to replace the traditional manufacturing in the fields of electronics, medical, and military industries [2]. For medical engineering, 3D printing can be used to print artificial tissues, scaffolds, bioprinted model, etc. For electronic engineering, it has been applied to the cavity-based components such as waveguides and horn antennas [3].

Although 3D printing technology can achieve high flexibility machining with reduced manufacturing costs, it still faces many problems and challenges [4]. First, the surface quality could be poor with unproper printing process so the quality might not satisfy user expectations. Second, it is not suitable for larger-scale projects due to the limitation of the 3D printer chamber volume. The third challenge is high-functional printing materials are still under development in terms of heat resistance, chemical resistance, flame resistance, and electrical properties, which restricts the application of 3D printing.

3D printing technologies can be divided into seven categories according to ASTM Standard F2792 [5]. All technologies follow the process from point to line, from line to surface, and from surface to solid [6]. Nevertheless, each technology has unique manner in the fabrication process, curing principle, and the initial material state. There are no debates on which mechanism or technology functions better because each technology has its specific applications. It is essential to understand the manufacturing principles, advantages, and disadvantages of each technology before choosing the appropriate one for a specific application. Table 6.1 lists seven categories and their characteristics.

Among these techniques, vat photopolymerization, material jetting, powder bed fusion, and material extrusion are suitable for microwave and millimeter wave (mmWave) antenna printing. Vat photopolymerization is widely applied for dielectric printed antenna. For metal printed antenna, material extrusion and powder bed fusion are usually applied to shape the metal powder or conductive ink with laser or electronic beam.

The rest of this chapter is organized as follows. Section 6.2 introduces five common 3D printing technologies for antenna designs. Their specific characteristics and printing processes are discussed and applied in different types of antenna prototypes. Section 6.3 gives two 3D printed antenna examples, one of them is dielectric antenna and the other is metal antenna. The comparison between two antennas is made to show the advantages and disadvantages of dielectric printed

Table 6.1 3D printing technology.

Technology	Materials	Precision	Speed	Cost	Representative method
Vat photopolymerization	Photoreactive polymers	High	Fast	High	SL (stereolithography)
Material jetting	Metals, polymers, ceramics, composite, biologicals, and hybrid	High	Fast	High	
Binder jetting	Metals, sands, polymers, ceramics, and hybrid	Medium	Medium	Low	
Powder bed fusion	Metals, ceramics, polymers, composite, and hybrid	Medium	Low	High	SLS (selective laser sintering)
Material extrusion	Metals and polymers	Low	Low	Low	FDM (fused deposition modeling)
Directed energy deposition	Metals and ceramics	Medium	Fast	Low	
Sheet lamination	Metals, ceramics, and polymers	High	Fast	Low	

antenna. Section 6.4 introduces 3D printed finger nail antennas for 5G application. The proposed nail antenna design is light, cheap, easy to install, and part of beauty accessory, which occupies a small surface area and is easy to wear. Finally, Section 6.5 gives remarks and summary of this chapter.

6.2 3D Printed Antenna

3D printed antenna is a representative work for 3D printing technologies that enables the fabrication of complex geometries in manifold materials. It provides a new paradigm for manufacturing antenna designed with 3D electromagnetic simulation software [1]. In this section, five types of 3D printing technologies that are suitable for antenna manufacturing are detailed with their definitions, pros and cons, mechanism, and printing process.

6.2.1 Vat Photopolymerization

Vat photopolymerization uses high-intensity light such as laser and projection to selectively cure liquid photopolymer materials. Classified by the curing method, it is categorized into stereolithography (SL), digital projection, digital light processing (DLP), and continuous digital light processing (CDLP)/continuous liquid interface production (CLIP). Vat photopolymerization has advantages of superior print resolution, relatively high speed, low cost, and flexibility in resin material design. Its applications include biomedical engineering and electronic engineering such as 3D printed antenna [7]. In [8], the SL method is applied in the fabrication of a dielectric reflect array antenna featuring one-shot deployability and wide-angle beam scanning. A bifocal phase distribution method is utilized to optimize the radio frequency (RF) performance of the array, realizing −30° to −10° and 10° to 30° beam scanning.

Figure 6.1 shows two typical SL machines with top-down and bottom-up architectures. They consist of three parts: (1) a high-intensity light source, such as laser or ultraviolet (UV) light; (2) a material cylinder and tray containing light-cured liquid resin; and (3) a control system for guiding the light source to selectively illuminate and cure the resin. At the same time, a plurality of galvanometers is distributed over the *X*-axis and *Y*-axis of the light source lens. The light beam is

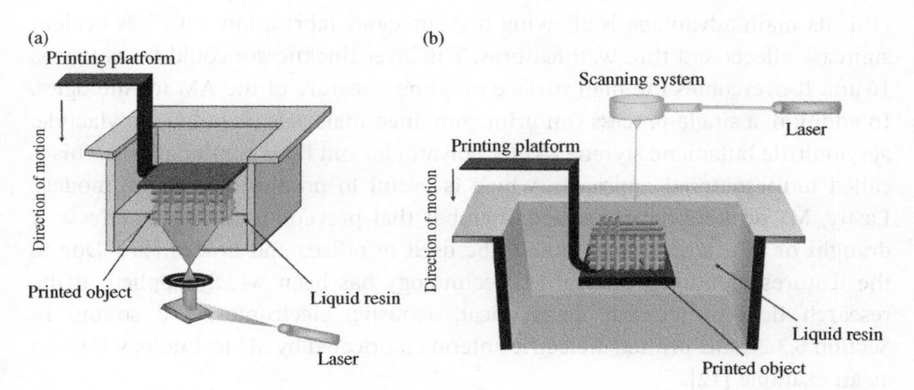

Figure 6.1 Schematic of typical SL printer. (a) Top-down architecture. (b) Bottom-up architecture.

quickly scanned by the deflection of the galvanometer, which is projected to the curing resin in the print slot.

The 3D printing process by SL technology is demonstrated as follows.

First, preprocessing, which prepares the desired mesh model. In this step, a 3D model is constructed by computer-aided design (CAD) software or 3D scanned from a physical object. The 3D model is subsequently converted to the standard triangulation language (STL) format, which has been the most frequently used format for preparing data for AM production since 1987.

Second, processing, where the model is printed. Based on the prepared optimal orientation of the model and the supports, the STL model along with the supports is sliced into layers with a plane parallel to the platform surface (*xOy* plane). Each layer is then built consecutively in the OZ direction. After a layer of new resin is cured, the printing platform rises or falls, changing the position of the object to be printed in the print slot and exposing a new layer of resin to be cured. The sliced model is subsequently sent to the printer until the entire model is printed. The layer thickness depends on the printer and the quality required.

Lastly, postprocessing, which removes the model from the platform and refines the printed model. In the case of photopolymerization, the as-built models are rinsed by wash solution such as isopropyl alcohol (IPA) to get rid of the liquid resin layer. Technical processing procedures can be used to enhance the mechanical properties [9].

6.2.2 Material Jetting

Material jetting (MJ), also named as poly-jet printing (PP), whose process is selectively deposited of dropping material. The printhead dispenses photosensitive material droplets, which is solidified under UV light in order to print layer-by-layer [10]. Its main advantage is allowing high precious fabrication with less evident staircase effects and thin wall features. The layer thicknesses could be as low as 16 μm. It overcomes the high surface roughness texture of the AM technologies. In addition, a single process can print combined materials including polylactide, acrylonitrile butadiene styrene (ABS), polyamide, and their combinations. This is called multi-material approach, which is useful to produce composite models. Lastly, MJ printers have a sealed chamber that prevents undesirable effects of draught or dirt, which is suitable to be used in offices and homes [11]. Due to the features mentioned above, MJ technology has been widely applied in the research field of medical, mechanical, acoustic, electronics, and so on. In Section 6.3.2, a 3D printed dielectric antenna fabricated by MJ technology is given as an example [12].

Figure 6.2 shows the schematic of a typical MJ printer. The printing steps are listed as follows. First, the print sprinkler sprays or sputters the material

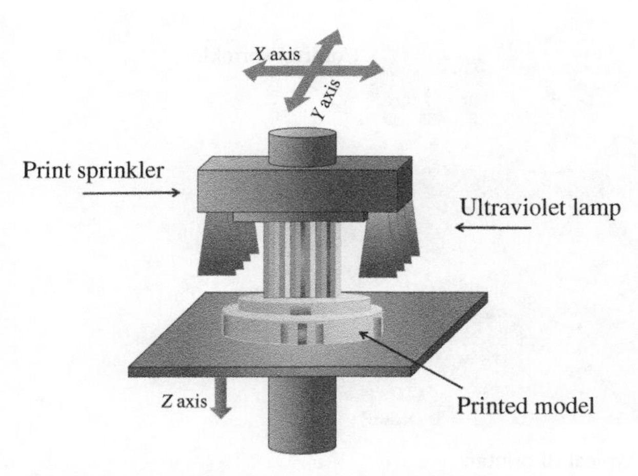

Figure 6.2 Schematic of a typical MJ printer.

horizontally along the X- and Y-axes of the print platform surface. Then, the UV lamp pumps the high-intensity UV light that solidifies the polymer layer from liquid to solid. After curing a layer, the building platform is lowered at certain layer thickness and new liquid material is jetted onto the previous layer until the print model is completed. Since liquid or molten material is initially used, a gel-like supporter is needed, especially in overhang regions. These supporters can be removed by applying other methods such as sonication in a bath of sodium hydroxide solution or using a high-pressure water jet.

6.2.3 Binder Jetting

Binder jetting (BJ) achieves rapid prototyping by using liquid binding agent to selectively deposit powder particles. It uses jet chemical binder to spread powder to form the printed layers. BJ offers many potential advantages: (1) realize color printing without supporters; (2) simple, fast, and cheap printing process that glues powder particles together; and (3) print various materials including metals, sands, polymers, hybrid, and ceramics. Ka-band circuits and antenna manufacturing examples using BJ are characterized in [13].

Figure 6.3 shows the schematic of a typical BJ printer. The BJ printer has a material chamber and a printing chamber. First of all, the piston plate of the material chamber rises by a certain height to eject the powder needed to cover a layer thickness of the printing platform. Then the roller rotates forward, pushing the powder ejected from the top of the material chamber onto the printing platform of the printing chamber to smooth and compact. Next, the sprinkler applies the adhesive to the powder and color the powder if the bonding is smooth. As the printing

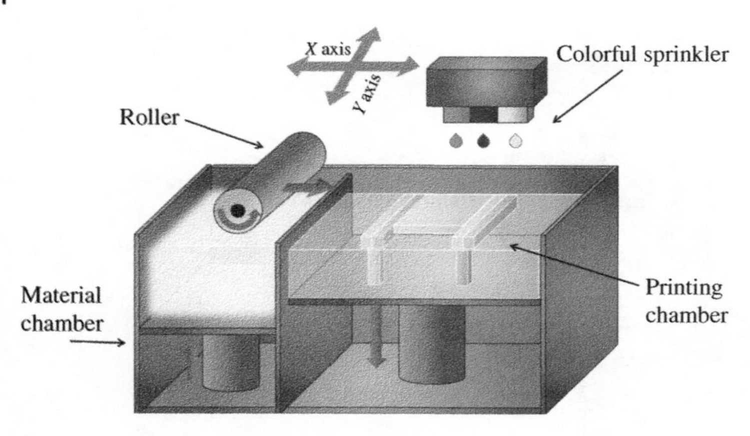

Figure 6.3 Schematic of a typical BJ printer.

material gets harder, the printing platform drops to enable a new layer of powder to lay along the printing platform for re-bonding. The printed object can be self-supporting in the unbonded powder bed, so there is no need to add supporter to the model before printing. After the printing process is completed, the remaining unbonded material powder around the part can be cleaned by vacuum, and all the residual powder can be blown away by air in the chamber. Finally, the surface of the printed object is hardened by infiltrating different materials such as cyano-acrylate, wax, and resin.

6.2.4 Powder Bed Fusion

Powder bed fusion (PBF) is realized by high-energy source (i.e. pulsed laser, electron beam, or UV light) to selectively bind or melt high-strength structures by sintering or melting polymers, pure metals, or metal alloys. It can be classified into selective laser sintering (SLS), selective laser melting (SLM), electron beam selective melting (EBM), direct metal laser sintering (DMLS), direct metal laser melting (DMLM), and multi-jet melting (MJF) depending on the source type and realization procedure. Metal powders can be printed by all categories, but the polymers can be printed by SLS and MJF only. The most common printed polymer is nylon, which has the characteristics of lightweight and high strength. Other printed polymers include polyether ketone, polyacryl ether ketone, polycarbonate (PC), polystyrene, and thermoplastic elastomer polyurethane. The metal powders include pure metals such as aluminum, gold, platinum, palladium, and pure titanium, as well as metal alloys such as cobalt, chromium, and titanium alloys and stainless steel. As example, an air-filled metallic waveguide antenna using SLS is detailed in Section 6.3.1 [14].

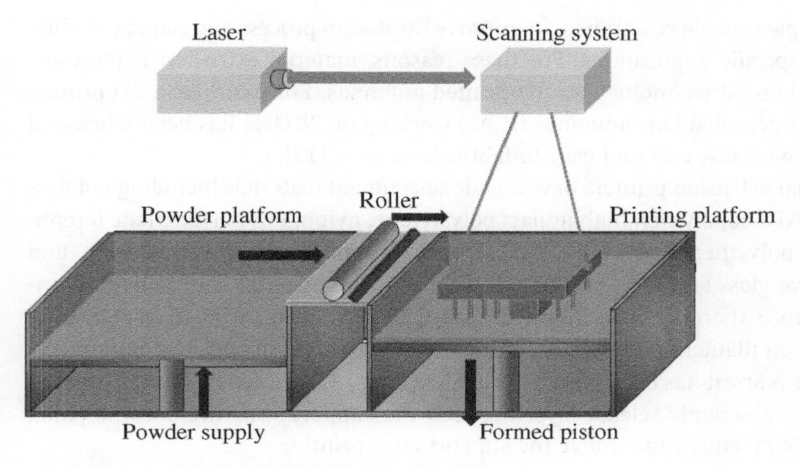

Figure 6.4 Schematic of a typical SLS printer.

SLS is chosen as a representative PBF technology to introduce here and its printing steps are presented as follows. Figure 6.4 shows the schematic of a typical SLS machine. First, place the powder material on the printing platform and spread the powder into a thin layer with the roller. As the printing platform is moving down, the laser beam selectively sinters to solidify the powder. Once one layer printing is completed, the formed piston drops to lay a new layer of powder material above the previous printed layer. Repeat this process and circulate layer by layer until the whole object is complete. What should be mentioned is that during construction, the supporter is not required because the unsintered/melted material retains the original position to support the print model. When the printing process ends, the printed parts need to be cooled before they are removed. After cooling, compressed air is sprayed over the surface of the object to remove the unmelted powder.

6.2.5 Material Extrusion

Material extrusion can produce porous materials, which is also known as fused filament fabrication (FFF) or fused deposition modeling (FDM). It features low cost, ease of operation, and variety in the choice of materials, but suffers from high surface roughness, imperfect sealing between layers and toolpaths, use of support material, and long processing time [15]. This technology can be used to extrude metallic materials, hydrogels, or cell-loaded suspensions in order to incorporate functional components such as antenna in microfluidic devices [16]. Unlike traditional microfluidic manufacturing methods (i.e. soft lithography) that require specialized fabrication skills and facilities, material extrusion is accessible and customizable to serve biology, chemistry, or pharma research. Open-source technologies enable researchers to improve the design process and reduce production for specific applications. For these reasons, material extrusion is the most commonly used technology for 3D printed antennas. For example, a 3D printed low-loss integrated lens antennas (ILAs) working on 28 GHz has been fabricated by FDM with low-cost and easy-to-fabricate process [17].

Material extrusion printers have a wide selection of materials including polylactic acid, ABS copolymer, high impact polystyrene, nylon, PC, polyethylene terephthalate, polyetherimide, polyvinyl alcohol, thermoplastic polyurethane, and innovative glass and fiber reinforced materials. The material is usually prefabricated into a thermoplastic filament core with a diameter of 1.75 or 3.00 mm. The unused filament cores are recommended to be stored in cool, dry, dark place. In recent years, it has been possible to realize metal 3D printed material extrusion by adding a ceramic release layer between the support structure and the parts, which is convenient to remove the support after printing.

Figure 6.5 shows the schematic of a typical material extrusion machine. The printing material roll is melted by thermal melting/softening and extruded by wheel to pass through the sprinkler under constant pressure, and then stacking molding layer by layer on the print platform. Supporters are required in the printing process, but they can be printed with the same material as the main part. If the supporters use different materials as the main part, these supporters can be dissolved in hot water or solvent (such as alkaline solution). If the supporters use the non-removable material or the same material with the main model, they can be designed with the removable shape and manually removed from the printed object in the final step. In some cases, the object can be oriented on the print platform without the use of supporters.

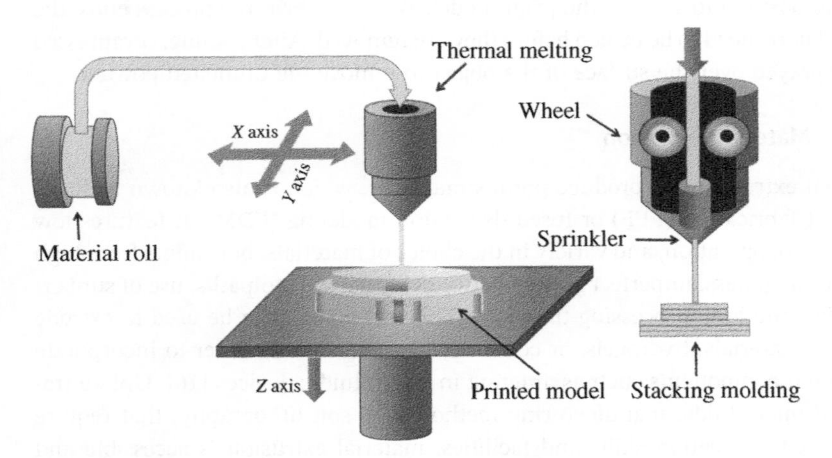

Figure 6.5 Schematic of a typical material extrusion printer.

6.3 3D Printed Antennas with Complementary Structure

In this section, a 3D printed metal antenna and a 3D printed dielectric antenna with complementary structure are given as examples. For each example, antenna configurations, working principles, fabrication prototypes, and tolerances are discussed for comparison. Both antennas are designed for mmWave communications. In recent years, mmWave technology has been widely used in communication, radar, remote sensing, astronomy, and many other advanced applications [18, 19]. Especially, 5G/6G antennas are currently under intensive research and development in both industry and academia. 3D printing technology has been proposed as an effective means to reduce antenna costs and improve the manufacturing efficiency [20].

6.3.1 3D Printed Air-Filled Metallic Waveguide Antenna (AFMWA)

The configuration and prototype of the proposed air-filled metallic waveguide antenna (AFMWA) is shown in Figure 6.6. It is designed in copper with hollow structure, simulated and optimized by Ansoft HFSS software, and manufactured by SLS technology. It is fed by WR-34 rectangular waveguide and designed with two symmetrical slots in taper shape. It was found that the taper shape is the key factor to achieve wideband performance and circular polarization radiation.

The operation of the proposed antenna is based on the transverse electric (TE) wave transmitted through the rectangular waveguide (WR-34), where the wide sides of rectangular waveguide have symmetrical slot structures. Figure 6.7 shows a cycle of distribution of the surface current at 28 GHz. When the current flows along the slot line, the electric field in the original vertical direction changes. The slots on wide side waveguide can produce horizontal electric field. Vertical electric field from the waveguide can couple with the horizontal electric field. By adjusting the position of the slot, the amplitude of the horizontal field and vertical field observed in the far-field are equal, thus achieving circular polarization. The circular polarization phase difference is also affected by the slot structure. Therefore, it is shown in Figure 6.8a that the simulated amplitude difference between horizontal field and vertical field is $\mathrm{dB}(E_h) - \mathrm{dB}(E_v)$, as the proposed antenna adopts the slot with W_2 of 13.5, 14.5, 15.5, 16.5, and 17.5 mm. It can be seen that W_2 alters the amplitude difference between the horizontal field and vertical field. As shown in Figure 6.8b, the increase of W_2 has significant effect on the phase difference between horizontal field and vertical field. The optimum value of W_2 is 15.5 mm.

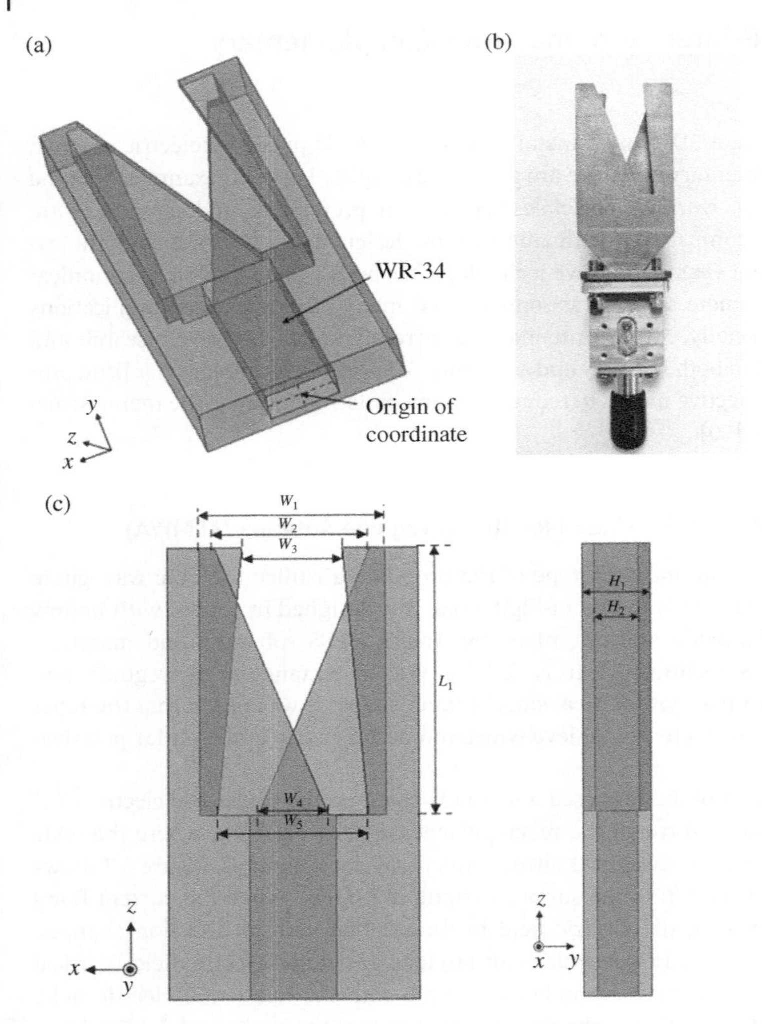

Figure 6.6 The proposed circularly polarized antenna. (a) Configuration. (b) Photos of the 3D printed antenna prototype. (c) The dimensions of the proposed antenna with W_1 = 18.5 mm, W_2 = 15.5 mm, W_3 = 10 mm, W_4 = 7 mm, W_5 = 15 mm, L_1 = 29.5 mm, H_1 = 7.2 mm, and H_2 = 4.7 mm.

Figure 6.9 shows the electric field of the proposed antenna at 24, 28, and 32 GHz. It can be seen that the slotted waveguide excites the mode of TE_{10}. Over a wide working frequency, the proposed antenna keeps the operation of TE_{10} mode. Therefore, it achieves a wide impedance bandwidth from 24 to 32 GHz.

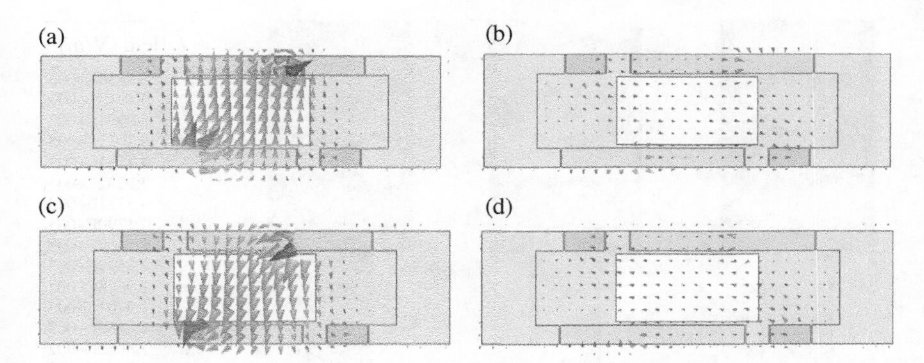

Figure 6.7 The distribution of electric field of the proposed antenna at 28 GHz with (a) 0 cycle, (b) 1/4 cycle, (c) 1/2 cycle, and (d) 3/4 cycle.

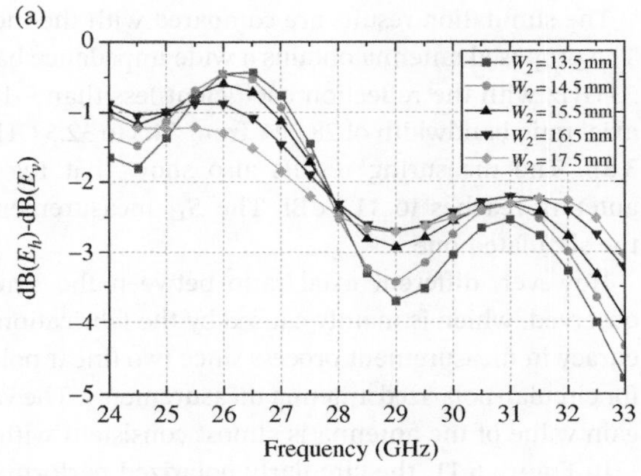

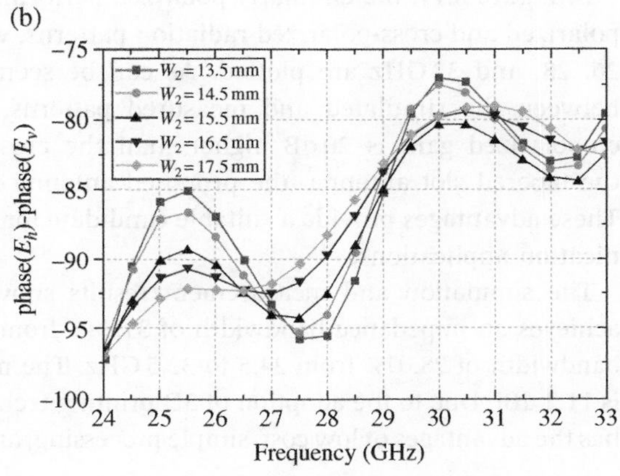

Figure 6.8 (a) Simulated dB(E_h)-dB(E_v) for different values of W_2. (b) simulated phase of E_h-phase of E_v for different values of W_2.

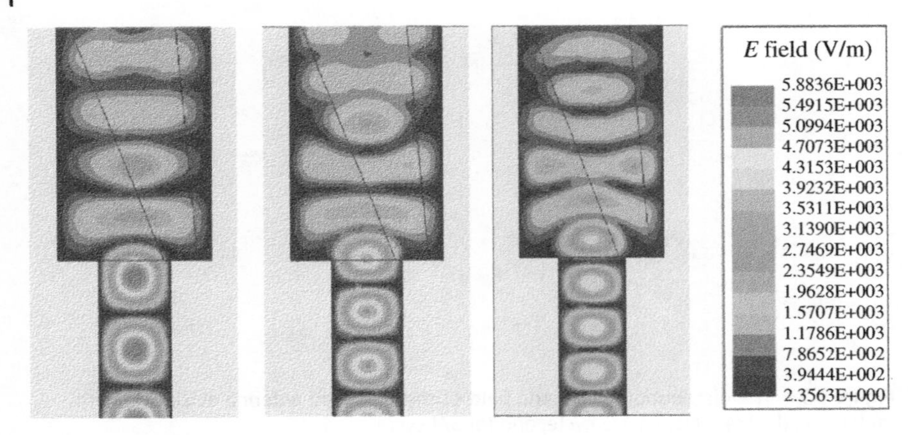

Figure 6.9 The electric field of the proposed antenna at 24, 28, and 32 GHz.

The simulation results are compared with the measured results in Figure 6.10. The proposed antenna obtains a wide impedance bandwidth of 31.58% from 24 to 33 GHz with the reflection coefficient less than −10 dB, which also achieves the axial ratio bandwidth of 28.07% from 24.5 to 32.5 GHz with the axial ratio less than 3 dB. The measuring results also show that the maximum gain of proposed antenna reaches to 11.2 dBi. The S_{11} measurement is in good agreement with the simulated one.

However, different axial ratio between the simulation and measurement is observed, which is mainly caused by the fabrication tolerance as well as the inaccuracy in measurement process since two linear polarized antennas are employed for circular-polarized antenna measurements. The variation trend of the measured gain value of the antenna is almost consistent with the simulation value.

In Figure 6.11, the circularly polarized performance is observed from the co-polarized and cross-polarized radiation patterns, where *E*-plane and *H*-plane at 26, 28, and 32 GHz are plotted. As can be seen, there is a good agreement between the simulated and measured patterns. In broadside direction, the co-polarized gain is 20 dB higher than the cross-polarized gain. Because of the tapered slot antenna, the proposed antenna has at least 10 dB back lobe. These advantages provide a suitable candidate for the millimeter-wave communication applications.

The simulation and measurement results show that the proposed antenna achieves an impedance bandwidth of 31.58% from 24 to 33 GHz and a 3 dB AR bandwidth of 28.07% from 24.5 to 32.5 GHz. The maximum gain of the antenna is 11.2 dBi. Due to the adoption of 3D printing technology, the proposed antenna has the advantages of low cost, simple processing, and environmental friendliness.

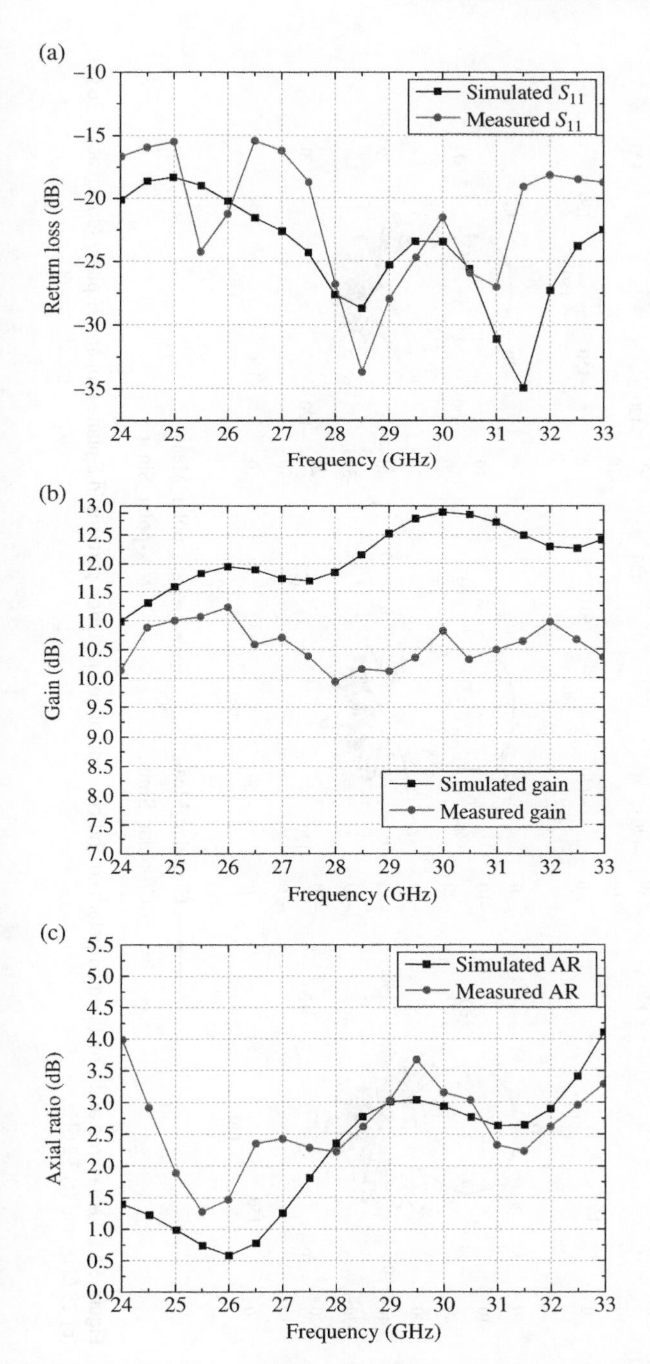

Figure 6.10 (a) $S_{(1,1)}$ of the proposed antenna, (b) gain of the proposed antenna, and (c) axial ratio of the proposed antenna.

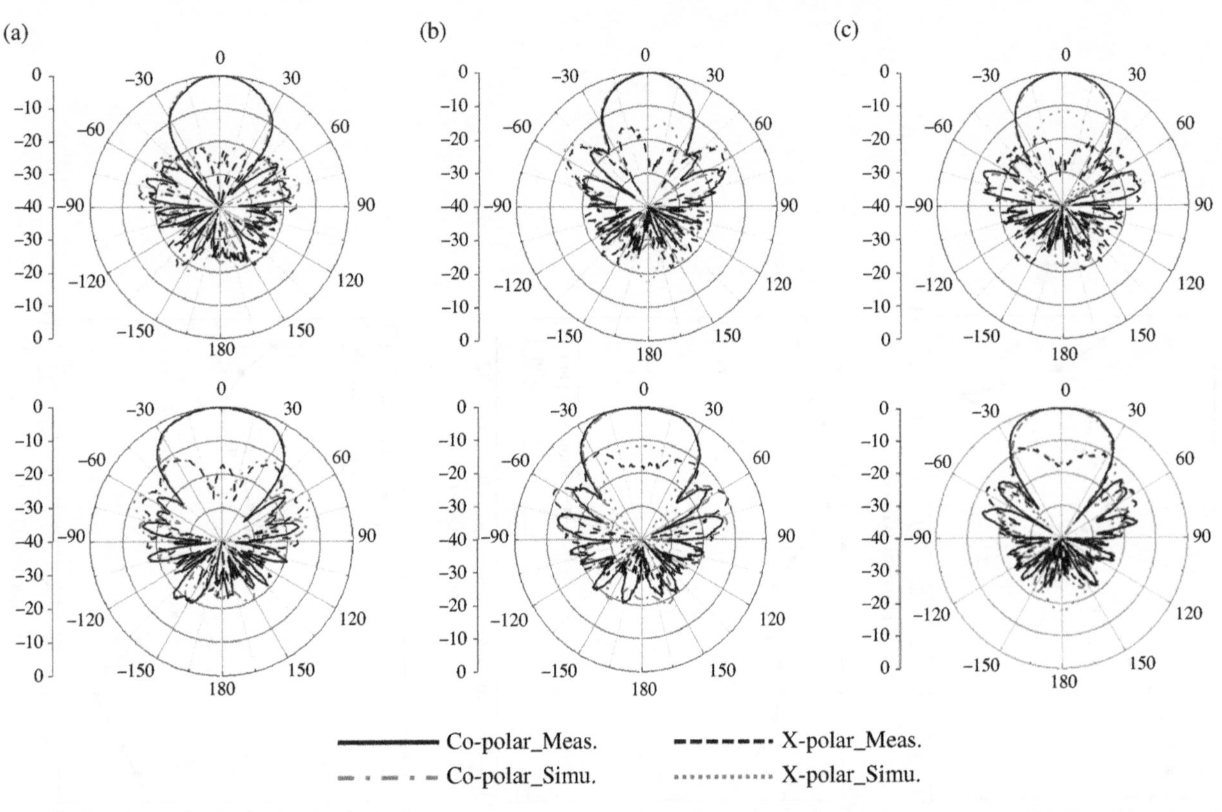

Figure 6.11 Measured and simulated co-polarized and cross-polarized radiation patterns in *E*-plane (up) and *H*-plane (down) at (a) 26 GHz, (b) 28 GHz, and (c) 32 GHz.

This indicates 3D printed antenna could be a competitive low-cost solution for 5G millimeter-wave communications.

6.3.2 3D Printed Dielectric Loaded Antenna (DLA)

A dielectric loaded antenna (DLA) fabricated by MJ technology is designed with complementary structure for comparison. As mentioned before, dielectric antenna shows its superior features on flexible design, compact structure, and high radiation efficiency. Recent advancement in 3D printing technologies has demonstrated their capability to overcome the drawbacks of high cost as well as the bulky nature of the ceramic dielectric, thus stimulating the drive to explore their use in printing dielectric antenna design. Most importantly, 3D printed dielectric also enables unconventional structures to realize some unique functions such as low cross-polarization [21] below and lower sidelobe level [22]. For these reasons, 3D printed dielectric antennas have aroused a growing interest in recent years.

To excite a solid dielectric antenna, transition would be required for guiding the input power into the dielectric structure. There are three excitation techniques, namely, the probe excitation, aperture excitation, and inset dielectric excitation. Probe excitation is commonly used for coaxial feed antennas, which are widely utilized in sub-6 GHz band but may suffer from high loss in mmWave band. Aperture excitation is usually adopted by the waveguide for substrate integrated waveguide (SIW) transition to feed the mmWave antenna and array. However, waveguide-to-SIW transition requires multilayered printed circuit boards (PCBs) for wideband operation and bulky flange for positioning [19]. The use of inset dielectric transitions alleviates the above problems by providing direct paths between the waveguide and dielectric structure [23].

In this work, a mmWave circular-polarized DLA incorporating with a novel inset dielectric transition is implemented by the 3D printing technology. This design has the advantages of low profile, low loss, and being able to clip onto the standard waveguide without the help of flange. This provides an effective means for excitation of the solid dielectric antenna.

The configuration of the proposed antenna is showed in Figure 6.12. It consists of a dielectric-loaded major waveguide, a dielectric-loaded minor waveguide, a pair of antipodal dielectric taper, and a pair of z-shaped inset dielectric transition.

They utilize the dielectric of relative permittivity of $\varepsilon_r = 2.9$ and loss tangent of $\tan\delta = 0.01$, which can be fabricated as a seamless integration. The shell over the dielectric is the thin metal film with 0.15 mm thickness. The inner size of the dielectric-loaded minor waveguide equals to that of a WR-34 waveguide (broad wall $a = 0.836$ mm, narrow wall $b = 4.318$ mm). The dielectric-loaded major waveguide has a larger size of $a = D_1$ and $b = H_1$. A pair of antipodal tapered slots is etched on the broad walls of the dielectric-loaded waveguide. The slots are

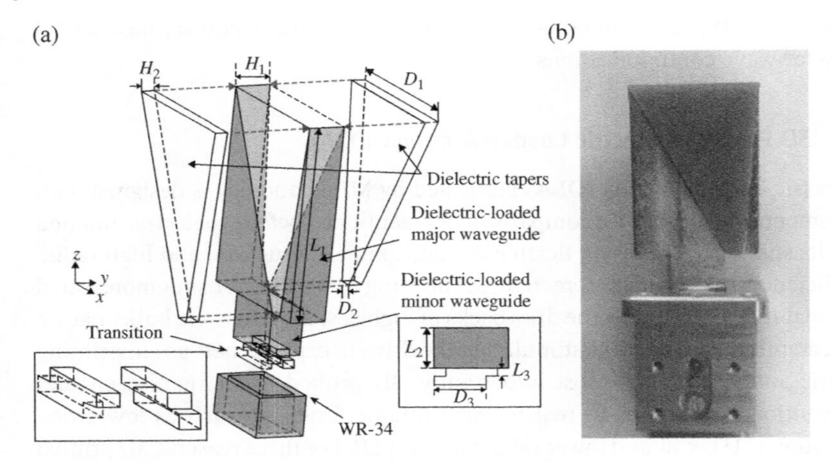

Figure 6.12 (a) Configuration of the proposed antenna with D_1 = 18.5 mm, D_2 = 3.73 mm, D_3 = 6.0 mm, H_1 = 4.7 mm, H_2 = 1.8 mm, L_1 = 25.0 mm, L_2 = 5.7 mm, and L_3 = 1.0 mm. (b) Photos of the 3D printed antenna prototypes.

covered with dielectric tapers with the same shapes. The inset transition has two z-shaped dielectrics, which are antipodal symmetrically placed on the opposite corners of the minor waveguide. Each z-shaped dielectric consists of two rectangular blocks with equal height but unequal thicknesses (top 1.7 mm and bottom 1.4 mm). This transition can be positioned and clipped onto the rectangular waveguide without the help of flange.

The power fed from the rectangular WR-34 waveguide is guided into the dielectric waveguide via the inset dielectric transition, and then radiates from the DLA. It is essential to design a low-loss transition to ensure good impedance matching and high antenna radiation efficiency. Moreover, DLA in proper design may present some unique functions from extra high-order mode excitations.

Figure 6.13 displays the mode dispersion curves of the air-filled WR-34 waveguide and dielectric-filled WR-34 waveguide. Over the band from 24 to 32 GHz, the air-filled WR-34 waveguide only supports TE_{10} mode, whereas dielectric-filled WR-34 waveguide could support eight modes. These results, in practice, the need in using a transition to address the issue on high transmission loss and unmatched mode dispersion.

As depicted in Figure 6.14, the comparison shows that without the transition, the transmission loss is in the range between 0.27 and 0.42 dB from 24 to 32 GHz. After adding a pair of rectangular blocks on the opposite dielectric corners, the transmission loss is reduced to less than 0.35 dB by improving the matching over the air-to-dielectric interface. The transmission loss can be

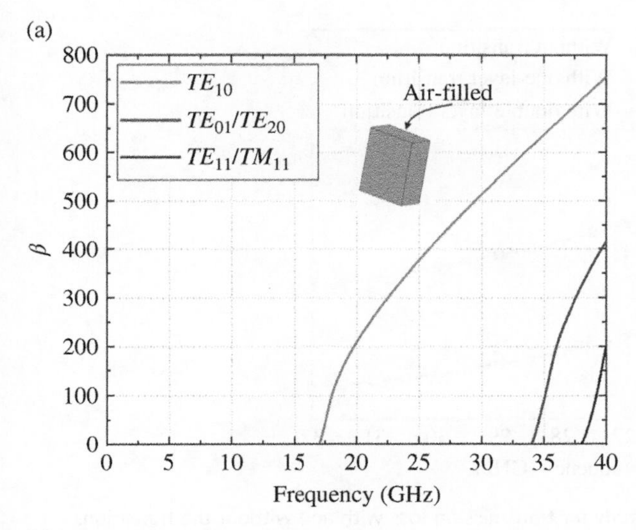

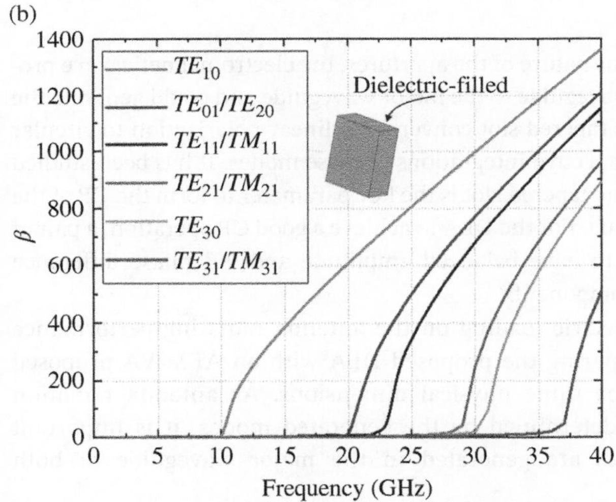

Figure 6.13 The dispersion curves of (a) air-filled waveguide. (b) Dielectric-filled waveguide with $\varepsilon_r = 2.9$.

further reduced by properly adding the dielectric blocks. In this design, a pair of z-shaped dielectrics is developed to achieve a low transmission loss of 0.04–0.19 dB. Their dimensions and positions are optimized by Ansoft HFSS software to achieve the minimum transmission loss and the lowest profile. In addition, this configuration is suitable to be clipped onto a standard waveguide on the antenna systems.

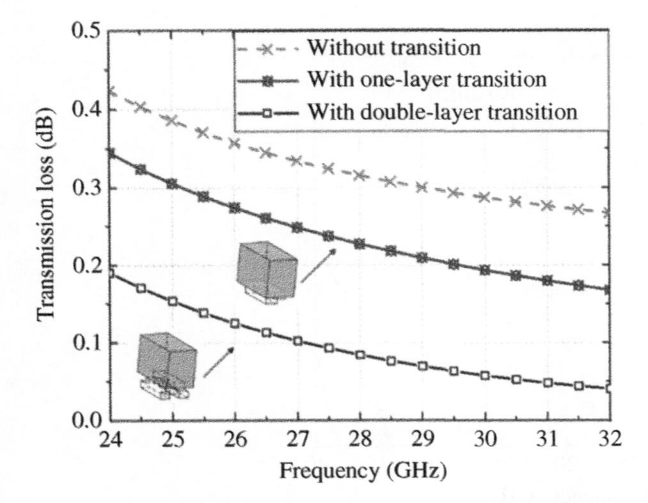

Figure 6.14 Parametric study for transmission loss with and without the transition.

Due to the discontinuous nature of the apertures, the electromagnetic wave propagates from the minor waveguide to the major waveguide and could generate the diverse modes. Then, the tapered slot converts the linear polarization to circular polarization (CP) with respective integrations of these modes. It has been studied in [14] that the shape of the tapered slot is the key parameter to form the CP of the metallic waveguide antenna. For the DLA to achieve a good CP operation, a pair of tapered blocks is added to have balanced amplitude and 90° phase difference between the E_x and E_y components.

The effects of the dielectric loading on the antenna radiating performance are investigated by comparing the proposed DLA with an AFMWA proposed in Section 6.3.1 with the same physical dimensions. As antenna radiation performance is mainly determined by the generated modes, it is important to explore which modes are generated in the major waveguide of both antennas.

It is well known that a certain high-order mode can propagate in the waveguide only if the cutoff frequency $f_{c,mn}$ is lower than the operating frequency f. The equation is demonstrated as [24],

$$f_{c,mn} = \frac{1}{2\pi\sqrt{\mu\varepsilon}}\sqrt{\left(\frac{m\pi}{a}\right)^2 + \left(\frac{m\pi}{b}\right)^2} < f \tag{6.1}$$

where ε and μ are the permittivity and permeability of the loading material, a is the dimension of the broad wall, and b is the dimension of the narrow wall. Both m and n are non-negative integers. Figure 6.15 shows the E-field distributions over

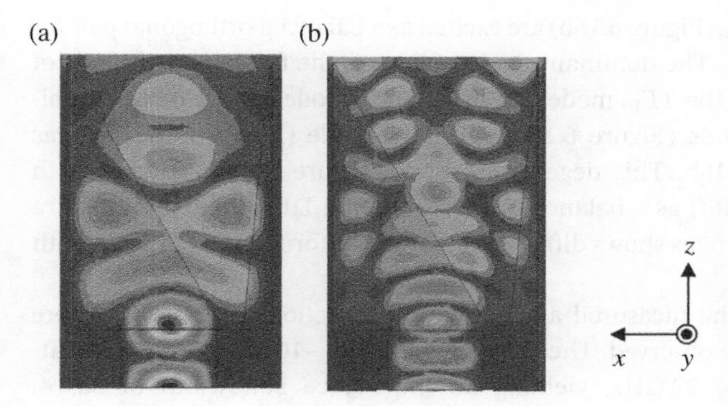

Figure 6.15 The *E*-field distribution over the *xoz*-plane (*y* = 0 mm from the antenna center). (a) AFMWA. (b) DLA.

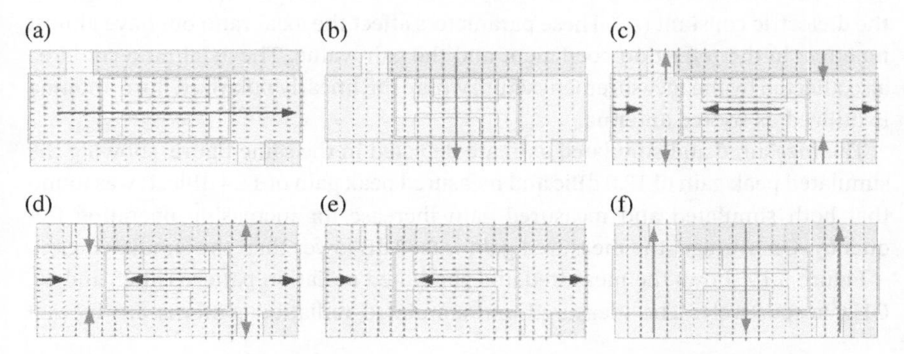

Figure 6.16 The mode field display over the antenna aperture. (a) TE_{01} mode. (b) TE_{10} mode. (c) TE_{21} mode. (d) TM_{21} mode. (e) Degenerate mode of TE_{21} and TE_{21}. (f) TE_{30} mode.

xoz-plane of the AFMWA and DLA. It is observed that the AFMWA is dominated by the fundamental modes, whereas DLA is excited with high-order modes. It is verified by (1) that the dielectric loaded waveguide supports extra higher modes due to the larger permittivity ε_r of the loading dielectric.

Figure 6.16 shows the fields of top five strongest modes over the antenna aperture. They include the fundamental modes of TE_{01} and TE_{10} and high-order modes of TE_{21}, TM_{21}, and TE_{30}. Other high-order modes can be also detected and calculated, but their amplitudes are 20 dB less than selected modes, hence they are omitted in this analysis.

The balanced amplitude and 90° phase difference of the E_x and E_y are essential for a good CP operation. For the AFMWA, fundamental modes of TE_{01}

(Figure 6.16a) and TE_{10} (Figure 6.16b) are excited as a balanced orthogonal pair for E_x and E_y, respectively. The dominant mode of DLA is the fundamental mode of TE_{10}. It is followed by the TE_{10} mode and degenerate mode, which is the combination of the TE_{21} mode (Figure 6.16c) and TE_{21} mode (Figure 6.16d) as was observed in Figure 6.16b. This degenerate mode (Figure 6.16e) matches with TE_{30} mode (Figure 6.16f) as a balanced pair. Therefore, DLA excited with extra high-order balanced modes shows different radiating performance compared with AFMWA.

Figure 6.17 shows the measured and simulated reflection coefficients, where good agreement can be observed. The $|S_{11}|$ is lower than −10 dB across all operating bands from 24 to 32 GHz, yielding an impedance bandwidth of 28.6%. Figure 6.18 shows the measured and simulated axial ratio and gain performance over the operating band. It was noted that the 3 dB axial ratio bandwidth is reduced from 28.6% to 8.2% in the measured results. A higher measured axial ratio is observed, which is mainly affected by the thickness of the taper plate (H_2) and the dielectric constant (ε_r). These parameters affect the axial ratio but have almost no effect to the reflection coefficient and the gain value. The axial ratio could be also affected by the measurement setup, where the linearly polarized horn antenna is utilized as source antenna.

The measured and simulated gain are depicted in the same figure, showing the simulated peak gain of 12.0 dBic and measured peak gain of 11.4 dBic. It was found that both simulated and measured gain increase for increasing operating frequency. On average, the measured gain is 0.82 dB lower than the simulated gain.

Figure 6.19 shows the measured and simulated radiation patterns at 25 and 30 GHz, respectively. The measured and simulated radiation patterns are in well

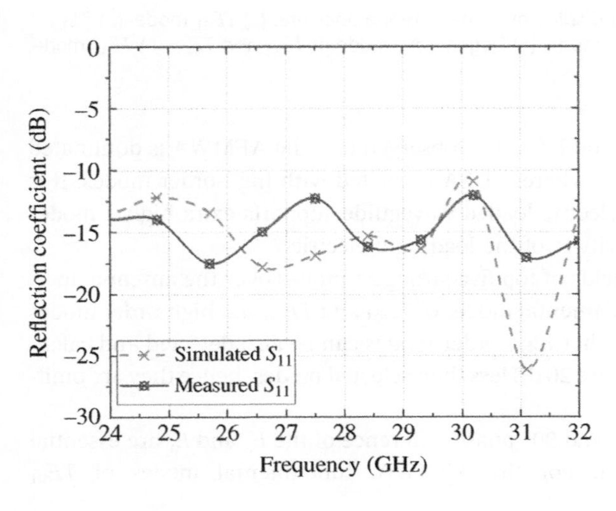

Figure 6.17 Measured and simulated reflection coefficient of the proposed DLA.

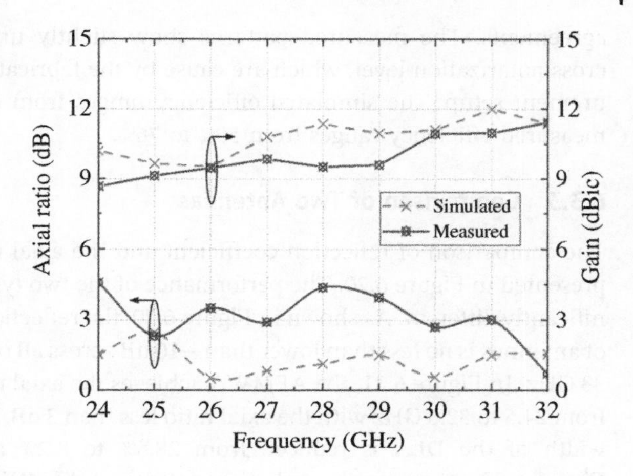

Figure 6.18 Measured and simulated axial ratio and gain of the proposed DLA.

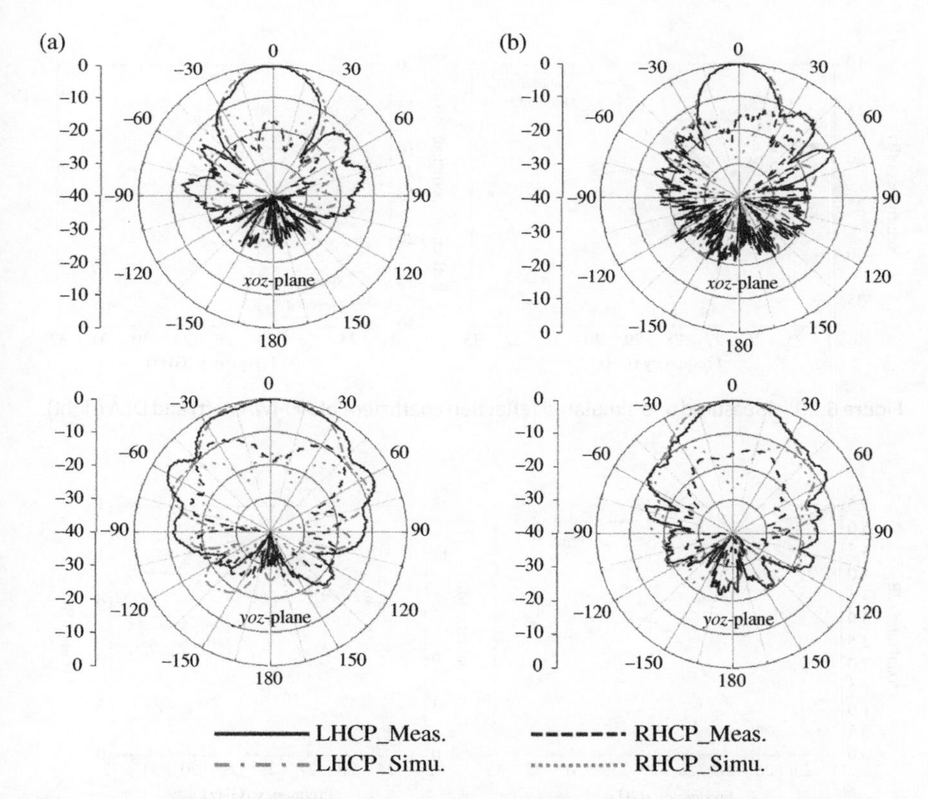

Figure 6.19 Measured and simulated LHCP and RHCP radiation patterns of the proposed DLA operating at (a) 25 GHz and (b) 30 GHz.

agreements. The measured patterns show slightly unsymmetrical and higher cross-polarization level, which are cause by the fabrication tolerances and measurement setup. The simulated efficiency ranges from 69% to 76%, whereas the measured efficiency ranges from 50% to 76%.

6.3.3 Comparison of Two Antennas

The comparison of reflection coefficient and the axial ratio of two antennas are presented in Figure 6.20. The performance of the two types of antennas is not significantly different. As shown in Figure 6.20, the reflection coefficient of two types of antennas is no less than lower than −10 dB across all operating bands from 24 to 33 GHz. In Figure 6.21, the AFMWA achieves the axial ratio bandwidth of 28.07% from 24.5 to 32.5 GHz with the axial ratio less than 3 dB. The 3 dB axial ratio bandwidth of the DLA is reduced from 28.6% to 8.2% in the measured results. Figure 6.22 compares the radiation pattern of AFMWA and dielectric antenna

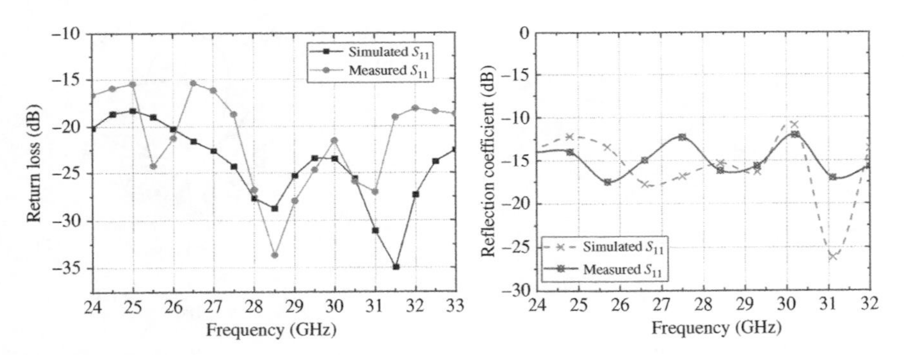

Figure 6.20 Measured and simulated reflection coefficient of AFMWA (left) and DLA (right).

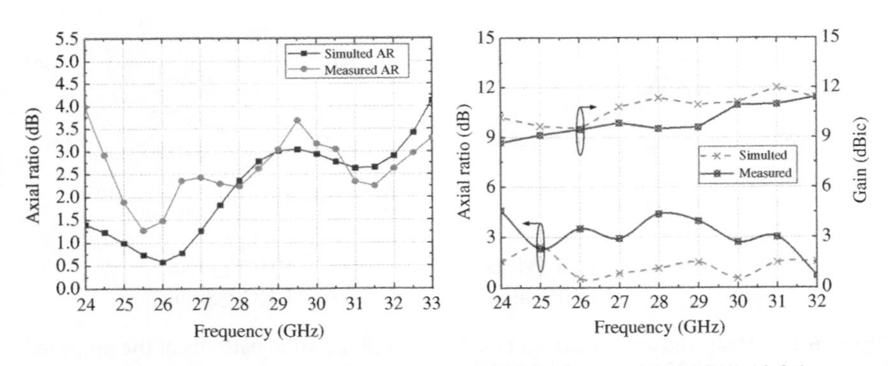

Figure 6.21 Measured and simulated axial ratio of AFMWA (left) and DLA (right).

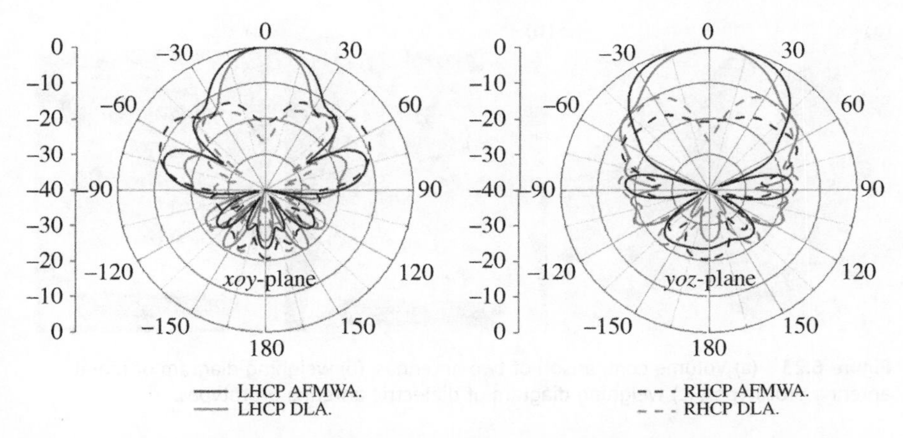

Figure 6.22 The LHCP and RHCP radiation patterns of the AFMWA and DLA at 30 GHz; *xoz*-plane (left); *yoz*-plane (right).

Table 6.2 Comparison of 3D printed tapered slot antennas.

Name	Volume (cm^3)	Weight (g)	0° Cx-polar @30 GHz (dBic)	Peak gain (dBic)
AFMWA [14] Sim.	$2.5 \times 0.72 \times 4.95$	41.585	−15.47	12.90
AFMWA [14] Mea.			−18.55	11.30
DLA Sim.	$1.87 \times 0.83 \times 3.27$	4.1739	−27.26	12.00
DLA Mea.			−16.47	11.42

at 30 GHz. It was noticed that the loading of dielectric reduces the 3 dB beamwidth from ±18 to ±14 in *xoz*- and ±33 to ±23 in yoz-plane. Moreover, the dielectric antenna shows around 5 dB lower cross-polarization level than that of AFMWA in broadside direction at 30 GHz. These characteristics are benefited from the excitation of extra high-order modes.

A comparison of the performance of tapered slot antennas is shown in Table 6.2. The proposed DLA has a very similar simulated radiation performance to that of a conventional tapered slot antenna without the dielectric loading [14]. As shown in Figure 6.23, it was found that antenna with dielectric loading can significantly reduce the antenna volume by 43% and the weight by 90%. In addition, with the help of transition, the proposed DLA can be clipped onto a standard waveguide without the need of flange. Note that narrower beamwidth is shown in theory

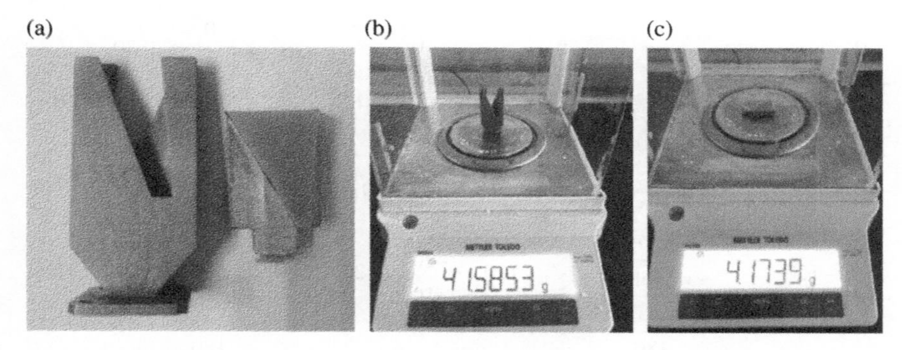

Figure 6.23 (a) Volume comparison of two antennas, (b) weighing diagram of metal antenna prototypes, (c) weighing diagram of dielectric antenna prototypes.

mainly because of the excitation of high-order modes. However, the fabrication error on the 3D printed dielectric antenna would degrade the axial ratio of the circular-polarized antenna. Nevertheless, due to advantages of low cost, compactness, and lightweight, the proposed DLA shows its optimistic prospect for the next generation.

6.4 3D Printed Finger Nail Antennas

In this section, we propose 3D printed antennas on a removable dielectric finger nail for on-body microwave and mmWave communications [25]. A dielectric-based patch antenna design is used to demonstrate this concept. Nanoparticle silver inks are dispensed using aerosol jet technology to produce the small features of the antennas. Flush curing was successfully employed to deposit layers of silver ink on the curved dielectric nails. The technique produced the high resolution required for the printed antennas as well as smooth and thin metallic layers. An additional copper layer was added to the 28 GHz through copper plating. Two antennas have been fabricated to operate at 15 and 28 GHz, respectively, providing good performance in terms of impedance match and bandwidth. The 15 GHz band antenna has been tested directly after printing and curing while an additional copper plating process is employed for the 28 GHz band antenna. The antennas' removability can allow portability of the electronic equipment it is attached to from one individual to another or the antenna from one device to another. This facilitates reusability of the device. These can find applications in future on-body sensing and communications though manicure electronics. The antenna designs, operation, and fabrication process of the two removable nail antennas operating at microwave and millimeter-wave are detailed as follows.

6.4.1 Microwave Removable Nail Antenna

In this section an antenna operating at 15 GHz was designed, simulated, and tested for the scenario of off- and on-finger. The fake finger nail substrate is made of an ABS of 0.5 mm thickness with a relative permittivity (ε_r) of about 2.7 and loss tangent of 0.0051 [26]. The designed patch antenna comprises of a rectangular radiating patch with a microstrip transmission line and a rectangular ground on the back plane. The designed antenna is illustrated in Figure 6.24a and Table 6.3 represents its dimensions. Figure 6.24b shows the antenna curved to the shape of a nail while Figure 6.24c depicts the angle, 55.38°, of the curvature. On-finger simulation was conducted to determine the effect of human tissue on the antenna's performance as shown in Figure 6.25. Bone, fat, skin, and nail tissues all of which have different ε_r were considered. Their estimated dimensions are skin (1 mm), fat (2.0 mm), nail (0.5 mm), and bone (2 mm). Their electrical characteristics at 15 GHz are given in Table 6.4 [27]. Figure 6.25a depicts the cross and Figure 6.25b the longitudinal sections of the patch antenna on a non-homogenous human tissues layer.

The simulation reflection coefficient, S_{11}, results of the flat, curved antenna and the curved antenna on the finger are shown in Figure 6.26. Flat and curved antenna S_{11} off-finger results are almost identical implying that bending had minimal effect on the S_{11}. However, with the antenna worn on finger, the resonant

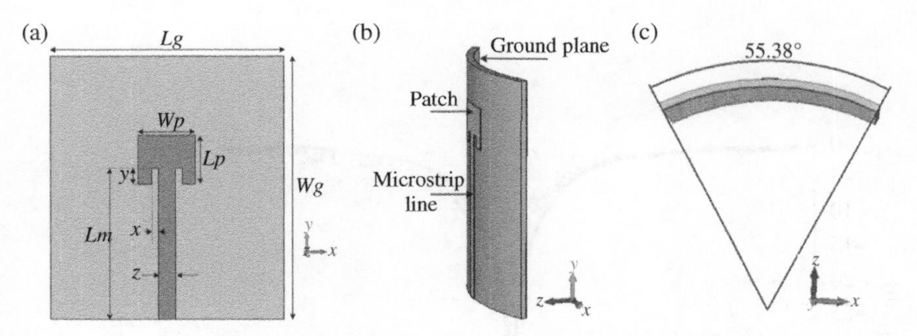

Figure 6.24 The designed (a) patch antenna design, (b) after curving into an arc, and (c) degree of curvature.

Table 6.3 Dimensions of the patch of the 15 GHz antenna (mm).

Wg	*Lg*	*Wp*	*Lp*	*Lm*	*x*	*y*	*z*
15	19	7.8	6.5	13	1.5	1.75	1.0

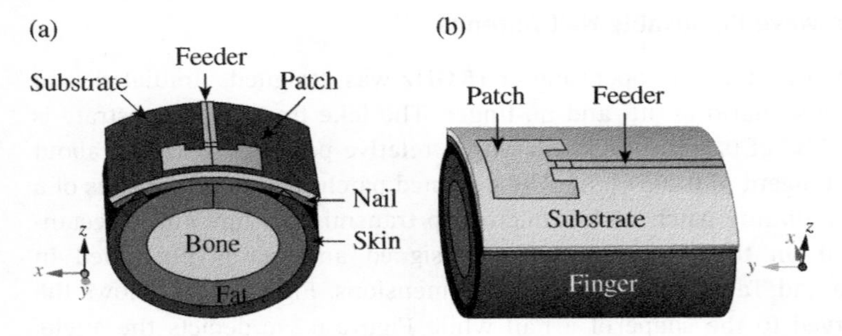

Figure 6.25 (a) Cross and (b) longitudinal section of the human finger model of the removable patch antenna.

Table 6.4 Electrical characteristics of human tissues at 15 GHz.

Tissue	Relative permittivity, ε_r	Conductivity, σ, (S/m)	Loss tangent, tan(δ)
Fat	4.2647	0.93625	0.26309
Skin	26.401	13.847	0.62855
Nail/bone	6.8698	3.1355	0.54695

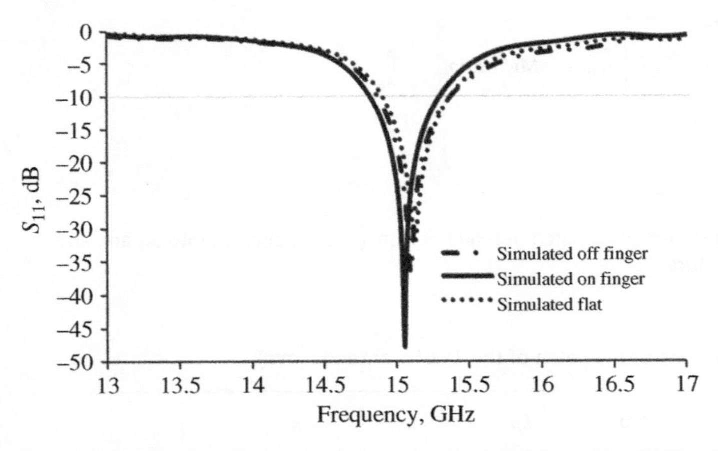

Figure 6.26 Simulated S_{11} results of flat, curved off, and on-finger simulated antenna at 15 GHz.

point slightly shifts to the left. The results indicate a −10 dB impedance bandwidth from 14.8 to 15.3 GHz (2.9%), 14.88 to 15.37 GHz (3.2%), and 14.82 to 15.30 GHz (3.2%) for the flat, curved off-finger and curved on-finger microstrip patch antennas, respectively. The results indicate bandwidth consistency for the three cases.

The microstrip patch nail antennas were fabricated by depositing the conductive ink using Optomec's aerosol jet technology. Aerosol jet printing manufacturing technology is emerging as a substitute for the traditional thick-film processes such as screen-print, photolithography, and micro-dispensing and has been described as superior to inkjet printing [28]. Figure 6.27 depicts working principle of the aerosol jet technology. The aerosol jet process uses aerodynamics to deposit functional material aerosolized droplets onto a substrate. The functional liquid is aerosolized into globules and then focused as collimated beams of diameter of around 10 μm after it has been passed through a deposition head. The deposition head sends out the aerosol beam that impinges the droplet on the substrate [30]. To print the features, the deposition head is translated in the *XYZ* and Theta directions with respect to the substrate. The CAD design file generated tool path guides the deposition head translation. This allows the deposition head to print in any orientation. Thus it can print on 3D surfaces and not just on smooth and flat surface.

To fabricate the antennas, the digital model was exported from CST Microwave StudioTM to an STL file. The metallic layers that constitute the radiator and the microstip transmission line were uniformly deposited onto the fake nail using the Optomec's aerosol jetting process that sprays Cabot CS-32 silver conductive ink.

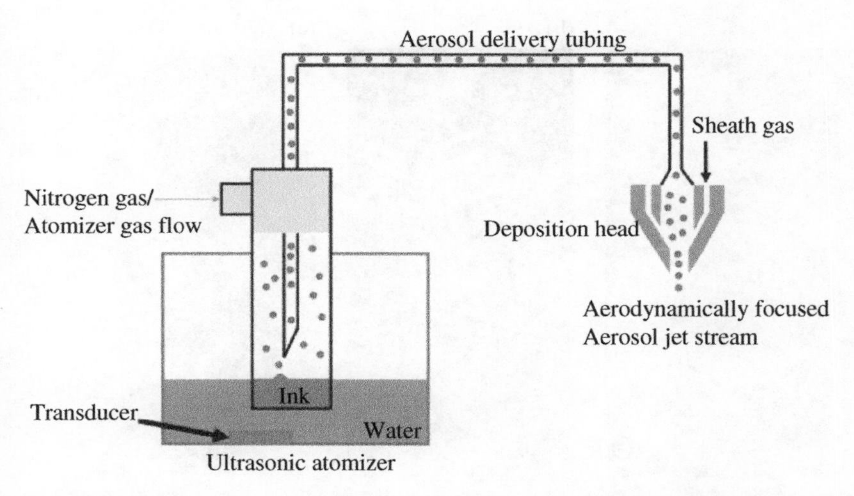

Figure 6.27 Schematic of aerosol jet printing process. *Source:* [29]/IOP Publishing/CCBY 3.0/Public domain.

The antennas were left to dry for about 24 hours before being transferred to a NovaCentrix PulseForge [31] machine to cure. The fabrication was done at the Centre for Process Innovation (CPI) [32]. Figure 6.28a shows the fake nail on which the antennas were printed; Figure 6.28b, the fabrication process; and Figure 6.28c, the fabricated antenna. An SMA End Launcher Connector was attached to the feedline of the antenna (Figure 6.28c). A digital multi-meter was used to measure the resistance of the print to test if the ink deposited on the substrate was cured as well as check for electrical continuity of the print. The resistance observed after curing was consistent with the expected resistivity of 2×10^{-7} Ωm [33] for silver ink. Figure 6.29 shows the surface profile of the silver ink layer of the patch element of antenna. The surface profile was analyzed using Talysurf CCI and showed a roughness of about 1 μm. The ground plane of the antenna was created using adhesive copper tape that was attached to the back of the ABS nail.

A graph of the measured and simulated S_{11} results is shown in Figure 6.30, where the S_{11} measurements were obtained using an Anritsu 37397C vector network analyzer for the nail antenna off- and on-finger. The measured S_{11} results show a better matching and wider −10 dB impedance bandwidth relative to the simulated results. This could be due to further resistive losses in the materials not accounted for, connectors, and errors in the fabrication. The resonant points also shifted slightly to the right. The −10 dB impedance bandwidths from 14.6 to 16.0 GHz (9.8%) and 14.5 to 15.9 GHz (9.1%) for the antenna off- and on-finger,

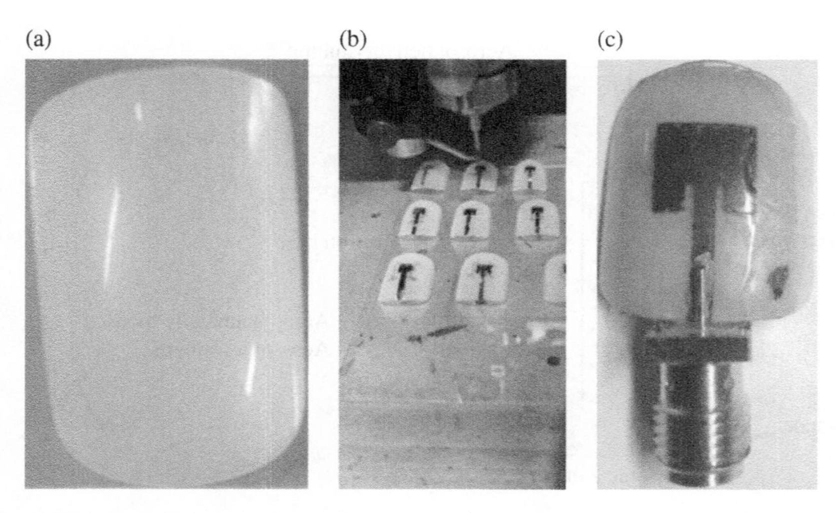

Figure 6.28 (a) The fake removable nail, (b) printing of the antennas using the Optomec machine, and (c) the fabricated antenna.

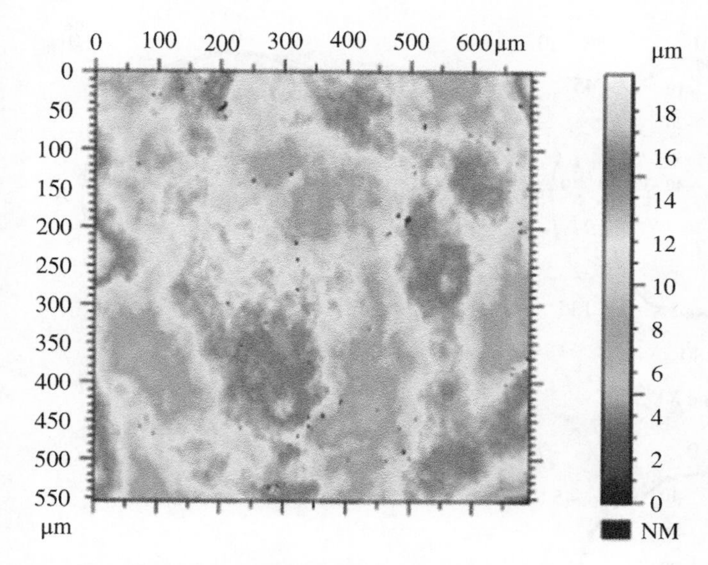

Figure 6.29 Surface profile of the antenna patch.

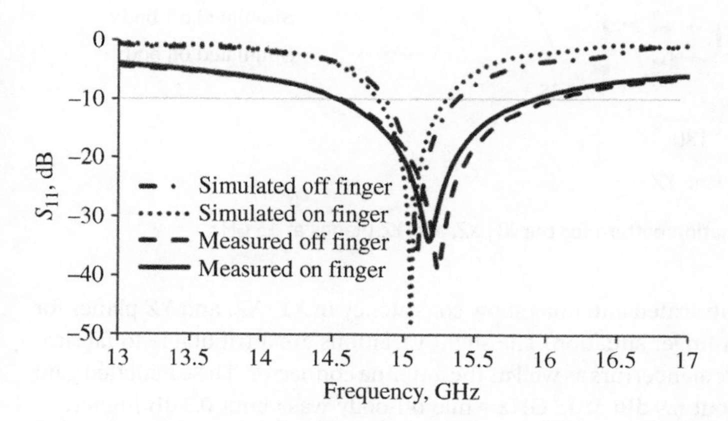

Figure 6.30 Reflection coefficient (S_{11}) of the optimized 15 GHz antenna.

respectively, are realized. A slight shift of the resonance point to the left is observed for the on-finger antenna. Far-field radiation pattern was performed in an anechoic chamber. Figure 6.31 shows the radiation patterns for both off- and on-finger for the simulated and fabricated antennas. Patterns are as expected for a patch antenna on a small curved ground plane with a main lobe out of the finger nail and lower radiation toward the finger and body. Radiation pattern for both

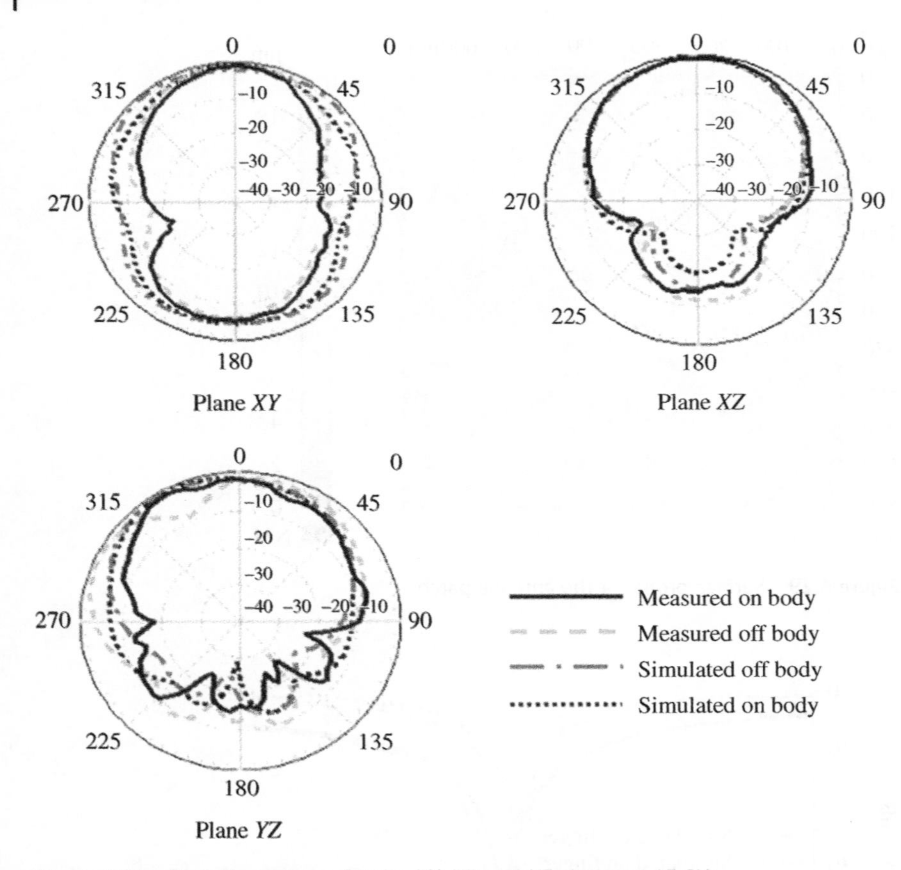

Figure 6.31 Radiation pattern for the *XY*, *XZ*, and *YZ* planes at 15 GHz.

simulated and fabricated antennas show consistency in *XY*, *XZ*, and *YZ* planes for both off- and on-finger situation. The slight variations are attributable to fabrication and measurement errors as well as the antenna connector. The calculated gain on body was about 6.9 dBi at 15 GHz while off-body was about 0.3 dB higher.

6.4.2 mmWave Nail Antenna

A mmWave nail antenna is introduced here to increase the communication bandwidth for higher data rate. For the higher frequency, a higher conductivity antenna surface material is preferable. To improve conductivity, a layer of copper can be added to the metallic tracks using an electroplating process. A mmWave frequency antenna with dimensions shown in Table 6.5 was designed, simulated, and tested

Table 6.5 Dimensions of the 28 GHz patch antenna (mm).

Wg	*Lg*	*Wp*	*Lp*	*Lm*	*x*	*y*	*z*
14.96	17.45	3.72	3.25	10.0	0.44	1.06	1.10

Table 6.6 Electrical characteristics of human tissues at 28 GHz.

Tissue	Relative permittivity, ε_r	Conductivity, σ (S/m)	Loss tangent, tan(δ)
Fat	3.6985	1.6979	0.29471
Skin	16.552	25.824	1.0016
Nail/bone	5.1671	4.9427	0.6141

for both off- and on-finger. On-finger worn antenna was simulated to determine the effect of human tissues on the antenna performance at the mmWave. Table 6.6 [27] shows the electrical characteristics of the same human body tissues considered for the microwave antenna for on-body antenna simulation at 28 GHz. Reflection coefficient and radiation pattern performance parameters were used to gauge the performance of the antennas. Figure 6.32 shows the simulated S_{11} of the flat, bent off-finger and on-finger antennas. The antenna resonances at 28 GHz have only minor frequency shifts for the three cases. The results indicate a −10 dB impedance bandwidth of 27.5 to 28.6 GHz (3.9%), 27.5 to 28.5 GHz (3.6%), and

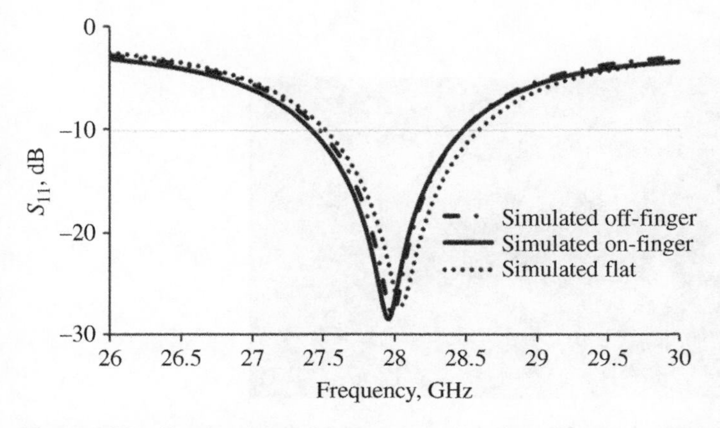

Figure 6.32 Simulated S_{11} of flat, curved off, and on-finger antennas at 28 GHz.

27.5 to 28.5 GHz (3.6%) for the flat, curved off-finger, and on-the-finger microstrip patch antenna, respectively.

The 28 GHz antenna was fabricated using the same fabrication process of the 15 GHz antenna. After the radiator and microstrip transmission line were printed and cured, a copper layer was added through an electroplating process.

A digital microscope from Keyence (UK) Limited was used to observe the antenna surface, measure it roughness, and photograph the surface. Figure 6.33 shows the surface of the copper plated radiator and its feedline at ×50 magnification. Surface roughness measurements of the radiator and feedline are shown in Figure 6.34. Figure 6.34a shows the conductor height while Figure 6.34b shows the two points at which the measurements were taken on either sides of the feedline inset. Figure 6.34c depicts the actual measurement readings of 9.4 and 10.9 μm at the edge of the tracks. The conductor height measured at the inset/patch point longitudinal to the feedline was also measured and found to be about 13.5 μm.

Figure 6.35a implies that though a viable antenna was produced, the fabrication process produced uneven surface. The surface roughness (R_a) was measured on the feedline section of the antenna and found to be 0.8 μm in the profile depiction as shown in Figure 6.35b. The patch antenna microstrip feedline matches the 50 Ω impedance of a low profile 2.92 mm SMA Jack (female) end launch connector from Southwest Microwave, Inc. as shown in Figure 6.36. Figure 6.36a shows the fabricated antennas after the electroplating process while Figure 6.36b depicts the antenna worn on a finger. Figure 6.37 shows the measured S_{11} of the fabricated antenna and its comparison with simulations for both the unworn and worn antennas. The measured antenna operates at 28 GHz with a minor difference off and on the body. The measured bandwidth was 27.0–29.8 GHz (10%) and 26.9–29.8 GHz (10.25%) for off- and on-finger antennas, respectively.

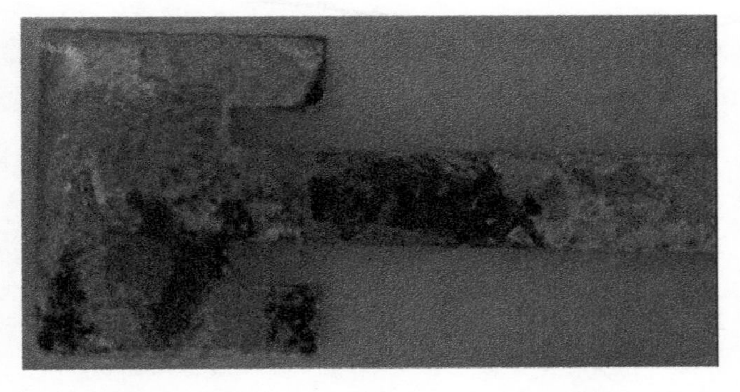

Figure 6.33 Blown out photo of the patch antenna and feedline.

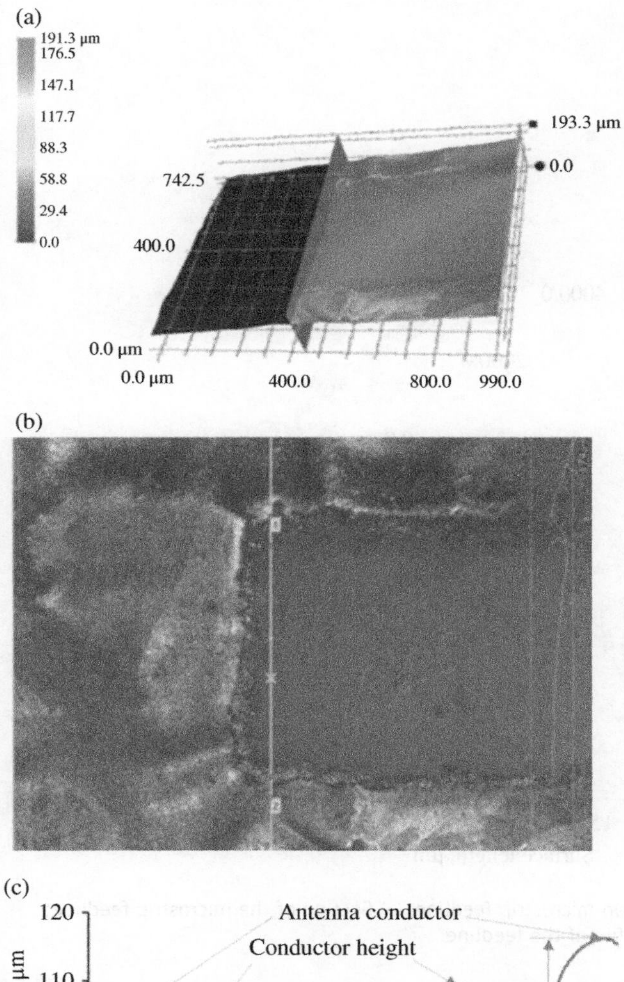

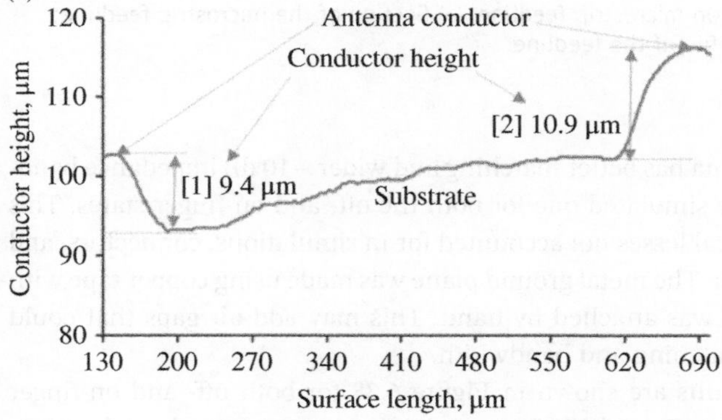

Figure 6.34 Evaluation of conductor height. (a) The antenna conductor height, (b) measurement point, and (c) the height at the points.

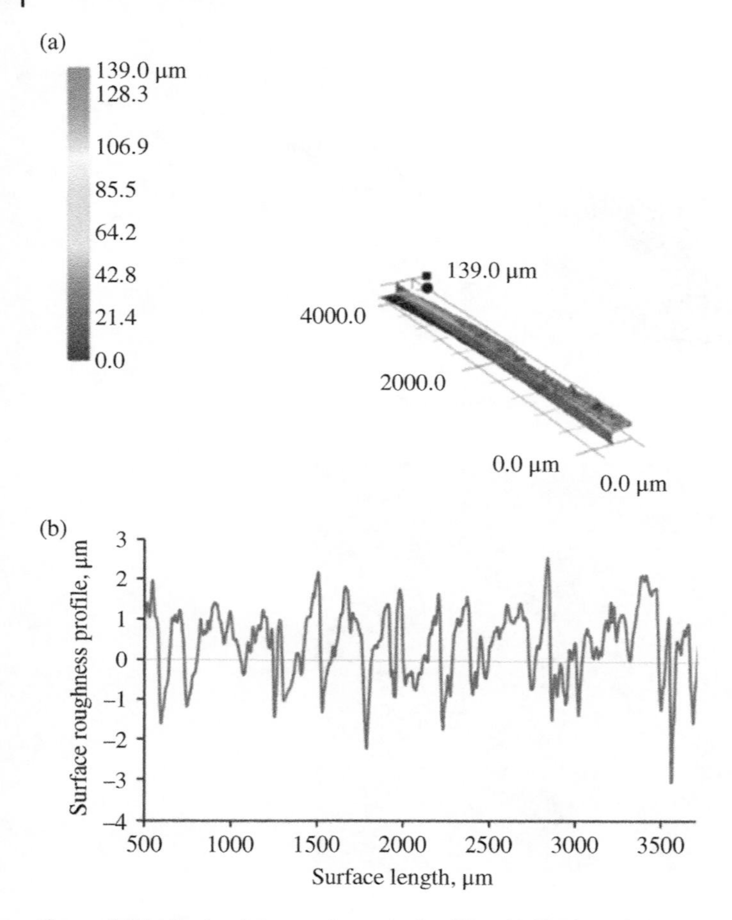

Figure 6.35 Evaluation on microstrip feedline. (a) Section of the microstrip feedline; (b) surface roughness profile of the feedline.

The fabricated antenna has better matching and wider −10 dB impedance bandwidth compared to the simulated one for both the off- and on-finger states. This could be due to electrical losses not accounted for in simulations, connectors, and errors in the fabrication. The metal ground plane was made using copper tape with an adhesive layer and was attached by hand. This may add air gaps that could potentially increase matching and bandwidth.

Far-field pattern results are shown in Figure 6.38 for both off- and on-finger antennas for planes *XY*, *XZ*, and *YZ*. The results show the expected patch antenna's hemispherical radiation pattern with moderate directivity. The main lobe

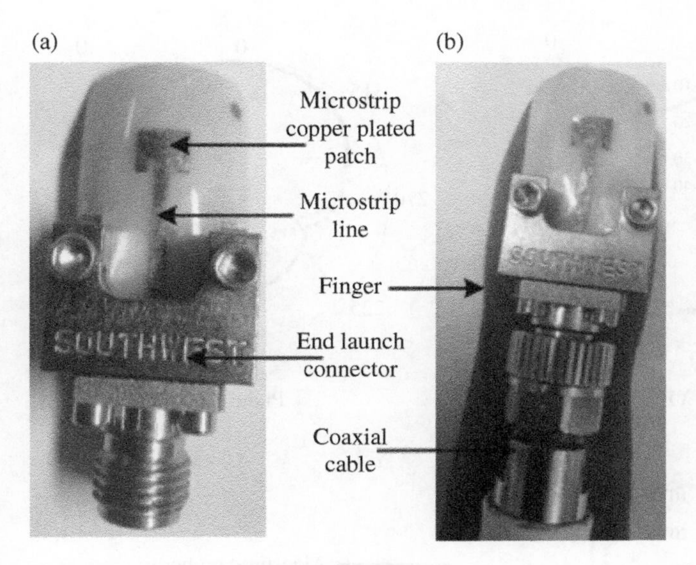

Figure 6.36 The fabricated antenna (a); antenna worn on a finger (b).

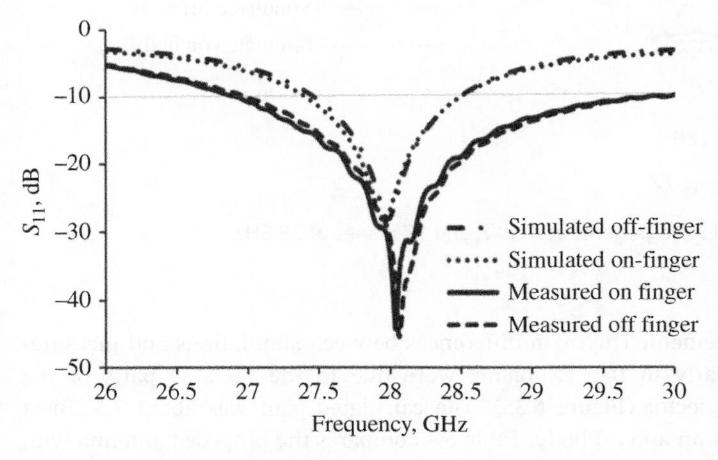

Figure 6.37 Reflection coefficient (S_{11}) of the optimized antenna.

points out of the finger nail while low back radiation is realized. Back radiation is also lower for the antenna at 15 GHz (Figure 6.30) due to the smaller size of the patch at 28 GHz in relation to the metallic ground plane. Radiation patterns were as expected for patch antenna with a main lobe out of the finger nail and low radiation toward the finger and body. The simulated and measured patterns were in

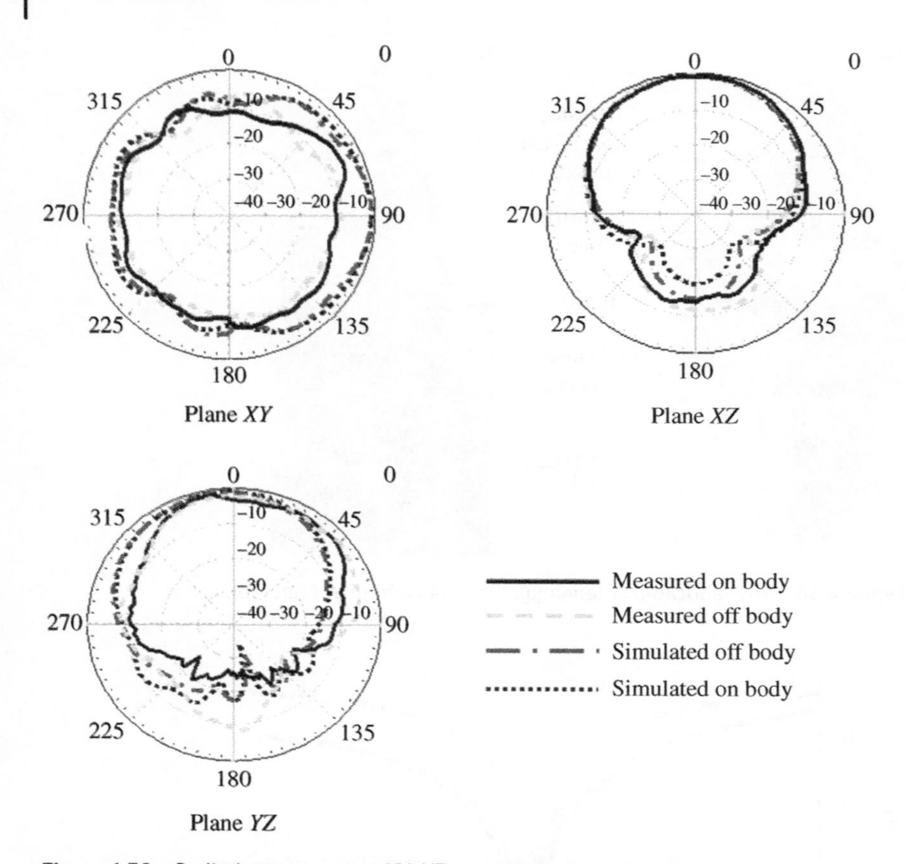

Figure 6.38 Radiation patterns at *XY*, *XZ*, and *YZ* planes at 28 GHz.

reasonable agreement. The main differences between simulations and measurements, particularly in the *YZ* plane, were due to the metallic parts of the end-launch connector (Figure 6.35). The calculated gain was about 7.5 dBi at 28 GHz for both on and off body. Table 6.7 compares the proposed antenna with previous wearable antennas. The benefit of this patch antenna is that it offers a highest gain and it will not be affected by body movements.

The antenna designs presented in this work can potentially be deployed in Internet of Things (IoT) solutions for 5G technology. The proposed nail antenna design is light, cheap, easy to install, and part of beauty accessory, which occupies a small surface area and is easy to wear. The requirement of different equipment at different stages presents a case for a production chain process. This could enable scaling up to a mass production industrial process. Further, multiple fingers on a human

Table 6.7 Comparison between proposed antenna and other wearable antennas at 28 GHz.

Reference	Frequency (GHz)	Bandwidth (GHz)	Gain (dB)	Substrate
[34]	28	1	7	PLA Medallion
[35]	28	15	3.5	Jeans
[36]	28	2.68	2.1	Rogers 5880
[37]	26/28	10	7	Woven polyester
This work	**28**	**2.87**	**8**	**ABS Nail**

hand can make antenna arrays and diversity systems feasible for signal reception improvement.

6.5 Summary

This chapter discussed the feasibility of 3D printing technologies on dielectric antenna and antenna with dielectric substrate. First, we introduce some basic knowledge of 3D printing technology, including its definition, advantages, application fields, classification, and characteristics of each kind of technology. Second, five common 3D printing technologies for antenna designs are discussed. It gives these technologies definition, characteristics, printing steps of various printers, and antenna examples. Then, a 3D printed dielectric antenna and a 3D printed metal antenna are proposed with complementary structure. Comparisons are made between them in terms of the size, weight, fabrication tolerance, and performance. After that, 3D printed finger nail antennas for 5G application are introduced to show their advantages of light, cheap, easy to install, and part of beauty accessory. Their fabrication process and fabrication tolerance have been discussed.

References

1 Tomassoni, C., Peverini, O.A., Venanzoni, G. et al. (2020). 3D printing of microwave and millimeter-wave filters: additive manufacturing technologies applied in the development of high-performance filters with novel topologies. *IEEE Microwave Magazine* 21 (6): 24–45.

2 Eid, A., He, X., Bahr, R. et al. (2020). Inkjet-/3D-/4D-printed perpetual electronics and modules: RF and mm-wave devices for 5G+, IoT, smart agriculture, and smart cities applications. *IEEE Microwave Magazine* 21 (12): 87–103.

3 Bahr, R., Tehrani, B., and Tentzeris, M.M. (2018). Exploring 3-D printing for new applications: novel inkjet- and 3-D-printed millimeter-wave components, interconnects, and systems. *IEEE Microwave Magazine* 19 (1): 57–66.
4 El-Sayegh, S., Romdhane, L., and Manjikian, S. (2020). A critical review of 3D printing in construction: benefits, challenges, and risks. *Archives of Civil and Mechanical Engineering* 20: 2.
5 ASTM F2792-12a (2012). *Standard Terminology for Additive Manufacturing Technologies*. West Conshohocken, PA: ASTM International.
6 Quan, H., Zhang, T., Xu, H. et al. (2020). Photo-curing 3D printing technique and its challenges. *Bioactive Materials* 5 (1): 110–115.
7 Zhang, F., Zhu, L., Li, Z. et al. (2021). The recent development of vat photopolymerization: a review. *Additive Manufacturing* 48, 102423.
8 Cui, Y., Bahr, R., Nauroze, S.A. et al. (2022). 3D printed "Kirigami"-inspired deployable bi-focal beam-scanning dielectric reflectarray antenna for mm-wave applications. *IEEE Transactions on Antennas and Propagation* 70 (9): 7683–7690.
9 Pagac, M., Hajnys, J., Ma, Q.P. et al. (2021). A review of vat photopolymerization technology: materials, applications, challenges, and future trends of 3D printing. *Polymers* 13 (4): 598.
10 Tyagi, S., Yadav, A., and Deshmukh, S. (2022). Review on mechanical characterization of 3D printed parts created using material jetting process. *Materials Today: Proceedings* 51 (1): 1012–1016.
11 Gülcan, O., Günaydn, K., and Tamer, A. (2021). The state of the art of material jetting—a critical review. *Polymers* 13 (16): 2829.
12 Chen, Z., Song, W., Wu, G. et al. (2020). Circular polarized 3-D-printed dielectric loaded antenna using inset waveguide-to-dielectric transition for 5G millimeter-wave application. *IEEE Antennas and Wireless Propagation Letters* 19 (11): 1929–1932.
13 Rojas-Nastrucci, E.A., Nussbaum, J.T., Crane, N.B., and Weller, T.M. (2017). Ka-band characterization of binder jetting for 3-D printing of metallic rectangular waveguide circuits and antennas. *IEEE Transactions on Microwave Theory and Techniques* 65 (9): 3099–3108.
14 Zhang, Z., Lu, Y., Liu, G. et al. (2020). 3D-printed high-gain circularly polarized antenna for broadband millimeter-wave communications over the power line. *International Journal of RF and Microwave Computer-Aided Engineering* 30 (3): e22094.
15 Gao, A.X., Qi, S., Kuang, X. et al. (2020). Fused filament fabrication of polymer materials: a review of interlayer bond. *Additive Manufacturing* 37: 101658.
16 He, Y., Wu, Y., Fu, J.Z. et al. (2016). Developments of 3D printing microfluidics and applications in chemistry and biology: a review. *Electroanalysis* 28: 1658–1678.
17 Malik, B.T., Doychinov, V., Zaidi, S.A.R. et al. (2019). Antenna gain enhancement by using low-infill 3D-printed dielectric lens antennas. *IEEE Access* 7: 102467–102476.

18 Zhu, S., Liu, H., Chen, Z., and Wen, P. (2018). A compact gain-enhanced Vivaldi antenna array with suppressed mutual coupling for 5G mmWave application. *IEEE Antennas and Wireless Propagation Letters* 17 (5): 776–779.

19 Chen, Z., Liu, H., Yu, J., and Chen, X. (2018). High gain, broadband and dual-polarized substrate integrated waveguide cavity-backed slot antenna array for 60 GHz band. *IEEE Access* 6: 31012–31022.

20 Zhang, B. and Zirath, H. (2016). Metallic 3-D printed rectangular waveguides for millimeter-wave applications. *IEEE Transactions on Components, Packaging and Manufacturing Technology* 6 (5): 796–804.

21 Hoel, K.V., Ignatenko, M., Kristoffersen, S. et al. (2020). 3-D printed monolithic grin dielectric-loaded double-ridged horn antennas. *IEEE Transactions on Antennas and Propagation* 68 (1): 533–539.

22 Zhang, S., Cadman, D., and Vardaxoglou, J.Y.C. (2018). Additively manufactured profiled conical horn antenna with dielectric loading. *IEEE Antennas and Wireless Propagation Letters* 17 (11): 2128–2132.

23 Cheng, X., Yao, Y., Yu, J., et al. (2017). Circularly Polarized Substrate-Integrated Waveguide Tapered Slot Antenna for Millimeter-Wave Applications. *IEEE Antennas and Wireless Propagation Letters* 16: 2358–2361.

24 Pozar, D.M. (2005). *Microwave Engineering*, 3e. Hoboken, NJ: Wiley Chapter 3.

25 Njogu, P., Sanz-Izquierdo, B., Elibiary, A. et al. (2020). 3D printed fingernail antennas for 5G applications. *IEEE Access* 8: 228711–228719.

26 Deffenbaugh, P.I., Rumpf, R.C., and Church, K.H. (2013). Broadband microwave frequency characterization of 3-D printed materials. *IEEE Transactions on Components, Packaging and Manufacturing Technology* 3 (12): 2147–2155.

27 Italian National Research Council. (2019). Calculation of the dielectric properties of body tissues in the frequency range 10 Hz - 100 GHz. Italian National Research Council. http://niremf.ifac.cnr.it/tissprop/htmlclie/uniquery.php (accessed 19 November 2019).

28 OPTOMEC. (n.d.). AEROSOL JET® printed electronics overview. OPTOMEC: Additive Manufacturing Systems-From Nano to MACROTM.

29 Agarwala, S., Goh, G.L., and Yeong, W.Y. (2017). Optimizing aerosol jet printing process of silver ink for printed electronics. *IOP Conference Series: Materials Science and Engineering*, Bangkok, Thailand.

30 Paulsen, J.A., Renn, M., Christenson, K., and Plourde, R.2012). Printing conformal electronics on 3D structures with Aerosol Jet technology. *2012 Future of Instrumentation International Workshop (FIIW) Proceedings*, Gatlinburg, TN, USA, pp. 1–4.

31 NovaCentrix. (2019). Metalon® conductive inks for flexible printed electronics. https://www.novacentrix.com (accessed 13 November 2019).

32 Centre for Process Innovation. (2019). https://www.uk-cpi.com (accessed 13 November 2019).

33 KrishnaRao, V., Abhinav, V., Karthick, S., and Singh, S.P. (2015). Conductive silver inks and their applications in printed and flexible electronics. *The Royal Society of Chemistry* 5: 77760–77790.

34 Fawaz, M., Jun, S., Oakey, W. et al. (2018). 3D printed patch Antenna for millimeter wave 5G wearable applications. *12th European Conference on Antennas and Propagation (EuCAP 2018)*, London, UK, pp. 1–5.

35 Sharma, D., Dubey, S.K., and Ojha, V.N. (2018). Wearable antenna for millimeter wave 5G communications. *IEEE Indian Conference on Antennas and Propogation (InCAP)*, Hyderabad, India, pp. 1–4.

36 Tong, X., Liu, C., Chen, Y. et al. (2019). A dual-mode multi-polarization millimeter wave wearable antenna for WBAN applications. *IEEE MTT-S International Microwave Biomedical Conference (IMBioC)*, Nanjing, China, pp. 1–3.

37 Wagih, M., Weddell, A.S., and Beeby, S. (2019). Millimeter-Wave Textile Antenna for on-Body RF Energy Harvesting in Future 5G Networks. *IEEE Wireless Power Transfer Conference (WPTC)*, London, UK, pp. 245–248.

7

Millimeter-Wave DRA and Array

CHAPTER MENU

7.1 Overview

5G, also named as the 5th generation mobile network, first emerged in 2018 to deliver home internet service to customers in selected cities. The 5G wireless communication enables various commercial applications such as autopilot, telemedicine, and Industrial Internet of Things (IIoT).

These applications have huge demand in communication transmission rate, resource utilization, wireless coverage and mobile interconnection capability. With large spectrum, millimeter-wave (mmWave) is considered as the key enabling technology for 5G and Beyond 5G (B5G). However, the propagation

Dielectric Resonator Antennas: Materials, Designs and Applications, First Edition.
Zhijiao Chen, Jing-Ya Deng, and Haiwen Liu.

Published 2024 by John Wiley & Sons, Inc.

characteristics of the mmWave are significantly different from microwave frequency bands in terms of path loss, diffraction and blockage, rain attenuation, atmospheric absorption, and foliage loss behaviors. In general, the overall loss of mmWave systems is significantly larger than that of microwave systems.

Fortunately, the short wavelengths of the mmWave frequencies enable the deployment of antenna array with a large number of antenna elements in the same occupied space. The antenna arrays offer high spatial processing gains that can theoretically compensate for at least the isotropic path loss. The fast development in mmWave technologies with antenna array significantly speed up its applications in numerous fields, such as automotive radars [1, 2], imaging sensors [3], and satellite communications [4, 5]. Moreover, low-cost and high-reliability mass production of the mmWave devices is required considering usability ratings and adoption by service providers and consumers, while most of the existing designs based on the high-cost bulky metal structure are infeasible for commercialization.

Driven by the demand for high gain, miniaturized and low-cost solutions for mmWave communications, dielectric resonator antennas (DRAs), antenna array, and dielectric-based fabrication have been investigated. DRA benefited from the dielectric with low-cost, lightweight, broad bandwidth, ease of circularly polarized (CP) excitation, and compact structure. A competitive solution for low-cost and high-reliability mass production of the mmWave antenna is offered by the dielectric-based fabrication. DRA has a low-profile structure but has limited gain enhancement, thus requires array feeding network for high gain radiation, which will be discussed in this chapter.

This chapter is organized as follows. Section 7.2 introduces a mmWave DRA element design, which is suitable for unmanned aerial vehicle (UAV) satellite application. This DRA element has the characteristics of dual-band dual circular polarization and broadened axial ratio bandwidth. In Section 7.3, substrate integrated waveguide (SIW), as a critical technique to achieve compact low-cost feeding network in mmWave band, is introduced to provide the merits of lightweight, high-gain, and high-efficiency of the antenna array. In Section 7.4, an SIW-fed DRA array is designed with wideband, high-gain, and enlarged dimensions for improved fabrication tolerance, which is suitable for 5G base station antenna. Finally, Section 7.5 gives remarks and summary of this chapter.

7.2 mm-Wave DRA for UAV Satellite

7.2.1 Background

For satellite communications in UAV, wideband antennas are demanded to support high data rate and high throughput exchange of massive data flows, whereas dual-band and dual-CPCP operations are required to reduce inter-channel interference and ensure high isolation full-duplex systems. The compact size of single-fed antenna design is also essential to be mounted on UAV. As a result, it is challenging to design a single-fed dual-CP antenna with relatively wide bandwidths to meet all requirements mentioned above. Usually, feeding networks with dual-/multi-fed result in a complex antenna structure and are hard to realize in mmWave band. Some dual-CP antennas [6, 7] have relatively wide bandwidths but are single band, which is unsuitable for full-duplex operation. An alternative approach is to integrate linear-polarized (LP) antennas with polarizers, but their volume and weight are significantly increased as was found in [8, 9].

Without the metal loss, DRA has advantages such as high gain and high efficiency, especially on the high-frequency band such as mmWave. DRA also has a compact lightweight structure with enhanced bandwidth and gain, which has the potential to be applied for mmWave communication with high throughput. Also, with the proper design, DRA is enabled with multi-functions such as dual-band dual-polarization, full-duplex with high isolation for antenna, which is desired by the UAV satellite applications. Here, a DRA [10] with wide bandwidths, dual-band, and dual-CP operation is designed for satellite communications in UAV. It is a stacked DRA backed by the cavity-slot aperture. Controlled by the independent phase compensation method, it achieves orthogonal CP waves that come from the slot in the lower band and the TE_{111} mode of DR stacking in the upper band. As a result, a single-fed dual-band dual-CP DRA is proposed for high throughput, high isolation, full-duplex, and small payload UAV low earth orbit satellite communications.

7.2.2 DRA Design and Working Principle

Figure 7.1 depicts the geometry of the proposed dual-band dual-CP DRA. It consists of three parts: a metal strip with a rotated angle of α from x-axis, a double-layer stacked rectangular DR, and an SIW cavity with a slot for feeding. The strip with dimension of $W_{st} \times L_{st}$ is adopted for CP. The top layer of the stacked DR, with the dimension of $a \times a \times h_3$, is made from Rogers 6010 substrate ($\varepsilon_r = 10.2$ and tan $\delta = 0.0023$). The bottom-layer DR, with the dimension of $a \times a \times h_2$, is made from Rogers 5880 substrate ($\varepsilon_r = 2.2$ and tan $\delta = 0.0009$). The SIW cavity is made from Rogers 5880 substrate ($\varepsilon_r = 2.2$ and tan $\delta = 0.0009$) with a thickness of $h_1 = 0.381$ mm.

Dual-band radiation of the proposed DRA is achieved by generating TE_{110} mode of the SIW cavity through the slot and TE_{111} mode of the stacked DR. With stacked DR, not only the impedance matching of two bands but also CP and gain performances of two bands (especially the upper band) are improved.

At lower operating frequency of 20 GHz, the antenna mainly operates as an SIW cavity-backed slot antenna with the dimension of $W_s \times L_s$. The TE_{mn0} modes of the SIW cavity can be excited for radiation through the slot, where m and n denote the

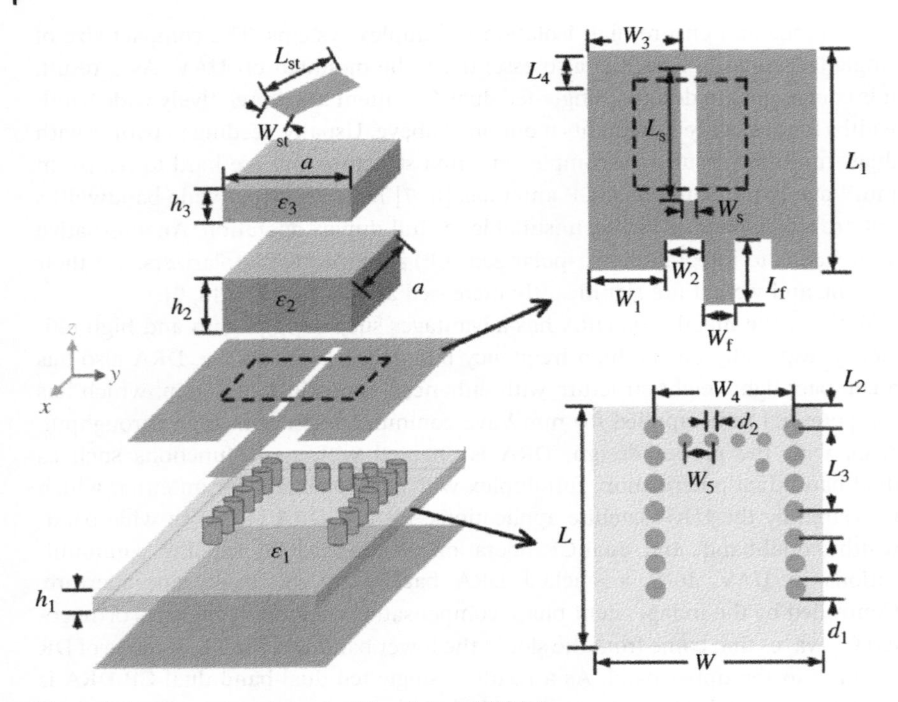

Figure 7.1 Geometry of the dual-band dual-CP DRA.

number of extrema in x- and y-directions in the cavity. The resonant frequency of the TE_{mn0} mode can be determined in [11]. Herein, TE_{110} mode of the SIW cavity is excited for radiation.

At higher operating frequency of 30 GHz, the radiation results from the TE_{111} mode of the stacked DR. The SIW cavity with the slot constitutes the feeding structure for the DR. The dimension of the DR is defined by the dielectric waveguide model (DWM), in which the modes can be divided into the TM_{mnl} and TE_{mnl} categories, where m, n, and l denote the number of extrema in the x-, y-, and z-directions in the waveguide [12].

To verify its operation, a single-layer DRA is given in Figure 7.2a as Ant. 1 to compare with the proposed double-layer DRA as Ant. 2. The single-layer DR is with a material constant of $\varepsilon_4 = 4.9$ while the double-layer DR is with $\varepsilon_2 = 2.2$ and $\varepsilon_3 = 10.2$. The overall dimensions of the single-layer and double-layer DRs are the same.

In Figure 7.2b, Ant. 1 shows a wideband radiation among 23.4–29.4 GHz, whereas Ant. 2 shows the desired dual-band radiation of 20/30 GHz. As is shown in Figure 7.2c, the axial ratio (AR) bands of Ant. 1 are not in accord with the impedance band, whereas the AR and impedance bands of Ant. 2 are overlapped. With the stacked DR, the impedance matching at 20 and 30 GHz are improved, leading

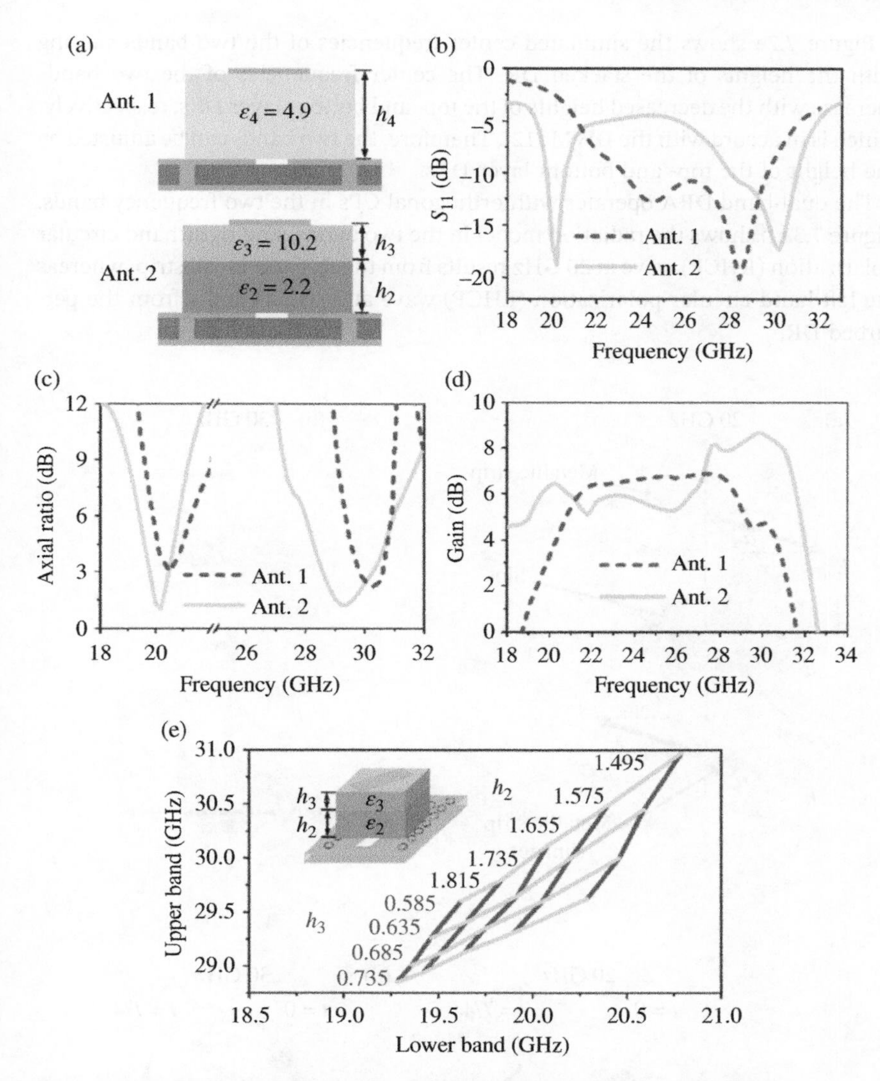

Figure 7.2 (a) Side-view geometries, simulated results of Ant. 1 and Ant. 2, (b) S_{11}, (c) AR, and (d) gain (h_2 = 1.575 mm, h_3 = 0.635 mm, h_4 = 2.210 mm) and (e) simulated frequencies of the two bands of Ant. 2 in terms of h_2 and h_3.

to the desired dual-band radiation. Furthermore, compared with Ant. 1, the 3-dB axial ratio bandwidth (ARBW) of Ant. 2 in the upper band is broadened from 2% to 4.3%, and the polarization purity of the two bands is improved. In Figure 7.2d, it is seen that compared to Ant. 1, the gain values of Ant. 2 in the upper band increase by more than 1.5 dBi averagely.

Figure 7.2e shows the simulated center frequencies of the two bands varying with the heights of the stacked DR. The center frequencies of the two bands increase with the decreased heights of the top- and bottom-layer DRs, respectively, which is in accord with the DWM [12]. Therefore, the two bands can be adjusted by the height of the top- and bottom-layer DRs.

The dual-band DRA operates with orthogonal CPs in the two frequency bands. Figure 7.3a,b shows the radiation model in the two bands. The right-hand circular polarization (RHCP) wave at 20 GHz results from the slot and metal strip, whereas the left-hand circular polarization (LHCP) wave at 30 GHz results from the perturbed DR.

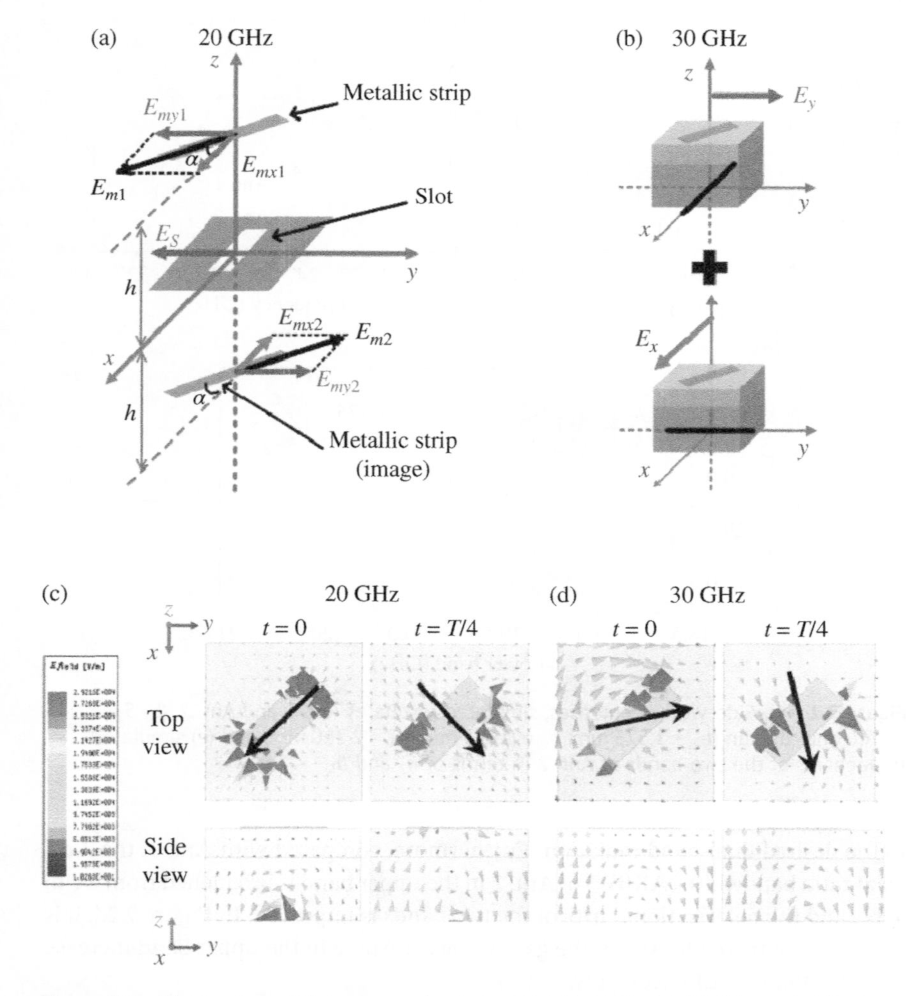

Figure 7.3 Mechanism of the proposed antenna at (a) 20 GHz, (b) 30 GHz, and *E*-field distributions of the antenna at (c) 20 GHz, (d) 30 GHz.

Figure 7.3c,d show the E-field distributions of the proposed antenna. At 20 GHz, the fields relatively concentrate around the slot, confirming the radiation source of the slot. Besides, the vector of major E-fields rotates counterclockwise as time increases, so an RHCP field is realized. At 30 GHz, the E-fields mainly concentrate inside the DR and show the distribution of TE_{111} mode, confirming the radiation source of the DR mode. The vector of major E-fields rotates clockwise as time increases, thus causing LHCP at 30 GHz. The CP waves in two bands can be controlled independently by a phase compensation method with coarse and fine tuning. The tuning principles are discussed below.

At 20 GHz, the CP wave results from the cavity-backed slot and the metal strip excited by the slot, which can be modeled as a slot-dipole antenna in Figure 7.3a. A slot is along x-axis and with the excitation amplitude and phase of E_s and ϕ_s. The radiation of the slot is modeled as a magnetic dipole along the x-axis, so the electric field only exists along y-axis ($\vec{E}_s$). Above the slot with a distance of h lies a metal strip with a rotated angle of α from x-axis, with excitation amplitude and phase of E_{m1} and ϕ_{m1}. The radiation of the strip is considered as an electric dipole with a rotated angle of α from x-axis, so the electric field has two components distributing along x- and y-axis ($\vec{E}_{mxl}$ and $\vec{E}_{myl}$). According to image theory, there is a metal strip under the slot with the distance of h. The field of it is $\vec{E}_{m2}$. It is noted that two strips are excited by the slot, so they are with the opposite excitation amplitude ($E_{m1} = -E_{m2}$) and same phase of ϕ_m. The field by the slot and the strips, traveling in the $+z$-direction, can be expressed as:

$$\vec{E}_S = -E_S \tilde{a}_y \tag{7.1}$$

$$\begin{aligned} \vec{E}_m = \vec{E}_{m1} + \vec{E}_{m2} &= E_{m1}(\cos\alpha\hat{a}_x - \sin\alpha\hat{a}_y) - E_{m2}(\cos\alpha\hat{a}_x - \sin\alpha\hat{a}_y) \\ &= E_s a_m (e^{jkh} - e^{-jkh})(\cos\alpha\hat{a}_x - \sin\alpha\hat{a}_y) \\ &= 2jE_s a_m \sin kh(\cos\alpha\hat{a}_x - \sin\alpha\hat{a}_y) \end{aligned} \tag{7.2}$$

The total field is composed of $\vec{E}_s$ and $\vec{E}_m$. RHCP wave can be realized by combining two orthogonal field components totally along x- and y-axis, which ideally have the same amplitude and −90° phase difference (PD). The desired amplitude can be adjusted by several parameters such as the length of the strip (L_{st}), rotation angle (α), and distance (h) [13]. The key condition of −90° PD is assured through the rotation angle (α) and distance (h) [14]. After being elaborately designed, RHCP wave at 20 GHz is generated.

Next, LHCP wave at 30 GHz can be generated by a phase compensation method. The CP wave at 30 GHz mainly results from the DR. The metal strip introduces perturbations to the DR, generating two near-degenerate orthogonal DR modes

(TE^x_{111} and TE^y_{111}). The radiation of two DR modes can be modeled as orthogonal magnetic dipoles along x- and y-axis in Figure 7.3b. LHCP wave at 30 GHz can be assured through 90° PD of the modes by the phase compensation method with coarse and fine tuning processes.

Coarse tuning the PD at 30 GHz can be achieved by adjusting the rotation angle (α) and the distance ratio (r). Figure 7.4a,b plot the simulated PDs of the two frequency bands varying with α and r, respectively, to illustrate the coarse tuning process. For example, to decrease the PD in the upper band, a two-step approach is practicable: increasing α and decreasing r. As shown in Figure 7.4a, increasing α obviously leads to the decrease in PD at 30 GHz, but also brings phase variation to the 20 GHz band. Next, to compensate the phase variation at 20 GHz, decreasing r can be adopted. On one hand, in Figure 7.4b, decreased r makes the PD decrease continuously at 30 GHz. On the other hand, decreased r decreases the PD at 20 GHz in Figure 7.4b, compensating the phase variation from increased α in Figure 7.4a. Therefore, decreased PD at 30 GHz and nearly stable PD at 20 GHz can be achieved. Similarly, a two-step approach of decreasing α and increasing r can be carried out to increase the PD at 30 GHz, while keeping the PD at 20 GHz nearly stable. Hence, coarse tuning the PD at 30 GHz can be realized.

Besides, fine tuning the PDs at 20 and 30 GHz can be obtained by adjusting the length and width (L_{st} and W_{st}) of the metal strip. Figure 7.4c,d plots the simulated PDs of two bands varying with L_{st} and W_{st} to illustrate the process. In Figure 7.4c, the PD at 20 GHz increases when increasing L_{st}, while it stays relatively stable at 30 GHz. This means minor phase correction at 20 GHz can be obtained by adjusting L_{st}. In Figure 7.4d, the PD at 30 GHz increases when increasing W_{st}, while it remains relatively stable at 20 GHz. Hence, slight phase correction at 30 GHz can be obtained by adjusting W_{st}. Hence, fine tuning the PDs in the two bands can be realized.

Figure 7.4e,f shows the mechanism of the phase compensation method. For coarse tuning, Figure 7.4e plots the relation of PDs in two bands with different α and r. For example, if the present PDs are at P1 (−90° at 20 GHz, 50° at 30 GHz), to achieve the ideal PDs at P2 (−90° at 20 GHz, 90° at 30 GHz), a two-step approach of increasing r and decreasing α is effective, thus ensuring increased PD at 30 GHz with constant PD at 20 GHz. For fine tuning, Figure 7.4f plots the relation of PDs in two bands with different L_{st} and W_{st}, showing that increasing L_{st} (W_{st}) is effective to increase the PD in the lower (upper) band with little effect to the other band. Similarly, decreased PDs can be attained by adjusting the parameters conversely.

7.2.3 Results and Comparison

To validate the proposed design, a dual-band dual-CP DRA was fabricated and tested. The fabricated antenna is connected to an SMA (SubMiniature Version

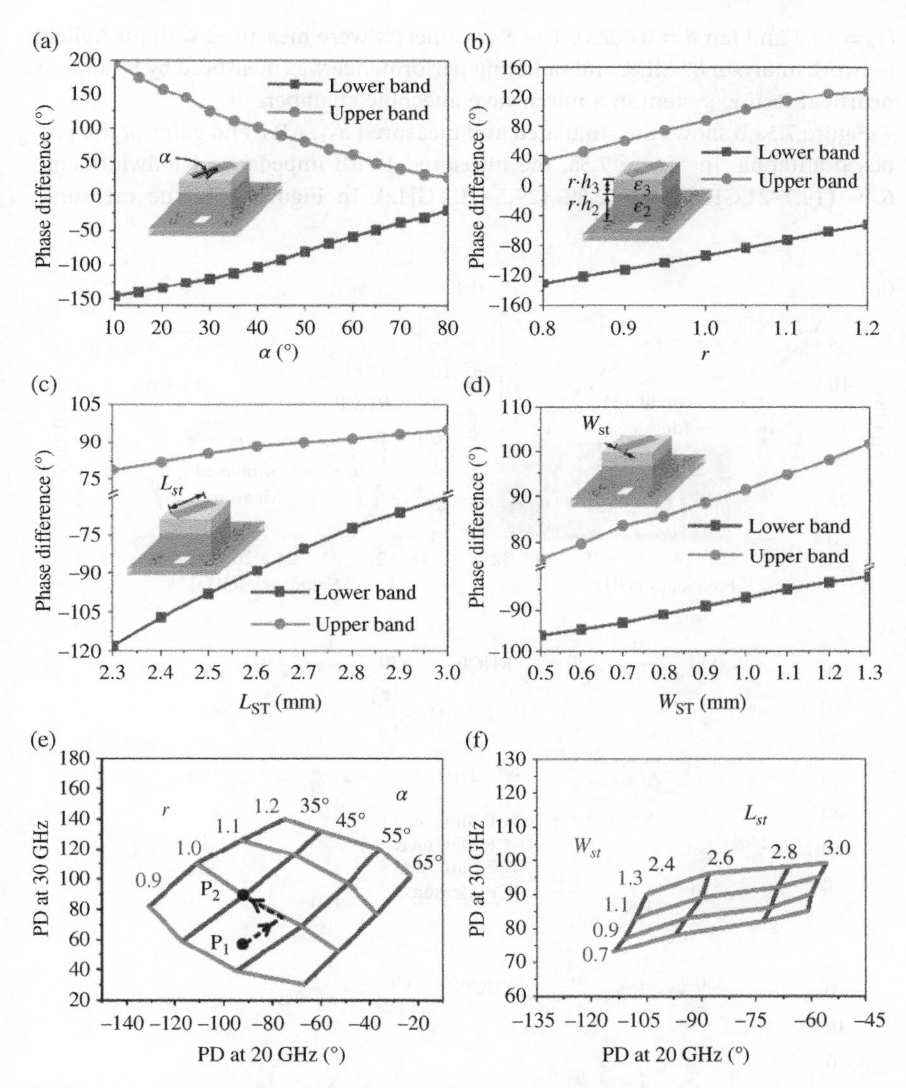

Figure 7.4 Simulated phase difference (PD) of the DRA at 20/30 GHz in terms of the (a) rotation angle α, (b) distance ratio r, (c) strip length L_{st}, (d) strip width W_{st}, and mechanism of the phase compensation method with (e) coarse tuning: simulated PDs in terms of α and r, (f) fine tuning: simulated PDs in terms of L_{st} and W_{st}.

A) connector (GWAVE KFD0830) as shown in Figure 7.5a. The printed circuit board (PCB) is made of 0.381 mm-thick Rogers 5880 substrate ($\varepsilon_r = 2.2$ and tan $\delta = 0.0009$). The stacked DR is fabricated with 1.575 mm-thick Rogers 5880 substrate ($\varepsilon_r = 2.2$ and tan $\delta = 0.0009$) and 0.635 mm-thick Rogers 6010 substrate

($\varepsilon_r = 10.2$ and tan $\delta = 0.0023$). The S-parameters were measured with an Agilent network analyzer 8722ES, and radiating performance was measured by SRainbow nearfield testing system in a microwave anechoic chamber.

Figure 7.5a,b shows the simulated and measured S_{11}, ARs, and gains of the proposed antenna. In Figure 7.5a, the measured 10-dB impedance bandwidths are 6.4% (19.7–21 GHz) and 12.8% (27.5–31.2 GHz). In Figure 7.5b, the measured

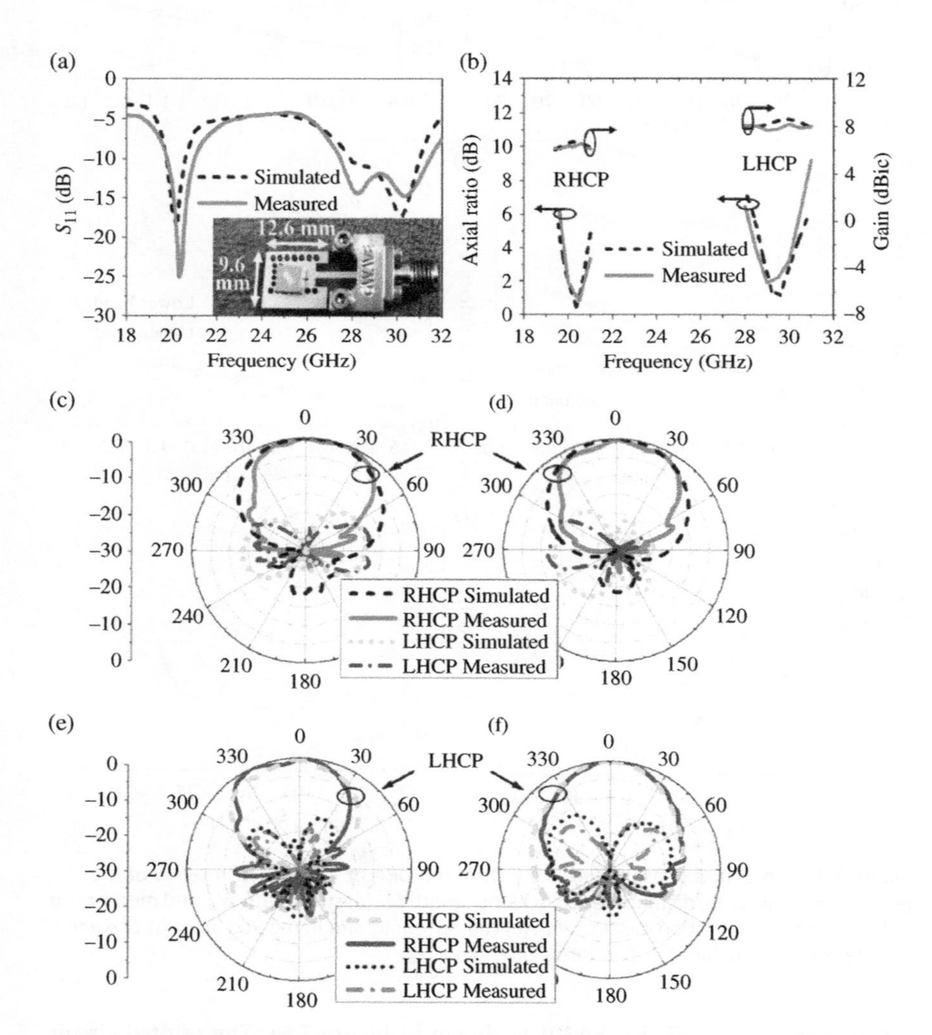

Figure 7.5 Simulated and measured (a) S_{11}, (b) ARs and gains, and radiation patterns (c) at 20 GHz in the *xoz* plane, (d) at 20 GHz in the *yoz* plane, (e) at 30 GHz in the *xoz* plane, (f) at 30 GHz in the *yoz* plane.

3-dB AR bandwidths are 5.2% (19.8–20.9 GHz) and 4.1% (28.7–29.9 GHz). The measured maximum gain of the antenna is 6.6 dBic for the RHCP in the lower band and 8.2 dBic for the LHCP in the upper band. The values of gain fluctuation within the operating bandwidths are 0.5 dBic and 0.7 dBic, which show good flatness in both two bands. It is seen that the simulated and measured results show good agreement.

Figure 7.5c–f plot the simulated and measured normalized radiation patterns at 20 and 30 GHz, respectively. Good agreement can be observed in both *xoz* and *yoz* planes. The metal strip provides the dominant RHCP at 20 GHz and LHCP at 30 GHz. The measured cross-polarization separation remains >29 and >26 dB in the lower and upper bands. Therefore, the DRA can radiate RHCP and LHCP waves with good cross-polarization in the boresight direction.

Table 7.1 tabulates a comparison between the proposed DRA and other previously reported dual-CP antennas that can be applied to the UVA satellite communications. Single-band dual-CP antennas were presented for satellites in [6, 7, 14], but they cannot meet the need of full-duplex for satellites. Dual-band antennas with orthogonal CPs have attracted great attention for satellites recently due to high capability and reduced size. However, it is challenging to realize dual-band dual-CP for mmW band satellites. The proposed antenna provides an approach to fulfil the task. It can be seen that the proposed antenna possesses the widest AR bandwidth that is superior to other dual-band dual-CP designs. Besides, enhanced and flat gain is achieved by the proposed DRA, which outperforms other reported

Table 7.1 Comparison between the proposed antenna and other antennas.

References	Operation frequency (GHz)	ARBW (%)	Gain (dBi)	Gain fluctuation (dBi)	Single-fed dual-band dual-CP
[15]	1.2/1.57	RH/RH	2.6/3.2	>1/>1	No
[16]	12.2/17.5	1.3/1.8	17.5/18.2	>2/>2.5	No
[8]	8.2/14.5	2.4/4.1	16.1/15.2	>1.5/>0.5	Yes
[17]	2.6/2.98	<2/<2	5.3/5.8	>1.5/>1.5	Yes
[6]	28	3.9	13.5	—	No
[7]	30	6.57	11.8	>3	No
[18]	30	7.6	31.8	>2.5	No
[19]	20/30	LP/LP	24.2/28.5	—	No
This work	**20/30**	**5.2/4.1**	**6.6/8.2**	**0.5/0.7**	**Yes**

works. Moreover, the proposed DRA provides a simplified design without complex feeding networks or bulky structures in [8]. Once connected with a diplexer, the proposed antenna can be employed for full-duplex systems on UAV. High isolation full-duplex, high throughput, and small payload UAV satellite communications can be achieved by the proposed antenna.

7.3 SIW Feeding Network for Antenna Array

A DRA element has a low-profile structure but has limited gain enhancement, thus requiring array feeding network for high-gain radiation. SIW is a good candidate for implementing complicated feed network of antenna array because it can be implemented by low-cost multilayered PCB laminates and machined by standard PCB facilities. The SIW is a waveguide-like structure in planar form by using periodic metallic via holes. It combines the advantages of high Q, high capacity of the waveguide and low cost, low-profile of the microstrip lines. It enables researchers to design and verify their cutting-edge ideas on mmWave antenna array designs in low cost. Based on the SIW technology, the planar arrays could be fabricated by various other means, such as complementary metal oxide semiconductor (CMOS), low-temperature co-fired ceramic (LTCC), and liquid crystal polymers (LCP). As a result, it is also preferred for industrial products, where fabrication complexity and costs have to be considered for high volume applications, like base stations and mobile devices.

7.3.1 SIW Working Principle

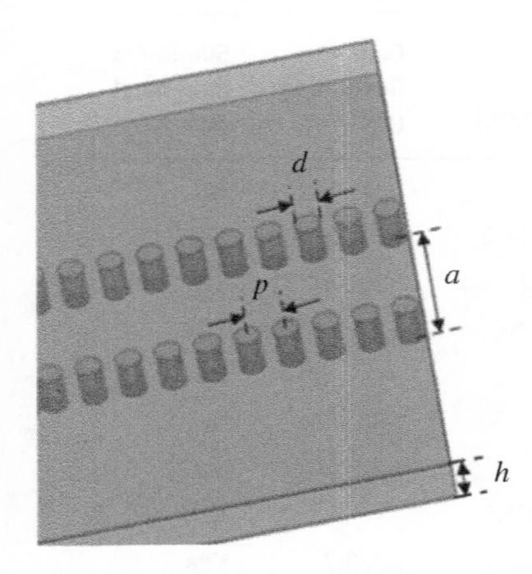

Figure 7.6 Configuration of the SIW structure.

Figure 7.6 shows the configuration of SIW structure with principle parameters. It is constructed by two rows of periodic metallic via holes, which are embedded into a dielectric substrate with double-side ground planes. In this case, a is the transverse spacing of the SIW metal via hole, d is the diametric of the SIW metal via hole, and p is the longitudinal spacing of the metal through hole. It should be noticed that SIW structure has similar propagation mode and dispersion characteristic to the dielectric loaded rectangular waveguide. The gaps between metalized vias prohibit the longitudinal current on the side wall of the SIW. Therefore, the SIW supports the TE_{n0} modes but not the TM modes. Several empirical relations about the geometrical dimension of SIW and rectangular waveguide have been proposed. For example, in [20], the effective width of SIW is formulated as,

$$w_{eff} = w - \frac{d^2}{0.95s} \tag{7.3}$$

where d is the diameter of the visa, w is the transverse spacing, and s is the longitudinal spacing. In [21], (7.3) is refined as,

$$w_{eff} = w - 1.08\frac{d^2}{s} + 0.1\frac{d^2}{w} \tag{7.4}$$

In [22], a comprehensive relation has been derived from an analytical method as,

$$w_{eff} = \frac{w}{\sqrt{1 + \left(\frac{2w-d}{s}\right)\left(\frac{d}{w-d}\right)^2 - \frac{4w}{5s^4}\left(\frac{d}{w-d}\right)^3}} \tag{7.5}$$

The effective width (w_{eff}) of the SIW structure is calculated from (7.3)–(7.5) equivalent to the air-filled waveguide. Then, the operation frequency and modes of SIW structure can be derived from the same size air-filled waveguide.

However, unlike the air-filled waveguide, SIW structure suffers from higher loss when unproperly designed. The loss in the SIW structure can be classified into conductor losses, dielectric losses, and radiation losses. The conductor loss is produced by the finite conductivity of the metal walls, which depends on the top and bottom metal surfaces of the SIW. It could be reduced by increasing the SIW substrate thickness because the surface integral is proportional to the square of the thickness. The dielectric loss is generated by the lossy dielectric material, which depends on the loss tangent of SIW dielectric material. For this reason, it is essential to use the PCB substrate with low dielectric loss to minimize the dielectric loss of SIW structure. The radiation loss is caused by the power leaking out between the through-hole vias, which usually takes a large portion of entire loss. The radiation loss could be minimized by proper design of SIW structure, which would be discussed as follows.

Figure 7.7 shows the E-field distributions within the SIW structures by using commercial simulation software, HFSS 15.0 [23]. These structures are designed with the same SIW width (a) and via diameter (d), but different via space (p), resulting in different d/a and p/d ratios. It can be seen that p/d should be as small as possible to avoid the power leakage between the vias. However, a smaller p/d would increase the fabrication difficulties, also yielding a fragile PCB structure.

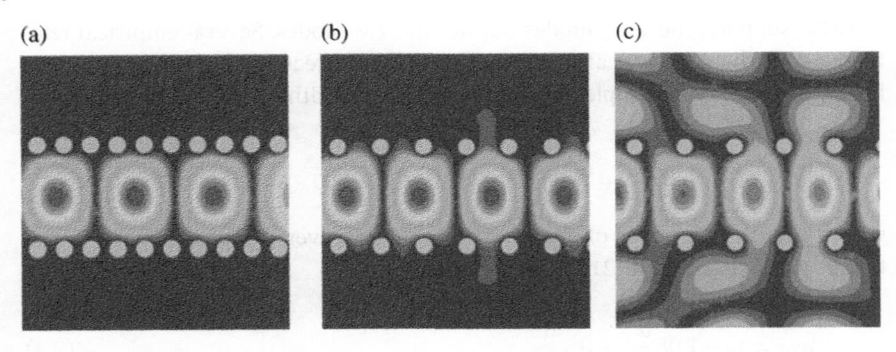

Figure 7.7 The *E*-field distributions in the SIW with different dimensions on *p*. (a) $d/a = 0.17$, $p/d = 1.5$. (b) $d/a = 0.17$, $p/d = 2.5$. (c) $d/a = 0.17$, $p/d = 3$.

For this reason, a tradeoff on the p/d is considered for avoiding power leakage while having the easy-to-fabricate prototype. Based on research experience, the SIW design guideline is

$$d/a < 0.2, p/d < 2 \tag{7.6}$$

When the a, d, and p meet the design criteria in (7.6), the propagation modes inside the SIW are close to the rectangular air-filled waveguide. Hence, under certain conditions, SIW can be equivalent to a rectangular dielectric-filled waveguide. The corresponding relationship between the SIW and the equivalent rectangular waveguide is demonstrated as [24],

$$\overline{a} = \xi_1 + \frac{\xi_2}{\dfrac{p}{d} + \dfrac{\xi_1 + \xi_2 - \xi_3}{\xi_3 - \xi_1}} \tag{7.7}$$

$$a_{RWG} = a \cdot \overline{a} \tag{7.8}$$

where $\overline{a}$ is the ratio of the rectangular waveguide width a_{RWG} to the SIW equivalent width a.

The values of ξ_1, ξ_2, ξ_3 are formulated as,

$$\xi_1 = 1.0198 + \frac{0.3465}{\dfrac{a}{p} - 1.0210} \tag{7.9}$$

$$\xi_2 = -0.1183 - \frac{1.2729}{\dfrac{a}{p} - 1.0210} \tag{7.10}$$

$$\xi_3 = -1.0082 - \frac{0.9163}{\dfrac{a}{p} + 0.2152} \tag{7.11}$$

The calculated equivalent width in (7.7) is close to the practical equivalent width, with less than 1% error. When $d \leq \lambda/10$, the formula (7.8) can be simplified to,

$$a_{RWG} = a - \frac{d^2}{0.95p} \tag{7.12}$$

For the case $p/d < 3$ and $d/h < 5$, this formula can be reduced into,

$$a_{RWG} = a - 1.08\frac{d^2}{p} + 0.1\frac{d^2}{a} \tag{7.13}$$

Based on this equivalent relationship, the SIW technique has been widely applied to the leaky wave antenna to replace the conventional leaky wave antenna based on the air-filled waveguide.

7.3.2 SIW Power Dividers

As mentioned, antenna elements have insufficient gain and are unsuitable for high-gain communications. To obtain a higher gain, an antenna array with a larger scale is designed and employed. The large-scaled antenna array can be constructed by using the parallel-fed network with power dividers. For example, in [25], a broadband, dual-polarized SIW-fed slot antenna array is demonstrated at 60 GHz to achieve a peak gain of 22.3 dB. In [26], a wideband, high-gain DRA array is proposed by using the SIW parallel feed network to achieve the peak gain of 14 dB. By adopting some special structures, SIW antenna array can further reduce the sidelobe level (SLL).

Figure 7.8 shows the configurations and dimensions of a T-junction and a H-junction. H-junction can be regarded as one to four power divider that consists

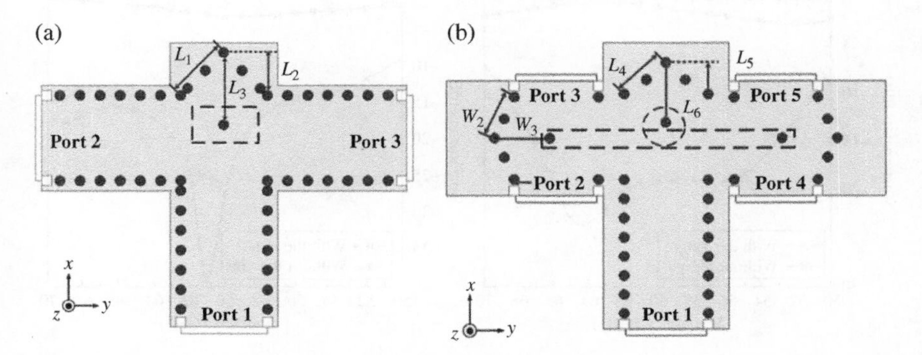

Figure 7.8 Configurations and dimensions of (a) T-junction. (b) H-junction. L_1 = 1.85 mm, L_2 = 1.31 mm, L_3 = 2.21 mm, L_4 = 1.72 mm, L_5 = 1.0 mm, L_6 = 1.9 mm, W_1 = 3 mm, W_2 = 3 mm, W_3 = 1.7 mm.

of three T-junctions, one in the center and two are connected on each side. As these three T-junctions are too close to each other, they should be considered as one structure by including the coupling effects between them. T-junction and H-junction are symmetrical with respect to their vertical axis. Hence the amplitudes of the output ports ($|S_{12}|$ to $|S_{13}|$ for T-junction, $|S_{12}|$ to $|S_{15}|$ for H-junction) are uniform. The phase error among the four output ports is within 1.4° from the ideal value.

It should be noted that offset vias are placed L_2 and L_5, respectively, from the center axis to obtain good impedance matching for input ports. Vias marked with circles play an important role for impedance matching of the T-junction. Figure 7.9 shows comparison of S_{11} of the T-junctions and H-junctions with and without the vias. It can be seen that the case without the via results in a poor impedance matching.

7.3.3 Waveguide to SIW Transition

The SIW-based antenna operates at mmWave band is mostly fed by standard rectangular air-filled waveguide. For this reason, waveguide to SIW transition is required to have a low-loss power conversion between the waveguide and SIW. The proper design of transition is also essential for the wideband impedance matching of mmWave antenna. Waveguide to SIW transition designs can be founded in many antenna designs, including leaky-wave antenna, dual-polarized antenna, CP antenna, reflect array antenna, and DRA array. SIW-based multi-beam antennas utilized a number of waveguides to SIW transitions to switch between different beams. The SIW antenna can be alternatively fed by

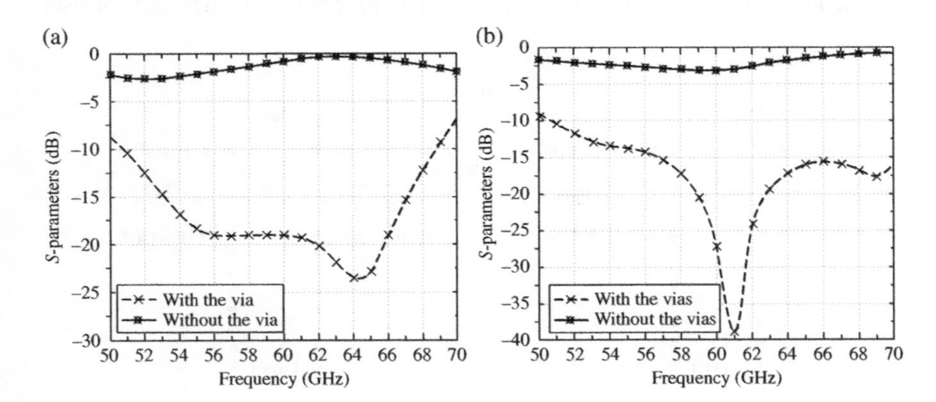

Figure 7.9 The comparison of S_{11} with and without the marked vias. (a) T-junction. (b) H-junction.

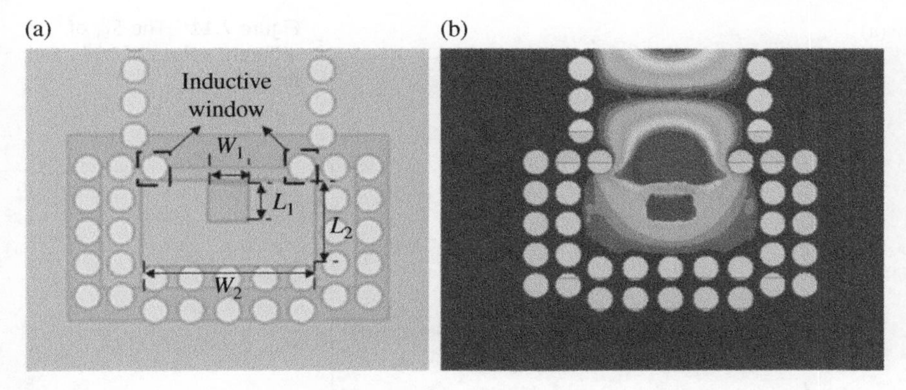

Figure 7.10 (a) The structure of the SIW with the parameters of W_1 = 0.3 mm, W_2 = 2.6 mm, L_1 = 0.6 mm, L_2 = 1.325 mm. (b) The electric field distribution of the SIW.

coaxial feed line. In this case, a coaxial to SIW transition is required for antenna feeding.

Figure 7.10a shows an example of waveguide to SIW transition design. It consists of an SIW resonant cavity with a central loaded microstrip patch. Figure 7.10b plots the electric-field distribution in the transition, which indicates the operation of the transition. The microstrip patch printed on the waveguide surface is excited by the waveguide to transfer the waveguide power into the substrate. Then the power in the substrate is guided by the surrounding metal via holes to propagate along the SIW. Therefore, the size of microstrip patch and the position of the inductive window vias have great influence on the impedance matching. These principle parameters are demonstrated in Figure 7.10a and their effects on impedance matching are studied.

In Figure 7.11, the parametric studies on the patch size are conducted by changing one parameter whereas keeping other parameter as the original value. When L_1 increases from 0.4 to 0.8 mm, the curves of S_{11} hardly changed, which means that the change of L_1 has little influence on the impedance matching performance of the SIW. However, when W_1 increases from 0.1 to 0.5 mm, S_{11} changed dramatically with the resonance continuously shifting to the low frequency. A new resonant will be introduced when the W_1 is increased to a certain value. It can be seen that the size of the microstrip patch is critical for impedance matching.

The inductive window also has a strong impact to the transition impedance matching [27]. Figure 7.12 compares the S_{11} with and without the inductive window. The case without the inductive window shows poor impedance matching and operates at narrow band. With the help of inductive window, the proposed transition covers a wide <−10 dB impedance matching over 50–65 GHz.

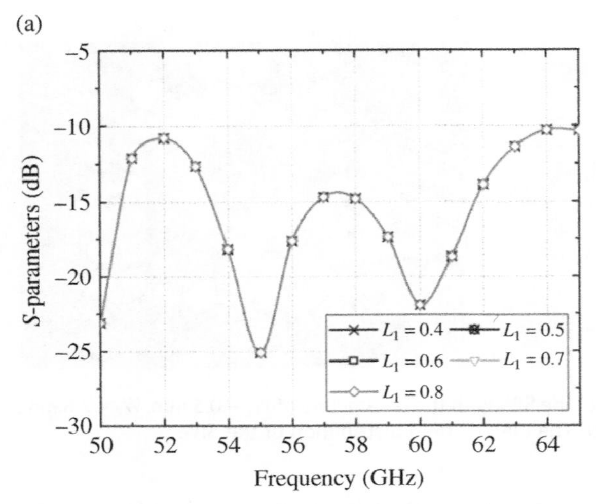

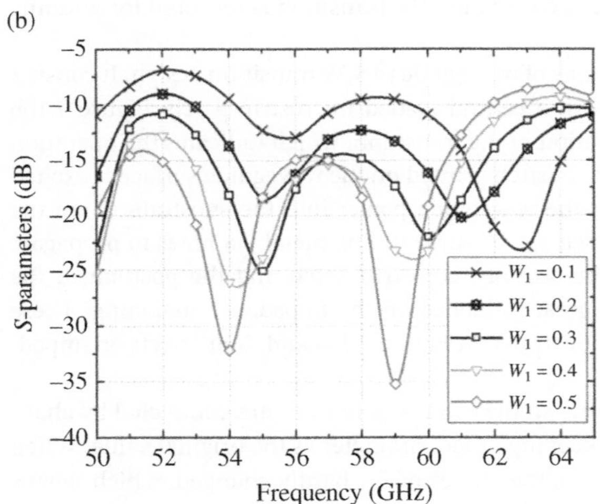

Figure 7.11 The S_{11} of different values of (a) L_1 and (b) W_1.

7.4 mm-Wave DRA Array for Base Station

7.4.1 Background

In recent years, the frequency range of 64–71 GHz was proposed by the US Federal Communications Commission as an unlicensed band to provide high-data-rate communication for next-generation 5G technologies [28]. DRA arrays, especially sDRA arrays, have the benefits of high-gain, high-efficiency mmWave communications.

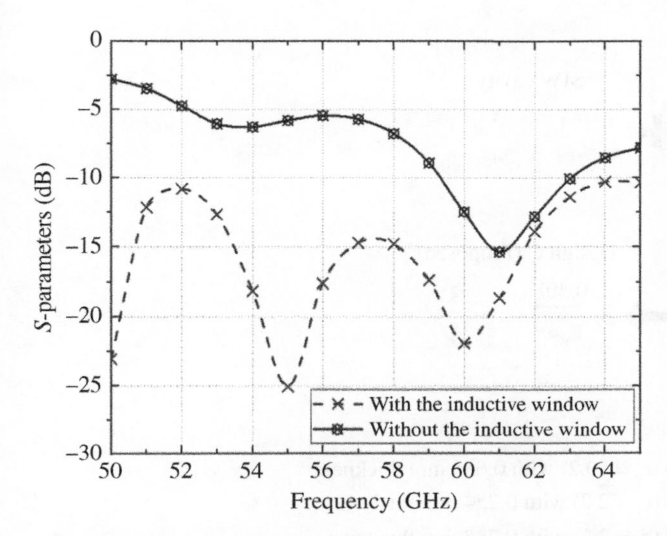

Figure 7.12 The S_{11} when the transition with and without the inductive window.

However, fabrication of the DRA array becomes challenging as the operating band increases. For instance, the side length of the rectangular DRA must be in the range of 1.5–3.0 mm to function over the Ka band [29, 30]. DRA arrays in this range of dimensions can be machined and positioned by standard PCB facilities. For DRA arrays operating over the V-band, however, the side length of the rectangular DRA is reduced to 1.05 mm [31], and the radius of the cylindrical DRA is reduced to 0.75 mm DRA [32]. To minimize fabrication errors at these sizes, DRAs are nested in a template [31] or designed in a perforated structure [32]. When the working frequency is increased to 67 GHz, the side length of the DRA shrinks to less than 1.0 mm and presents challenges to standard PCB technology. DRAs can be machined by CMOS technology to work at higher bands [33]. Nevertheless, the dimensions of the DRA must be as large as possible for easier fabrication, especially for DRAs working over the terahertz band. In addition, the fabrication tolerance of the DRA must be as accommodating as possible to minimize the effects of fabrication error.

7.4.2 Working Principle

In this design, an SIW cavity is employed as a high-pass filter to allow a larger DRA. In Figure 7.13, the geometries of a standard DRA array (Design A), a cavity-backed DRA array (Design B), and the proposed sDRA array (Design C) are demonstrated. By appropriately integrating the SIW-backed cavity, the DRAs in Design B and Design C have an enlarged side length of $a = b = 0.40\lambda$, which is

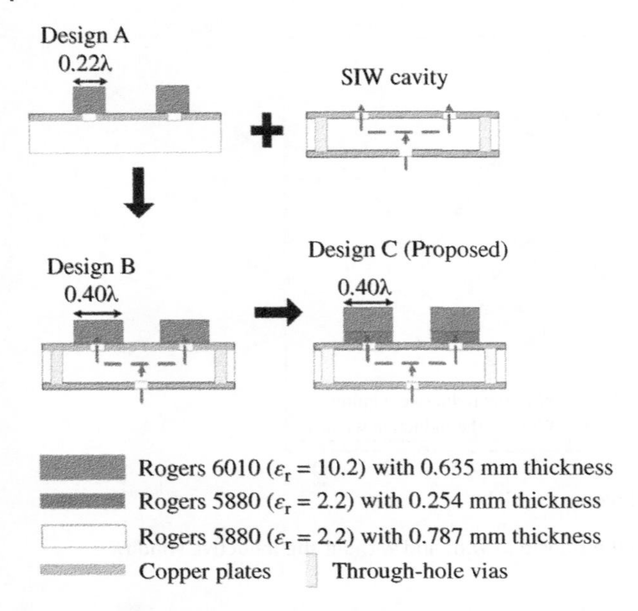

Figure 7.13 Side-view geometries of a standard DRA array (Design A), a cavity-backed DRA array (Design B), and the proposed sDRA array (Design C).

1.8 times that of Design A, which is $a = b = 0.22\lambda$. In Design C, the sDRA is constructed by two substrate layers to improve the antenna bandwidth and gain.

This study provides a comparison of the physical operation of the designs. It is known that the SIW cavity features the high-pass characteristics of a conventional waveguide [34]. In this work, the SIW cavity is used as a high-pass filter and a 1–2 × 2 power divider. As shown in Figure 7.14a, the input port is a rectangular slot located at the center of the square SIW cavity. The output ports are composed of four rectangular slots that are placed over the adjacent E-field resonance locations. Within the cavity, the TE_{14} mode is excited at resonance, as observed from the E-field and H-field in Figure 7.14b,c. The E-field distribution for the 2 × 2 output ports is relatively weak, whereas H-field distribution around the slot regions is strong. Magnetic couplings are generated around the 2 × 2 output slot with the same amplitude and phase, therefore constructing a 1–2 × 2 power divider.

The dimensions of the DRA are defined by the DWM [12]. The DWM introduces a dramatic simplification that enables one to obtain a closed-form solution [12]. In this model, the modes of a rectangular dielectric waveguide can be divided into the TM^y_{mnl} and TE^y_{mnl} families of modes, where m, n, and l denote the number of extrema in the x-, y-, and z-directions, respectively, inside the dielectric waveguide [35].

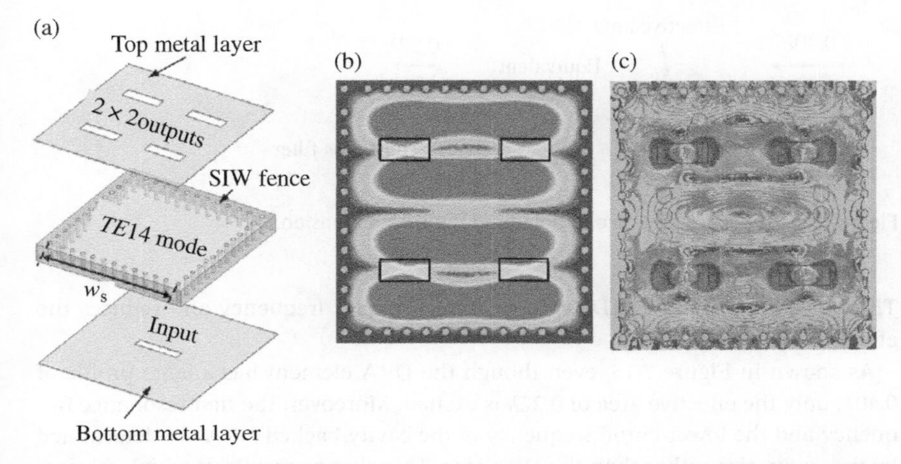

Figure 7.14 Proposed SIW cavity. (a) Schematics. (b) *E*-field and (c) *H*-field plots of the TE_{14} mode.

The fundamental resonant frequency of the rectangular DRA, TE^y_{111} mode, is defined as,

$$f_{DRA} = \frac{c}{2\pi\sqrt{\varepsilon_r}}\sqrt{k_x^2 + k_y^2 + k_z^2} \tag{7.14}$$

$$k_y = \frac{\pi}{b} = \frac{\pi}{a}, k_z = \frac{\pi}{2h}, \tag{7.15}$$

$$\tan\left(\frac{ak_x}{2}\right) = \left(\frac{k_{xa}}{k_x}\right), k_{xa} = \sqrt{k_y^2 + k_z^2} \tag{7.16}$$

where k_x, k_y, and k_z are the wavenumbers along the *x*-, *y*-, and *z*-directions, respectively. Here, *c* is the speed of light. Note that (7.16) is a transcendental equation.

For a standard DRA (Design A), the dimensions are calculated as $a = b = 0.22\lambda$ by setting a material constant of $\varepsilon_r = 10.2$ and using $h = 0.635$ mm in (7.14)–(7.16). At 67.0 GHz, 0.22λ is equivalent to 0.98 mm, which is challenging to machine in a standard PCB facility. Additionally, a fabrication error of ±0.1 mm would result in an f_{DRA} shift of 7.6%, as calculated from (7.14) to (7.16).

It is possible to scale the DRA size by lowering the ε_r of the material. However, the stored electric energy inside the DRA is proportional to ε_r [12]. A high ε_r guarantees an approximate magnetic boundary on the DRA edges to maintain the resonance mode, which is critical for providing a high boresight gain.

In Design B, the side length of rectangular DRA is enlarged to 0.40λ by properly integrating the SIW cavity. The high-pass filtering characteristic of the SIW cavity suppresses the fundamental mode of the DRA for the 0.40λ side length.

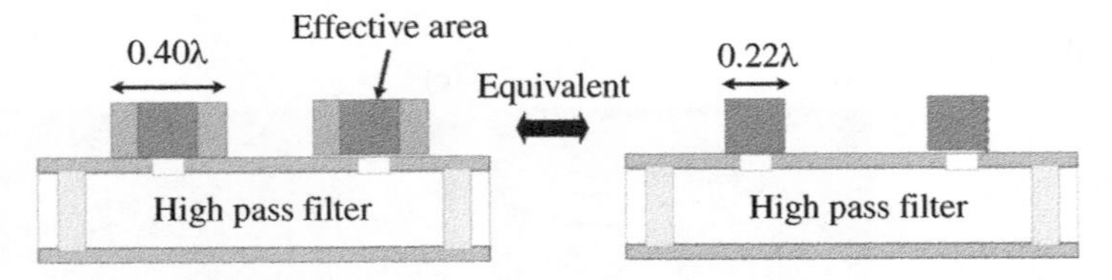

Figure 7.15 Physical principle of enlarging the DRA dimension.

This condition causes the DRA to work at a higher frequency and reduces the effective area of the DRA.

As shown in Figure 7.15, even though the DRA element has a large profile of 0.40λ, only the effective area of 0.22λ is excited. Moreover, the first resonance frequency and the lower cutoff frequency of the cavity-backed DRA are determined by the cavity size rather than the DRA size. This change results in a high fabrication tolerance for the DRA dimension. The mutual coupling between the enlarged DRA is not increased because the effective area of the DRA is fixed by the cavity.

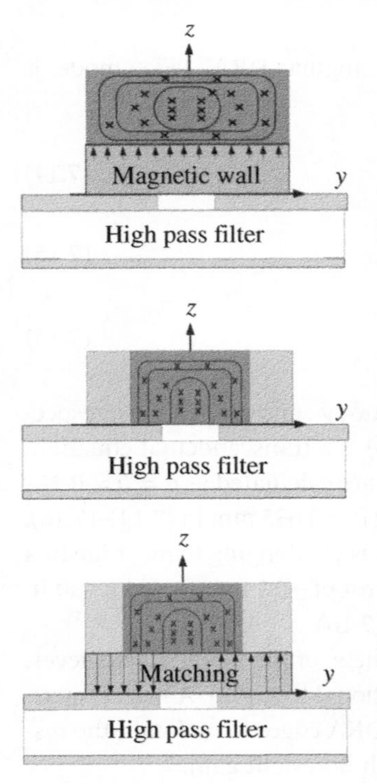

Figure 7.16 Physical operation of the TE^x_{111} and TE^x_{112} modes in the DRA and sDRA.

Design C of the sDRA array is expected to achieve a wider bandwidth and better gain than the single-layered DRA in Design B. The basic concept behind this approach is that an additional dielectric space allows the generation of an additional TE^x_{112} mode, as indicated in Figure 7.16.

In the first row of Figure 7.16, Design B shows a half circle resonance above the ground plane. Due to the mirror effect of the ground plane, a full loop is generated on the z-axis that is equivalent to the TE^x_{111} mode. However, Design B cannot excite the TE^x_{112} mode because this mode is eliminated by the ground plane (electric field boundary) [36].

In the second row of Figure 7.16, a space with lower permittivity is added between the DRA and the ground plane. The approximate magnetic boundary of the high-dielectric DRA is maintained because the permittivity of the upper dielectric is higher than that of the lower dielectric. Therefore, a full circle resonance mode is generated within the magnetic field boundary, which is equivalent to the TE^x_{112} mode, as indicated in Figure 7.16.

In comparison to the standard DRA array in Design A, the proposed DRA array in Design C can be characterized as follows:

1) The use of the SIW-backed cavity allows an enlarged DRA dimension.
2) A higher fabrication tolerance exists for the DRA dimension.
3) An improved bandwidth and enhanced gain are obtained by exciting the TE^x_{112} mode in the sDRA structure.

7.4.3 DRA Array Design

In this section, Design C operating at 67.0 GHz is designed with sDRA for improved bandwidth. Figure 7.17a shows the geometry of the proposed sDRA array, which comprises a shared backed cavity (Sub 1 + Sub 2) and a 2 × 2 sDRA array (Sub 3 + Sub 4).

The backed cavity consists of two PCB layers, which are indicated as Sub 1 and Sub 2 in Figure 7.17. Both PCB layers use Rogers 5880 laminates ($\varepsilon_r = 2.2$ and $\tan\delta = 0.002$) with a thickness of 0.787 mm. Sub 1 is used to interconnect the input SIW feed network with the input coupling aperture of Sub 2. Sub 2 works as the high-pass filter and the 1 : 4 power divider.

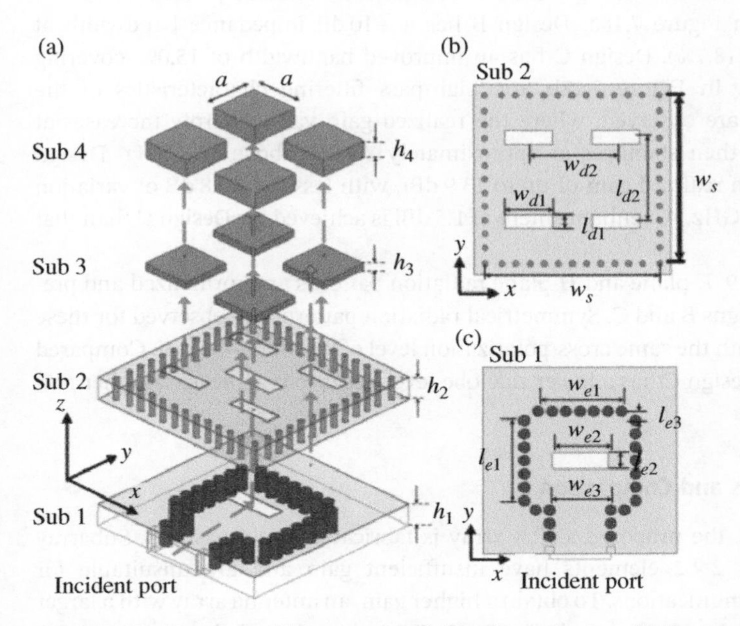

Figure 7.17 The 2 × 2 sDRA subarray. (a) 3D schematic view. (b) Top view of Sub 2. (c) Top view of Sub 1.

Table 7.2 Dimensions of the element array (units: mm).

Parameters	a	h_1	h_2	h_3	h_4	
Values	1.8	0.787	0.787	0.254	0.635	
Parameters	w_s	l_{d1}	w_{d1}	l_{d2}	w_{d2}	
Values	6.2	0.5	1.7	3.1	3.1	
Parameters	$le1$	$le2$	$le3$	$we1$	$we2$	w_{e3}
Values	3.0	0.6	0.35	3.0	2.0	2.2

The topology of each PCB layer is illustrated in Figure 7.17b,c, and the dimensions are listed in Table 7.2. With the use of this vertical power divider, the distance between the slots is reduced to 3.1 mm or to 0.69λ at 67.0 GHz in both the x- and y-directions.

The DRAs are made from commercial Rogers 6010 laminates ($\varepsilon_r = 10.2$, $\tan \delta = 0.0023$) with a thickness of 0.635 mm. The DRA operating at 67.0 GHz is calculated to have a side length of $a = b = 0.98$ mm, which is enlarged to $a = b = 1.80$ mm as shown in the final design in Figure 7.17. The inserted Sub 3 layer is made of Rogers 5880 laminate with $\varepsilon_r = 2.2$ in the thickness dimension of h_3. The DRAs are centrally placed over the slot to generate a symmetrical radiation pattern.

As shown in Figure 7.18a, Design B has a −10 dB impedance bandwidth at 63.6–69.4 GHz (8.7%). Design C has an improved bandwidth of 15.0%, covering 63.1–73.4 GHz. In Figure 7.18b, the high-pass filtering characteristics of the backed cavity are observed, where the realized gain value sharply increases at 60.0 GHz and then stabilizes at approximately 8.0 dBi above 62.8 GHz. Design C shows a high realized gain of up to 13.9 dBi, with less than 0.8 dB of variation over 61.4–74.2 GHz. An enhancement of 1.5 dBi is achieved for Design C than that of Design B.

In Figure 7.19, E-plane and H-plane radiation patterns are normalized and presented for Designs B and C. Symmetrical radiation patterns are observed for these two designs, with the same cross-polarization level of less than −35 dBi. Compared to Design B, Design C has a lower sidelobe level because it achieves a higher realized gain.

7.4.4 Results and Comparison

In this section, the proposed sDRA array is fabricated and measured. Subarray antennas with 2 × 2 elements have insufficient gain and are unsuitable for high-gain communications. To obtain a higher gain, an antenna array with a larger scale is designed and employed. The 2 × 2 sDRA array is scaled to a 4 × 4 sDRA array by integration with the full-corporate feed network.

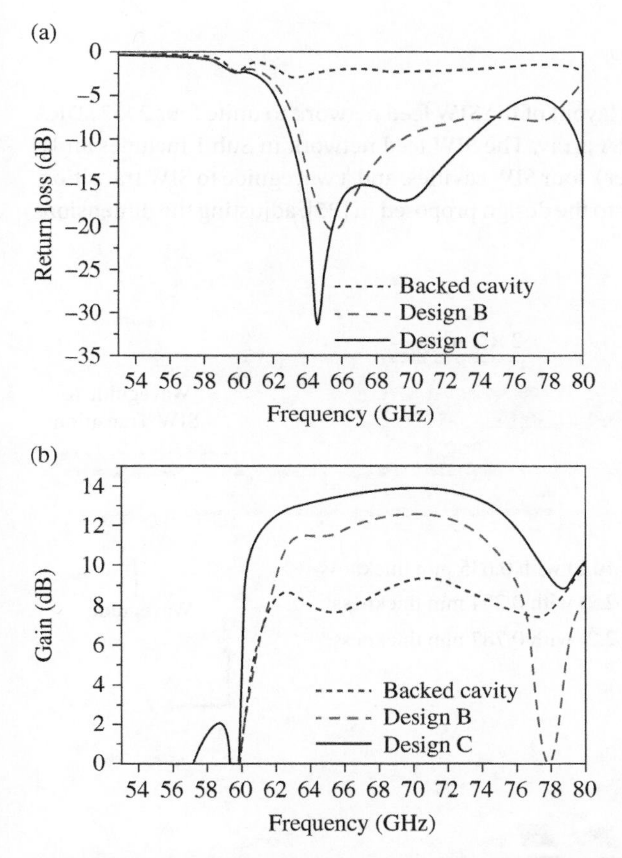

Figure 7.18 Comparisons of the simulated results for the backed cavity; Design B and Design C. (a) Return loss. (b) Gain.

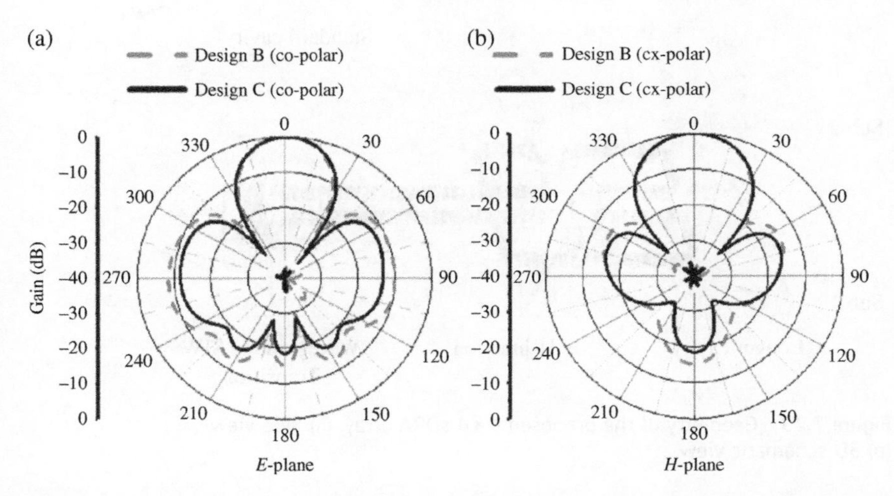

Figure 7.19 Radiation patterns for the backed cavity; Design B and Design C at 67.0 GHz. (a) *E*-plane. (b) *H*-plane.

Figure 7.20 presents the layout of the SIW feed network to unite four 2 × 2 sDRA arrays to form a 4 × 4 sDRA array. The SIW feed network in Sub 1 includes an H-junction (1–4 power divider), four SIW cavities, and a waveguide to SIW transition. The Sub 1 layout is similar to the design proposed in [25], adjusting the dimensions

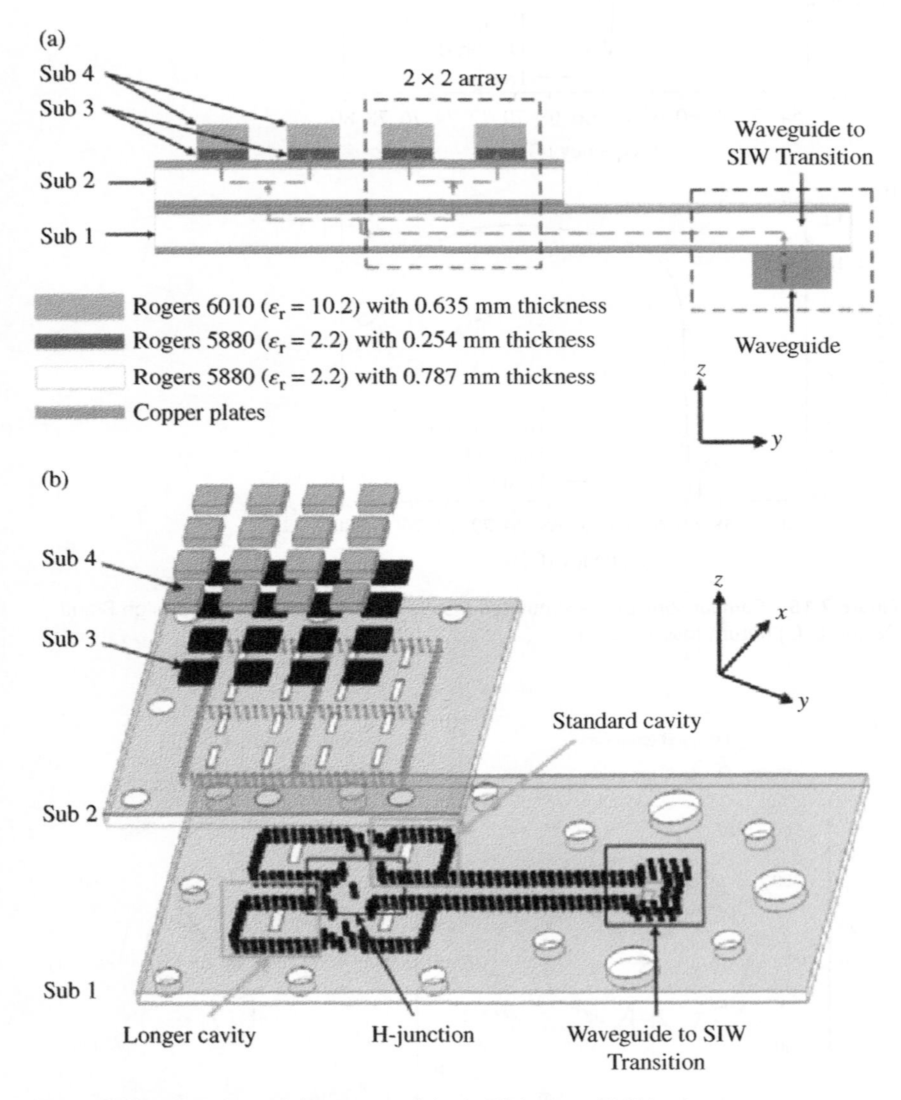

Figure 7.20 Geometry of the proposed 4 × 4 sDRA array. (a) Side view. (b) 3D schematic view.

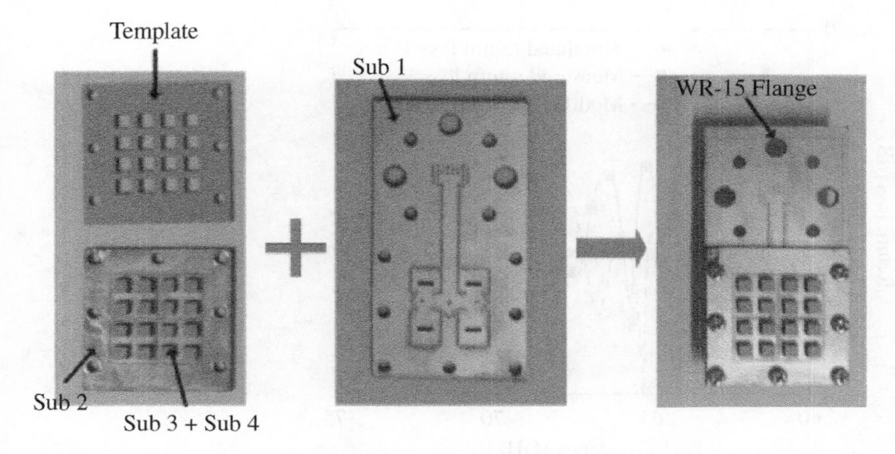

Figure 7.21 Photographs of the 4 × 4 mmW sDRA array.

to shift its operating band from 57 to 64 GHz to 64–71 GHz. For brevity, details about the waveguide to SIW transition and H-junction are not given here but have been detailed in Section 7.3. The prototype of the proposed 4 × 4 sDRA array is presented in Figure 7.21. Compared to the single-layered DRA array fabrication process, the fabrication process of the sDRA array has only one additional step, which is to bond two dielectric boards with a nonconductive adhesive. Then, the bonded stacked dielectric boards are cut into blocks as usual. Due to the large DRA profile of 1.8 × 1.8 mm, the proposed DRA in the stacked structure can be machined by standard PCB processing. The integration of the sDRA element is positioned with the help of dielectric templates.

The size of the array is 13.2 × 13.2 × 2.55 mm, containing 4 × 4 elements with 3.3 mm or $0.69\lambda_{67\text{GHz}}$ spacing. The input impedance of the DRA array was measured by an Agilent N5247A network analyzer. The radiation characteristics were measured by a compact antenna test range (CATR) measurement system.

In Figure 7.22, the simulated and measured return losses are compared for the proposed 4 × 4 sDRA array. The measured bandwidth is 16.4% (ranging from 62.7 to 73.9 GHz) for $|S_{11}| < -10$ dB, apart from a slight increase at 63.8 GHz with −9.0 dB. These results are similar to those of the simulated bandwidth that covers 62.0–73.2 GHz. The return loss of the 4 × 4 sDRA array is slightly wider than that of the 2 × 2 sDRA array and generates more resonances on the operating band. Resonances arise from the interactions between the sDRA, SIW feed network, and SIW transition. Therefore, it is normal that some resonances are not exactly matched in the simulated and measured return losses.

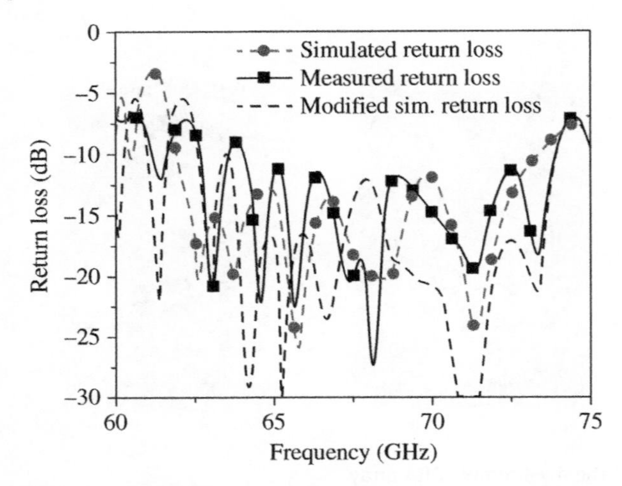

Figure 7.22 Measured and simulated return loss of the proposed array.

Figure 7.23 exhibits the simulated and measured array gains and directivity values. The comparison between the simulated and measured gains shows good agreement at the lower cutoff frequency at 60.0 GHz and the higher cutoff frequency at 73.0 GHz, indicating that the cavity and the DRA are fabricated to be the right size. However, between the lower and higher cutoff frequencies, the simulated gain is as high as 18.7 dBi, with a variation of 0.9 dB, whereas the measured

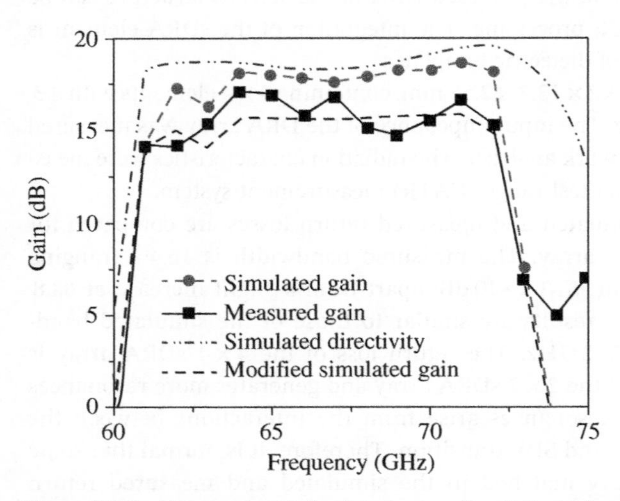

Figure 7.23 Measured and simulated array directivity and gain for the proposed array.

gain is as high as 17.2 dBi with a variation of 2.5 dB. This degradation is not unusual and could be attributed to the uncertain dielectric loss. The simulated model is modified by increasing the tan δ value of the Rogers 5880 laminate from 0.002 to 0.005 or by increasing the tan δ value of the Rogers 6010 laminate from 0.0023 to 0.005. The modified simulated gain and return loss match the measured gain, as shown in the black dashed lines in Figures 7.22 and 7.23. The additional dielectric loss also decreases the efficiency of the antenna array. The simulated maximum efficiency of the antenna array decreases from 89.2% to 72.3% when the tan δ value is modified.

Figure 7.24 shows the measured and simulated radiation patterns of the proposed sDRA array at 64.0, 68.0, and 72.0 GHz. In general, all depicted radiation patterns are symmetrical with the main beams pointed in the boresight direction.

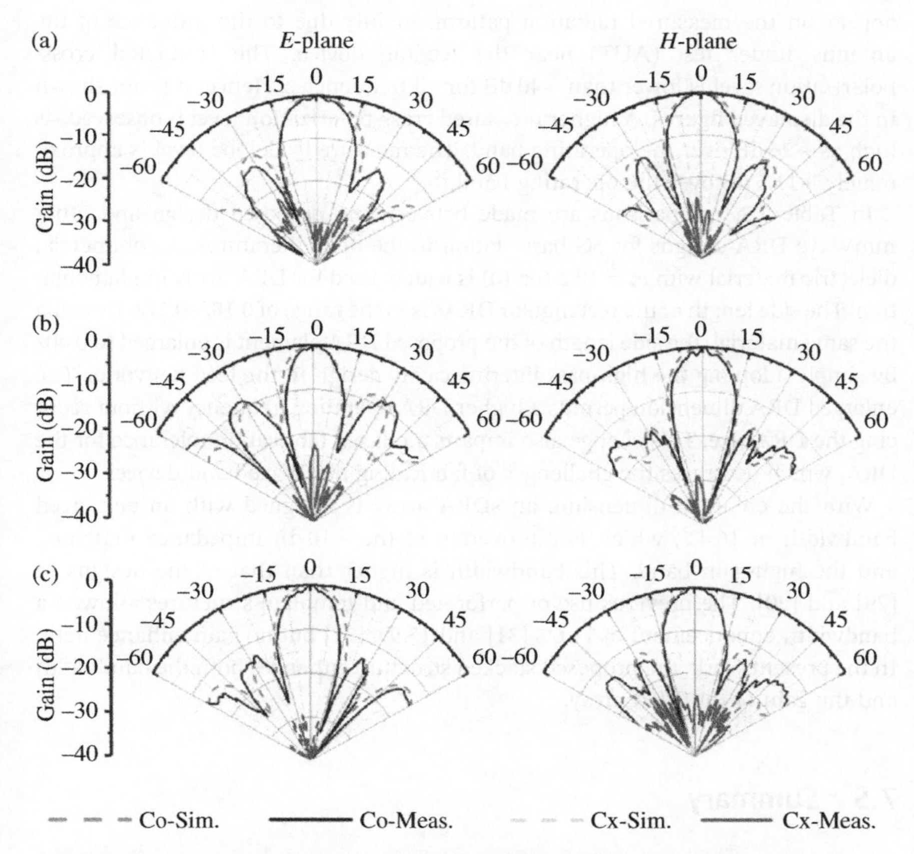

Figure 7.24 Measured and simulated radiation patterns for the proposed array. (a) 64.0 GHz, (b) 68.0 GHz, and (c) 72.0 GHz.

Table 7.3 Comparison between the proposed and reported MMW DRA arrays.

References	ε_r	DRA dimension (mm)	BW	Gain (dBi)
[29]	10.2	2.54 × 2.7 × 1.5 (0.25λ × 0.27λ × 0.15λ)	1.7%	13.6
[30]	10.2	3.0 × 1.5 × 0.787 (0.26λ × 0.18λ × 0.09λ)	6.1%	22.2
[31]	10.0	1.05 × 1.05 × 0.5 (0.21λ × 0.21λ × 0.10λ)	11.6%	10.5
[32]	10.2	Radius: 0.75 (0.25λ) Height: 0.63 (0.13λ)	15.9%	11.4
This work	**10.2/2.2**	**1.8 × 1.8 × 0.889 (0.40λ × 0.40λ × 0.20λ)**	**16.4%**	**17.2**

Good agreement can be observed for both the E-plane and H-plane results. Ripples appear on the measured radiation pattern, mainly due to the influence of the antenna under test (AUT) near the feeding device. The simulated cross-polarization level is lower than −40 dB for all frequencies. Hence, it is not shown in the displayed figures. A higher measured cross-polarization level is observed, as high as −26 dB over the operating band. The measured sidelobe level is approximately −11.6 dB over the operating band.

In Table 7.3, comparisons are made between the proposed design and other mmWave DRA designs for 5G base station in the open literatures. A commercial dielectric material with $\varepsilon_r = 10.2$ (or 10) is widely used for DRA array implementation. The side length of the rectangular DRAs is in the range of 0.18λ–0.27λ. By using the same material, the side length of the proposed DRA element is enlarged to 0.40λ by simply adopting the high-pass filtering cavity design in the feed networks. The enlarged DRA dimension permits a higher DRA operating frequency without reducing the DRA size. This change also imparts a relaxed fabrication tolerance for the DRA, which decreases the challenges of fabricating mmWave-band devices.

With the enlarged dimension, an sDRA array is designed with an enhanced bandwidth of 16.4%, which is the overlap of the −10 dB impedance matching and the high-gain band. This bandwidth is higher than that of the designs in [29] and [20]. The previous use of perforated and template structures allowed a bandwidth enhancement of 11.6% [31] and 15.9% [32] but no gain enhancement. In the present work, the proposed stacked structure enhances both the bandwidth and the gain of the DRA array.

7.5 Summary

In this chapter, DRA and DRA array are proposed for 5G mmWave applications. First, the background of the UAV satellite communication is introduced. As an example, a DRA element is designed with characteristics of dual-band dual-CP and broadened axial ratio bandwidth. The fabrication and measure results verified the proposed design. Then, SIW technology is introduced to provide the merits of lightweight, high gain, and high efficiency of the antenna array in mmWave band. SIW power dividers are designed for large-scaled antenna array, while waveguide to SIW transition is introduced to have a low-loss power conversion between the waveguide and SIW. After that, an SIW-fed DRA array is designed with wideband, high gain, and enlarged dimensions for improved fabrication tolerance, which is suitable for 5G base station antenna. The results and comparison are discussed, showing DRA as promising candidates for 5G mmWave applications.

References

1 Yeap, S.B., Qing, X., and Chen, Z.N. (2015). 77-GHz dual-layer transmit-array for automotive radar applications. *IEEE Transactions on Antennas and Propagation* 63 (6): 2833–2837.

2 Dokhanchi, S.H., Mysore, B.S., Mishra, K.V. et al. (2019). A millimeter-waveave automotive joint radar-communications system. *IEEE Transactions on Aerospace and Electronic Systems* 55 (3): 1241–1260.

3 Yanik, M.E. and Torlak, M. (2019). Near-field MIMO-SAR millimeter-wave imaging with sparsely sampled aperture data. *IEEE Access* 7: 31801–31819.

4 Rappaport, T.S., Xing, Y., MacCartney, G.R. et al. (2017). Overview of millimeter wave communications for fifth-generation (5G) wireless networks—with a focus on propagation models. *IEEE Transactions on Antennas and Propagation* 65 (12): 6213–6230.

5 Zhou, F., Li, W., Meng, L. et al. (2019). Capacity enhancement for hotspot area in 5G cellular networks using mmWave aerial base station. *IEEE Wireless Communications Letters* 8 (3): 677–680.

6 Park, S.J. and Park, S.O. (2017). LHCP and RHCP substrate integrated waveguide antenna arrays for millimeter-wave applications. *IEEE Antennas and Wireless Propagation Letters* 16: 601–604.

7 Yang, Y.H., Sun, B.H., and Guo, J.L. (2019). A low-cost, single-layer, dual circularly polarized antenna for millimeter-wave applications. *IEEE Antennas and Wireless Propagation Letters* 18: 601–604.

8 Zhu, J., Yang, Y., Li, S. et al. (2019). Dual-band dual circularly polarized antenna array using FSS-integrated polarization rotation AMC ground for vehicle satellite communications. *IEEE Transactions on Vehicular Technology* 68 (11): 10742–10751.

9 Yue, T.W., Jiang, Z.H., and Werner, D.H. (2019). A compact metasurface-enabled dual-band dual-circularly polarized antenna loaded with complementary split ring resonators. *IEEE Transactions on Antennas and Propagation* 67 (2): 794–803.

10 Xu, H., Chen, Z., Liu, H. et al. (2022). Single-fed dual-circularly polarized stacked dielectric resonator antenna for K/Ka-band UAV satellite communications. *IEEE*

Transactions on Vehicular Technology 71 (4): 4449–4453. https://doi.org/10.1109/TVT.2022.3144414.

11 Cassivi, Y. and Wu, K. (2003). Low cost microwave oscillator using substrate integrated waveguide cavity. *IEEE Microwave and Wireless Components Letters* 13: 48–50.

12 Mongia, R.K. and Ittipiboon, A. (1997). Theoretical and experimental investigations on rectangular dielectric resonator antennas. *IEEE Transactions on Antennas and Propagation* 45 (9): 1348–1355.

13 Min, K.S., Hirokawa, J., Sakurai, K. et al. (1997). Phase control of circularly polarized waves from a parasitic dipole mounted above a slot. *Proceedings of IEEE Antennas Propagation Society International Symposium*, pp. 1348–1351.

14 Rocher, M.F., Herruzo, J.I.H., Nogueira, A.V. et al. (2016). Circularly polarized slotted waveguide array with improved axial ratio performance. *IEEE Transactions on Antennas and Propagation* 64 (9): 4144–4148.

15 Zhong, Z., Zhang, X., Liang, J. et al. (2019). A compact dual-band circularly polarized antenna with wide axial-ratio beamwidth for vehicle GPS satellite navigation application. *IEEE Transactions on Vehicular Technology* 68 (9): 10742–10751.

16 Zhang, J.D., Wu, W., and Fang, D.G. (2016). Dual-band and dual-circularly polarized shared-aperture array antennas with single-layer substrate. *IEEE Transactions on Antennas and Propagation* 64 (1): 109–116.

17 Wang, S., Zhu, L., Zhang, G. et al. (2020). Dual-band dual-CP all-metal antenna with large signal coverage and high isolation over two bands for vehicular communications. *IEEE Transactions on Vehicular Technology* 69 (1): 1131–1135.

18 Pham, K.T., Clemente, A., Blanco, D. et al. (2020). Dual-circularly polarized high-gain transmitarray antennas at Ka-band. *IEEE Transactions on Antennas and Propagation* 68 (10): 7223–7227.

19 Hao, R.S., Cheng, Y.J., and Wu, Y.F. (2020). Shared-aperture variable inclination continuous transverse stub antenna working at K- and Ka-bands for mobile satellite communication. *IEEE Transactions on Antennas and Propagation* 68 (9): 6656–6666.

20 Cassivi, Y., Perregrini, L., Arcioni, P. et al. (2002). Dispersion characteristics of substrate integrated rectangular waveguide. *IEEE Microwave and Wireless Components Letters* 12 (9): 333–335. https://doi.org/10.1109/LMWC.2002.803188.

21 Feng, X. and Ke, W. (2005). Guided-wave and leakage characteristics of substrate integrated waveguide. *IEEE Transactions on Microwave Theory and Techniques* 53 (1): 66–73. https://doi.org/10.1109/TMTT.2004.839303.

22 Salehi, M. and Mehrshahi, E. (2011). A closed-from formula for dispersion characteristics of fundamental SIW mode. *IEEE Microwave and Wireless Components Letters* 21: 4–6.

23 Ansoft Corporation. (n.d.). HFSS: high frequency structure simulator based on the finite element method. Canonsburg, PA. http://www.ansoft.com

24 Yan, L., Hong, W., Wu, K. et al. (2005). Investigations on the propagation characteristics of the substrate integrated waveguide based on the method of lines. *IEE Proceedings - Microwaves, Antennas and Propagation* 152 (1): 35–42. https://doi.org/10.1049/ip-map:20040726.

25 Chen, Z., Liu, H., Yu, J. et al. (2018). High gain, broadband and dual-polarized substrate integrated waveguide cavity-backed slot antenna array for 60 GHz band. *IEEE Access* 6: 31012–31022.

26 Chen, Z., Shen, C., Liu, H. et al. (2020). Millimeter-wave rectangular dielectric resonator antenna array with enlarged DRA dimensions, wideband capability, and high-gain performance. *IEEE Transactions on Antennas and Propagation* 68 (4): 3271–3276.

27 Gong, K., Chen, Z.N., Qing, X. et al. (2012). Substrate integrated waveguide cavity-backed wide slot antenna for 60-GHz bands. *IEEE Transactions on Antennas and Propagation* 60 (12): 6023–6026. https://doi.org/10.1109/TAP.2012.2213060.

28 Federal Communications Commission. (2016). Fact sheet: Spectrum Frontiers rules identify, open up vast amounts of new high-band Spectrum for next generation (5G) wireless broadband. *Online*. 14. https://apps.fcc.gov/edocs_public/attachmatch/DOC-340310A1.pdf; https://www.fcc.gov/document/rules-facilitate-next-generation-wirelesstechnologies.

29 Chu, H. and Guo, Y.X. (2017). A novel approach for millimeter-wave dielectric resonator antenna array designs by using the substrate integrated technology. *IEEE Transactions on Antennas and Propagation* 65 (2): 909–914.

30 Abdel-Wahab, W.M., Wang, Y., and Safavi-Naeini, S. (2016). SIW hybrid feeding network-integrated 2-D DRA array: simulations and experiments. *IEEE Antennas and Wireless Propagation Letters* 15: 548–551.

31 Qureshi, A., Klymyshyn, D.M., Tayfeh, M. et al. (2017). Template-based dielectric resonator antenna arrays for millimeter-wave applications. *IEEE Transactions on Antennas and Propagation* 65 (9): 4576–4584.

32 Sun, Y. and Leung, K.W. (2018). Circularly polarized substrate-integrated cylindrical dielectric resonator antenna array for 60 GHz applications. *IEEE Antennas and Wireless Propagation Letters* 17 (8): 1401–1405.

33 Hou, D., Xiong, Y., Goh, W. et al. (2012). 130-GHz on-chip meander slot antennas with stacked dielectric resonators in standard CMOS technology. *IEEE Transactions on Antennas and Propagation* 60 (9): 4102–4109.

34 Hao, Z., Hong, W., Chen, J. et al. (2005). Compact super-wide bandpass substrate integrated waveguide (SIW) filters. *IEEE Transactions on Microwave Theory and Techniques* 53 (9): 2968–2977.

35 Marcatili, E.A.J. (1969). Dielectric rectangular waveguide and directional coupler for integrated optics. *The Bell System Technical Journal* 48 (7): 2071–2102.

36 Leung, K.W., Lim, E.H., and Fang, X.S. (2012). Dielectric resonator antennas: from the basic to the aesthetic. *Proceedings of the IEEE* 100 (7): 2181–2193.

8

Duplex Filtering DRA

CHAPTER MENU

8.1 Overview

The diverse wireless communication scenarios stimulate the development of miniaturized antennas with multiple functions [1]. For example, antennas are enabled with filtering functions to achieve specific frequency response for both the reflection coefficient and antenna gain. These antennas, which are called filtering antenna (or filtenna), efficiently integrate the antenna and the filter into one module to achieve the miniaturized size and lower loss. Besides, antennas with duplex function provide high-level isolation between the transceiver links for dual-band operation. In diplex/duplex antennas, component number and circuit footprint are highly reduced and the interfaces between them are avoided, thus reducing the loss in devices.

Dielectric Resonator Antennas: Materials, Designs and Applications, First Edition.
Zhijiao Chen, Jing-Ya Deng, and Haiwen Liu.

Published 2024 by John Wiley & Sons, Inc.

The filtering antennas have been implemented by patch antennas [2], slot antennas [3], and dielectric resonator antennas (DRAs) [4, 5]. The diplex/duplex antennas can be realized with microstrip [6], patch [7], or slot design [8, 9]. As a 3D configured antenna, DRA possesses several attractive advantages such as small size, lightweight, ease of excitation, low dissipation, and high degree of design flexibility [10–12]. With proper design, DRAs can fulfill various requirements of multifunctional antennas such as reconfigurable DRA [13], filtering DRA (FDRA) [14], and duplex DRA [15]. The ease of excitation enables DRA to be easily integrated with filtering or duplexing devices such as baluns and couplers, enjoying simplicity in structural design. Also, its high degree of freedom in shape and multi-mode nature allows simple design of filtering structures inside the DRA.

The rest of the chapter is organized as follows. In Section 8.2, the working principle of filtering antenna is demonstrated and classified. Antenna examples are given and the merits of FDRA are clarified. Section 8.3 defines and differentiates the concepts of diplexer and duplexer, along with their applications. The types of diplexer are categorized and presented with examples. Section 8.4 introduces two examples, including an FDRA and a duplex filtering DRA (DFDRA). The proposed antennas are preferable candidates for highly integrated wireless millimeter-wave communication system applications. Finally, remarks are concluded in Section 8.5.

8.2 Filtering Antenna

As essential components in the radio frequency (RF) front ends, antennas and band-pass filters (BPF) work separately and require matching circuits or devices for interconnection, leading to increased physical size and extra systematic loss. Filtering antenna is a co-design of the antenna and the filter, which has been proposed to improve the integration of passive components in the RF front ends by combining both filtering and radiating functions into one single passive structure. They have been developed to meet the ever-increasing performance requirements of high performance and miniaturization of wireless systems. The main challenge with filtering antenna is how to effectively reduce the loss caused by interconnection and matching networks while miniaturizing its overall size.

Figure 8.1 shows a typical filtering antenna with band-pass-type filtering radiation response, which provides good skirt selectivity with flat antenna gain in the passband and high suppression in the stopband. A filtering antenna will have at least one radiation null at the lower and upper band edges, respectively. A radiation null is a frequency point where the antenna barely radiates any energy.

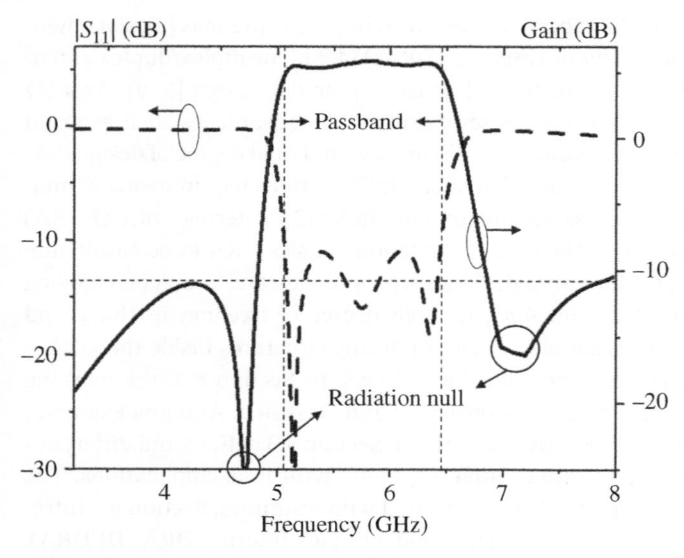

Figure 8.1 Frequency response of a typical filtering antenna.

In this way, it will cause a dip at the corresponding frequency point in the gain plot. This dip in the gain plot can be determined as an effective radiation null only when the value of its S_{11} is close to zero. This definition is to make sure that the radiated power is none not only in the boresight direction but also in any other radiating directions at this specific frequency point.

8.2.1 Filtering Antenna Classifications

Filtering antennas are classified here as Types A, B, and C as indicated in Figure 8.2. In general, Type A cascades filters or filtering devices with the antenna; Type B replaces filter's last-order resonator with an antenna; Type C designs filtering structures inside the antenna.

Type A cascades filters or filtering devices with the antenna. This cascaded method is comparably simple because the filter and the antenna are designed separately and then integrated together. Extra lines or matching network are usually needed to achieve impedance matching between the filter and the antenna. Therefore, this cascaded methodology tends to introduce additional loss or occupy a large area. Figure 8.3 illustrates a filtering antenna employing the cascaded design by combining a stub-loaded resonator (SLR) filter and a four-element patch array through an additional 50 Ω interface. It has simple structure but occupies a large area since the filter and the antenna array are placed on the same substrate.

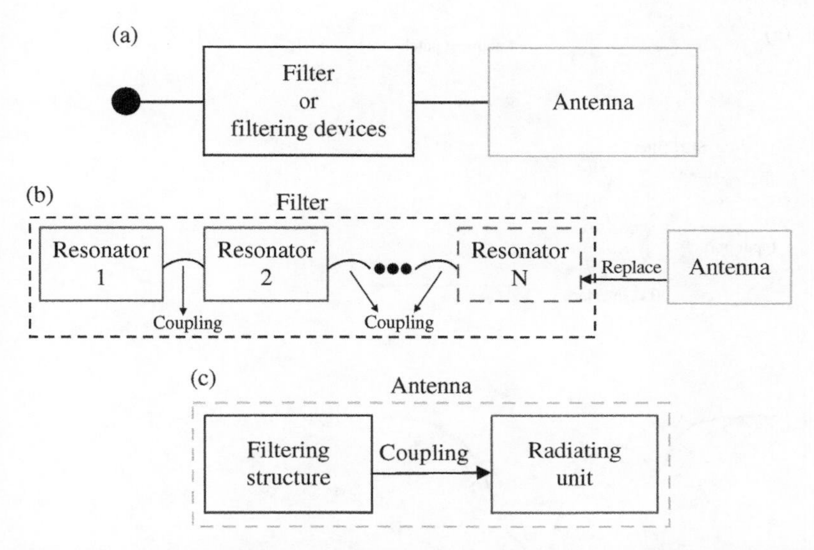

Figure 8.2 Diagrams of three design methodologies of filtering antenna. (a) Type A: Combine filters or filtering devices with antenna. (b) Type B: Replace filter's last-order resonator with antenna. (c) Type C: Design filtering structure inside antenna.

Besides, the overall bandwidth is restricted by the bandwidth of the patch, feeding network, and filtering device.

Type B is designed based on the synthesis approach of the filter. The last-order resonator of the traditional BPF is replaced with the antenna, which makes the antenna act as the load impedance of the filter and the radiator at the same time. This kind of cascaded design can reduce the external loss caused by extra matching network. However, multiple resonators used in filter synthesis occupy large area or high profile. Also, the internal insertion loss cannot be avoided and will ultimately deteriorate antenna's gain performances. Figure 8.4 shows an example of Type B filtering antenna. A microstrip rectangular capacitively loaded loop (CLL) BPF with two transmission zeros is first designed before substituting the last CLL element for the radiator. In this design, skirt selectivity is not sharp because it is a second-order filter. The sharper edge of the filter can be generated by increasing the order of the filter, but the design will occupy much larger area that is not suitable for practical use.

In Type C, the filter is designed inside the antenna to achieve band-stop characteristics in the gain plot. It eliminates the presence of the filter and thus minimizes the insertion loss and the overall size. This method has been explored extensively in recent years for it addresses the problematic issues of the cascaded

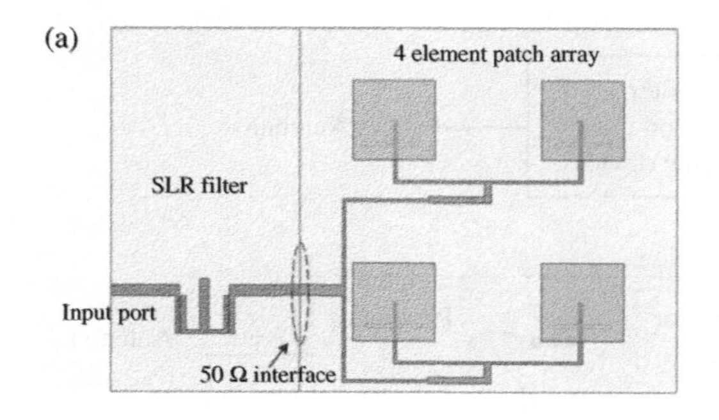

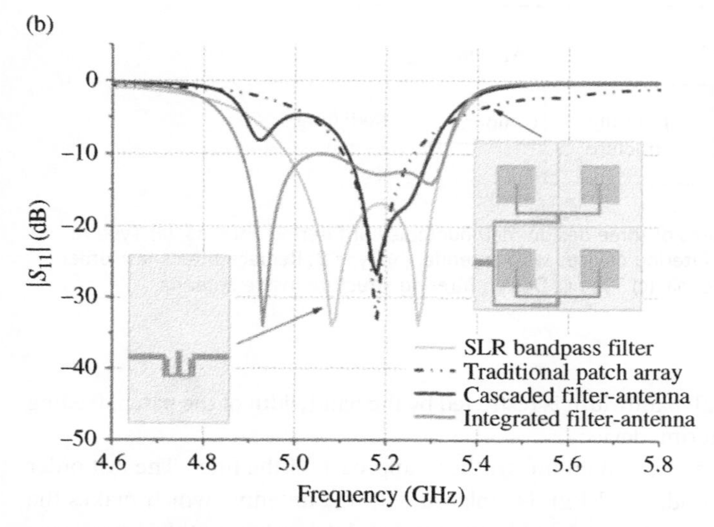

Figure 8.3 Example of Type A filtering antenna. *Source:* Adapted from Mao et al. [16]. (a) Configuration. (b) Reflecting coefficient.

design presented above. Without any complex filtering circuits involved, in-band antenna performances can be ensured. The antenna consists of filtering structures and radiating units, usually interacting with each other through coupling. This design realizes filtering response by introducing radiation nulls at band edges and enhancing suppression in the stopbands.

The most common way to inspire radiation nulls is to design unique structures to produce electric fields with opposite phase distribution at specific frequencies. The two out-of-phase electric fields will cancel each other and result in radiation nulls in the gain plot. An alternative way to introduce radiation nulls is utilizing unexcited non-radiating mode [18]. Figure 8.5a illustrates a Type C filtering

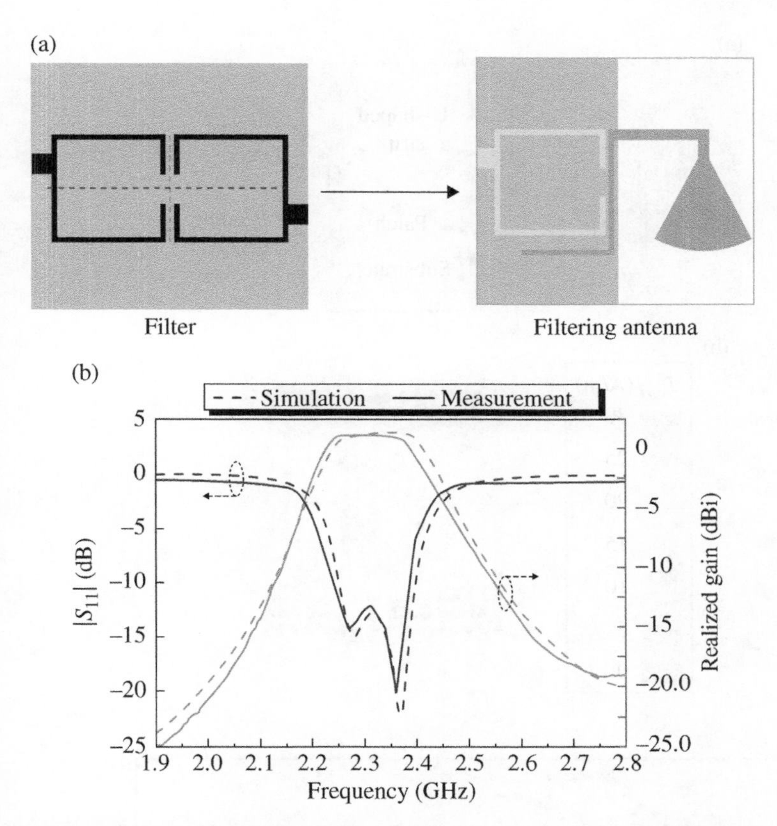

Figure 8.4 Example of Type B filtering antenna. *Source:* Tang et al. [17]/IEEE. (a) Designing process. (b) Reflecting coefficient and realized gain response.

antenna example. A U-shaped slot is etched on the patch to cause opposite currents at 4.2 GHz as shown in Figure 8.5b,c where a radiation null is generated. The other radiation null at 6.6 GHz is caused by the U-shaped strip that functions as a pair of $\lambda/4$ open-ended transmission lines.

Apart from generating radiation nulls, stopband suppression can prevent power from being transmitted into the radiator. Stopband suppression can be achieved by suppressing unwanted resonances in the stopbands. For example, harmonic suppression is implemented in filtering antenna design in [19] to improve out-of-band rejection, as shown in Figure 8.6. In this design, the U-slot patch antenna is coupled with a filtering SLR, which eliminates the strong harmonics at 8.75 and 11.7 GHz produced by traditional patch antenna.

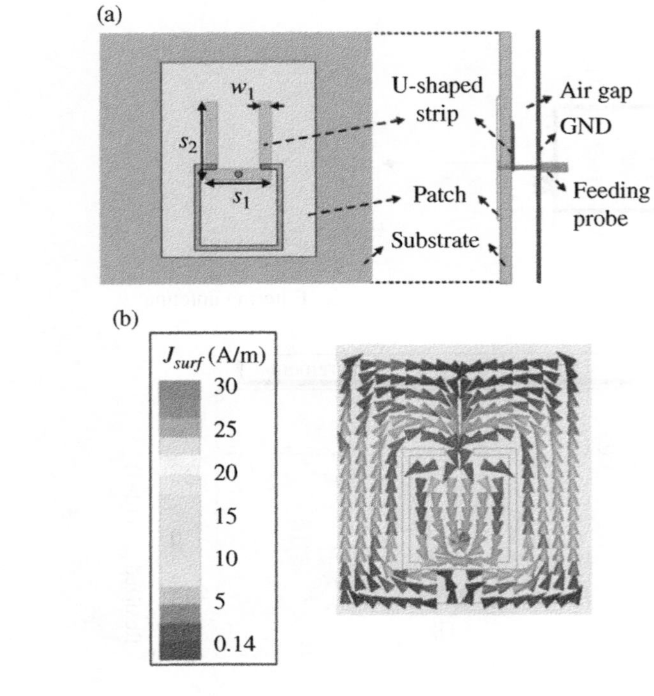

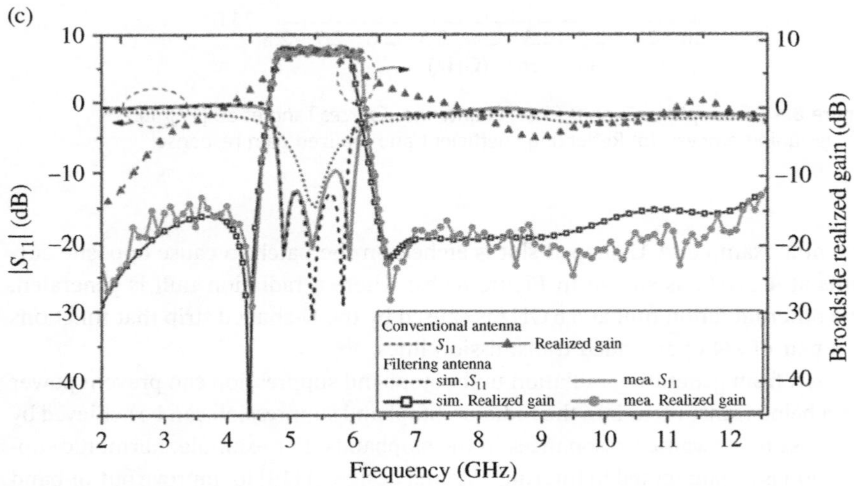

Figure 8.5 Example of Type C filtering antenna. *Source:* Yang et al. [18]/IEEE. (a) Configuration. (b) Electric current distributions at 4.2 GHz. (c) Reflecting coefficient and realized gain response.

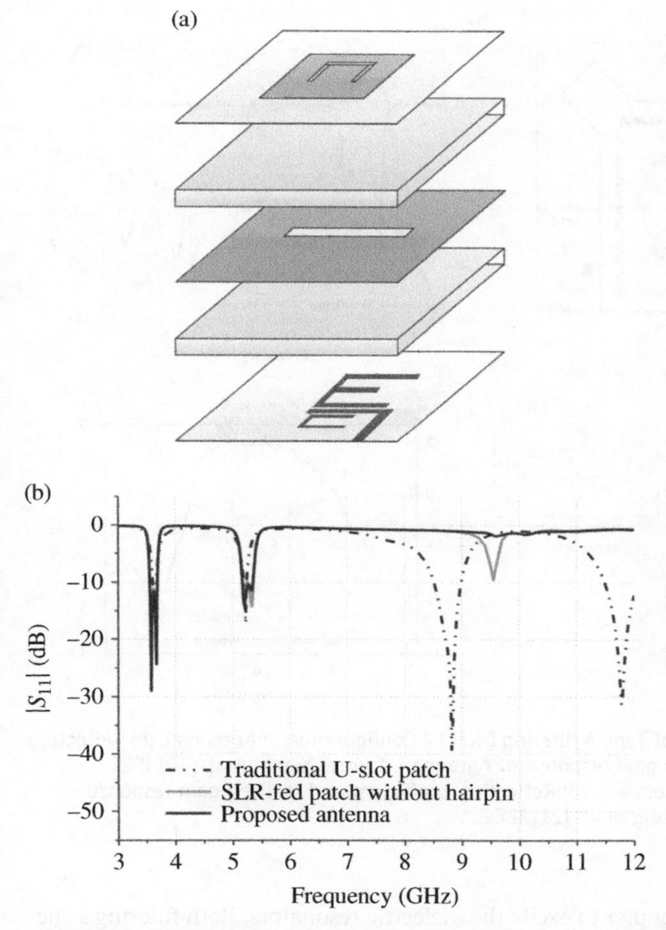

Figure 8.6 Example of Type C filtering antenna. *Source:* Mao et al. [19]/IEEE. (a) Configuration. (b) Harmonic responses.

8.2.2 Filtering DRA

DRA has great potential to work as filtering antennas, especially for Type A and C designs. For Type A filtering antenna, DRA with the merits of small size, low dissipation loss, and ease of excitation is easier to be integrated with components such as filters, couplers [20], and baluns [21]. Compared with all-metal integrated antennas such as microstrip or slot antennas, DRA with dielectric radiating structure enjoys simplicity in being integrated with metal feeding networks. Figure 8.7 illustrates two examples of Type A FDRA. The former uses filtering balun and the

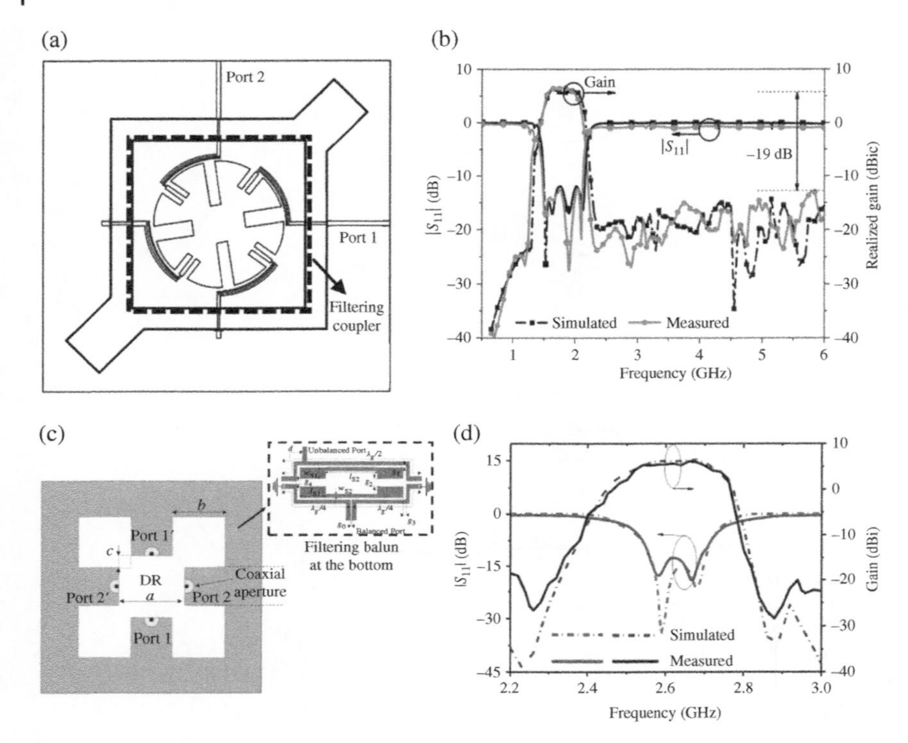

Figure 8.7 Examples of Type A filtering DRA. (a) Configuration of Antenna I. (b) Reflecting coefficient and realized gain response of Antenna I. *Source:* Xiang et al. [20]/IEEE. (c) Configuration of Antenna II. (d) Reflecting coefficient and realized gain response of Antenna II. *Source:* Tang et al. [21]/IEEE.

latter uses filtering coupler to excite the dielectric resonators. Both filtering structures are printed on the back of the substrate without extra space on the same plane, resulting in a compact structure.

For Type C design, DRA enjoys high degree of design flexibility to become a suitable candidate for filtering antenna. Dielectric resonators can be easily cut or etched to introduce electric field with opposite phase distribution, creating radiation nulls. Besides, higher-order modes in dielectric resonators can be used to generate natural radiation nulls with out-of-phase electric field distributions. Figure 8.8 illustrates two Type C FDRA examples. The former one employs hybrid microstrip line and conformal strip feeding, exciting electric fields inside the DRA that are out of phase with each other. These two electric fields are canceled out and two radiation nulls are produced at 1.56 and 2.25 GHz for a sharp roll-off rate. The latter example uses a probe and a disk to excite out-of-phase electric fields inside

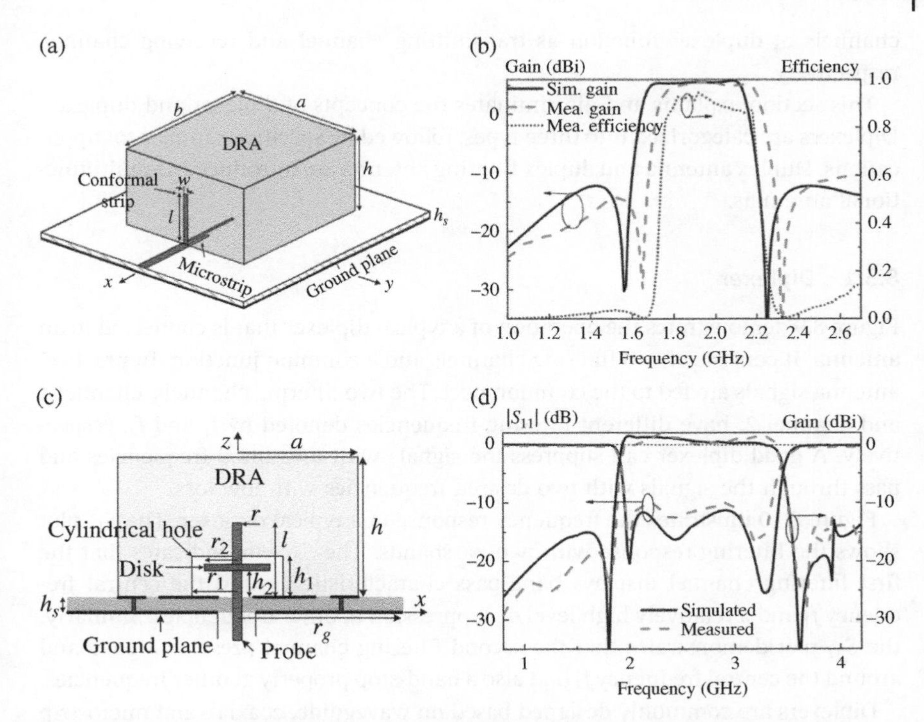

Figure 8.8 Examples of Type C filtering DRA. (a) Configuration of Antenna I. (b) Efficiency and gain response of Antenna I. *Source:* Pan et al. [4]/IEEE. (c) Configuration of Antenna II. (d) Reflecting coefficient and gain response of Antenna II. *Source:* Hu et al. [5]/IEEE.

the DRA. It forms a radiation null at 3.6 GHz, providing upper stopband filtering performance.

8.3 Diplexer and Duplexer

Diplexers/duplexers are key components in the RF front end. They provide two channels connecting the antenna and the transceiver links for signal extraction. Diplexers/duplexers are three-port device with one input port and two output ports, usually consisting of two filtering channels. Signals with specific frequencies are filtered out through the channels for frequency selection. The distinction between diplexer and duplexer can be clarified as follows. Diplexer can only process signals with two different frequencies. The two channels of diplexer both function as transmitting channels or receiving channels simultaneously. In contrast, duplexer can process signals with either different or same frequencies. The two channels of duplexer function as transmitting channel and receiving channel, respectively.

This section explains and discriminates the concepts of diplexer and duplexer. Diplexers are categorized into three types, followed by specific examples for applications. Duplex antenna and duplex filtering antenna are introduced as multifunctional antennas.

8.3.1 Diplexer

Figure 8.9 demonstrates the operation of a typical diplexer that is connected to an antenna. It consists of two filtering channels and a common junction. In practice, antenna signals are fed to the common port. The two filtering channels, channel 1 and channel 2, have different filtering frequencies denoted by f_1 and f_2, respectively. A good diplexer can suppress the signals with unwanted frequencies and pass through the signals with two desired frequencies with low loss.

Figure 8.10 illustrates the frequency response of a typical diplexer. The S_{11} plot shows the filtering response with two passbands. The S_{21} plot indicates that the first filtering channel displays band-pass characteristic around the central frequency f_1 and a relatively high level of suppression at other frequencies. Similarly, the S_{31} plot demonstrates that the second filtering channel presents a passband around the central frequency f_2 and also a band-stop property at other frequencies.

Diplexers are commonly designed based on waveguide, coaxial, and microstrip structures. They can be classified into the following three types based on the operations of the common junction: Type A with T-type matching network, Type B with distributed coupling structure, and Type C with common resonator, as illustrated in Figure 8.11.

Type A diplexer uses T-type matching network with 1/4 λ_g line to create open circuits, realizing isolation between two separated channels. To specify, we denote the resonant wavelengths of two filtering channels by λ_1 and λ_2, respectively. The

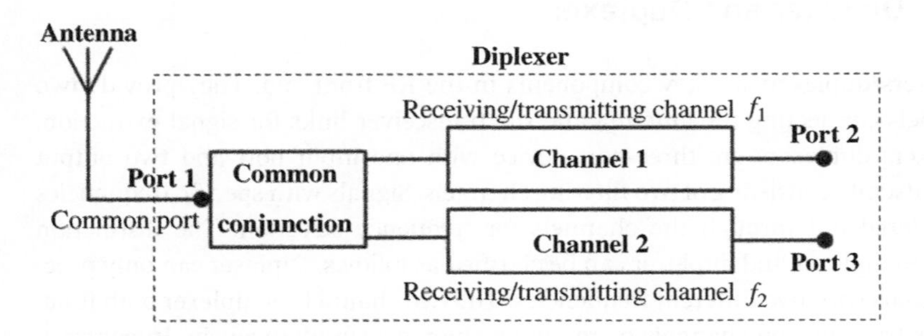

Figure 8.9 Transceiver front end including antenna and diplexer.

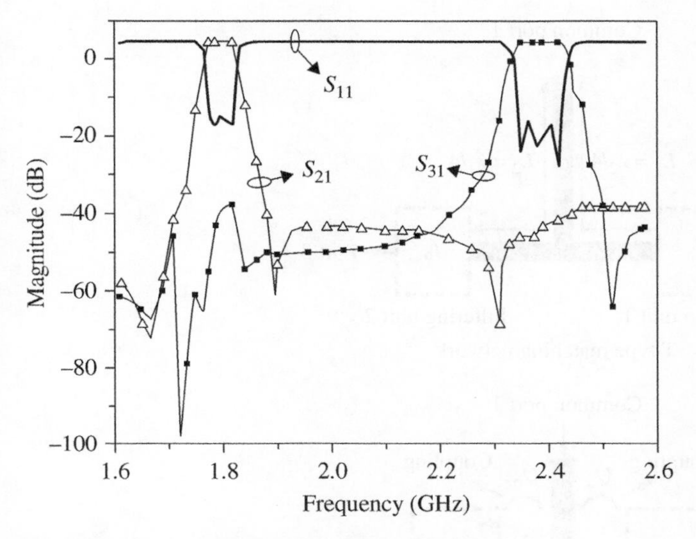

Figure 8.10 Typical frequency response of diplexer.

linear length between common port and the input port of channel 1 is designed to be the equivalent of 1/4 λ_2. This is to make sure that channel 1 will be an open circuit when channel 2 is working. In this way, no signal with operating frequency f_2 will flow into channel 1, avoiding interference between two channels. Similarly, the linear length between common port and the input port of channel 2 is designed to be 1/4 λ_1, making channel 2 an open circuit when channel 1 is functioning. Signals with operating frequency f_1 will be prevented from flowing into channel 2 and isolation between channels is achieved. A classic Type A diplexer is proposed in [22]. Its matching network is composed of only two sections of transmission lines featuring different characteristic impedances. The diplexer in [23] is also a Type A diplexer with an optimized T-junction that utilized corner cut to improve impedance matching. In [24], the diplexer T-junction uses stepped-impedance lines instead of uniform-impedance lines for optimization.

Type B diplexer employs distributed coupling structure. A common feeding line connected with the input port is directly coupled with two filtering channels. A classic Type B diplexer is proposed in [25]. The eight BPFs that form two filtering channels are deployed along the feeding lines for direct coupling. The diplexer in [26] is also a Type B diplexer, employing source-loaded coupling lines as direct coupling structures. Similar Type B diplexer designs can be found in [27, 28].

Type C diplexer uses common resonators that are connected to the input port directly for impedance matching. The common resonator possesses at least two

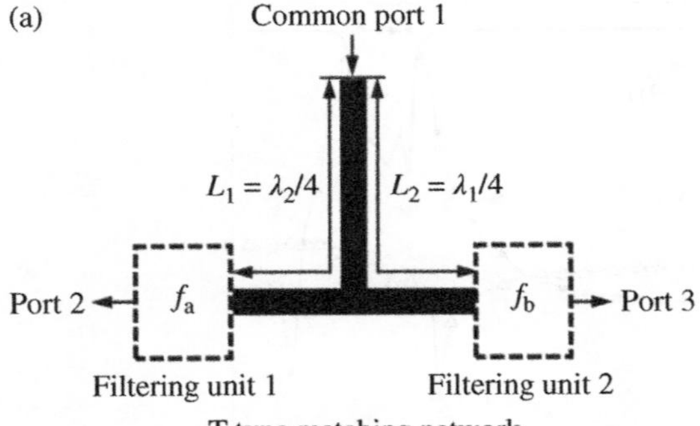

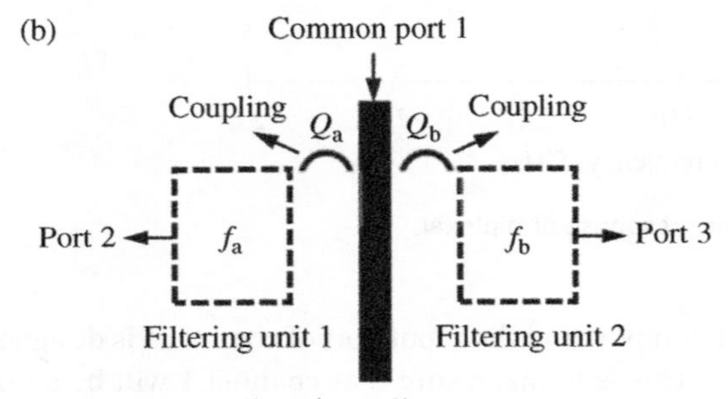

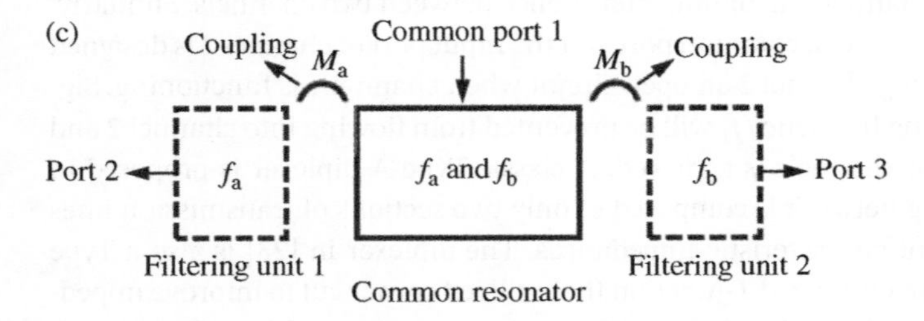

Figure 8.11 Diagrams of three design methodologies of diplexers. (a) Type A: Diplexer with T-type matching network. (b) Type B: Diplexer with distributed coupling structure. (c) Type C: Diplexer with common resonator.

operating frequencies, so that it can function as a filter resonator in both channels at the corresponding frequencies. It usually interacts with two filtering channels through coupling. The design in [29] is a classic Type C diplexer, using a dual-mode resonator at the input port as the common resonator. Besides input common resonator, additional common resonators placed in the middle of filtering channels can also be employed as in [30].

The pros and cons of these three types of diplexers are demonstrated here. Type A diplexer with T-junction has the simplest design and thus enjoys a wide range of applications. With matching networks, the input port can be well matched with the external 50 Ω circuits. Also, high-level isolation between two filtering channels can be easily achieved. However, the additional matching networks lead to bulky circuit layout, making it challenging for wireless communication applications. Type B diplexer eliminates the matching networks by using direct coupling between common feeding lines and filtering channels. In this way, compact structures are realized, which contributes to device miniaturization. Nevertheless, direct coupling will lead to deterioration in isolation level between two channels. Type C diplexer employs common resonators to get rid of additional matching networks, realizing compact size. The common resonators also help reduce the number of resonators needed by filtering channels, which further shrinks the overall size. However, since signals with two frequencies can both be transmitted through the common resonator, Type C diplexer suffers from low level of isolation as well. Besides, free control of multi-frequency is needed with the common resonator, which increases design complexity.

8.3.2 Duplexer

The concept of duplexer and diplexer can be easily confused. They are both three-port device and can be applied to different-frequency operation to single out signals with specific frequencies. Diplexer enables one antenna to use two signal paths by means of the frequency selectivity of filters. In this way, transceiver of two different frequencies can be realized using only one antenna. The two channels of diplexer can function only as receiving channels or transmitting channels alternatively at one time. Diplexers are usually used when multiple modulation modes and carriers need to be applied to one antenna. For example, neighbor cell sites often suffer from limited offsite space for additional antennas. As a result, different kinds of signals need to be transmitted and received by one antenna. For this reason, diplexers are employed to separate signals with different frequencies.

Duplexers are also capable of signal separation. Furthermore, duplexers enable one antenna with two-way communications, i.e. allocating the receiving signals and transmitting signals at the same operating frequencies. For example, radar transceiver modules usually have extremely close transceiver frequencies and can be separated properly only by duplexers. Therefore, diplexers cannot replace duplexers in common circuits. In comparison, duplexers enjoy a wider application in future communication for realizing higher spectrum efficiency.

The conventional transceiver system that uses different frequency bands leads to complicated system design. For example, the integration of one broadband antenna and multiple filters operating at different frequencies will result in a cascaded structure with bulky circuit layout and higher insertion loss. With the rapid development of mobile communications, full duplex or simultaneous transmit and receive (STAR) mode has been widely used. The STAR mode has attracted attention for its advantage of double data throughput with improved communication efficiency [31–33]. Thus, diplex/duplex antennas have been numerically investigated in recent years [9, 34–36].

There are generally two ways to achieve diplex/duplex function in antennas: using two resonators with different modes or exciting two modes within one resonator. Designs using the latter method share similarities with Type C diplexers. The resonator is shared by both channels as the common resonator, with each channel using one resonant mode. So far, most of the diplex/duplex antennas are implemented by substrate integrated waveguide (SIW)-based patch antennas. There are very few designs based on DRA. The potential of duplex DRA is worthy to be explored for the multi-mode resonance and ease of excitation characteristics of DRA.

Furthermore, in recent research, multiple functions such as radiation, filtering, and duplexing can be merged into one device. This reduces component number and circuit footprint without using extra transmission lines and interfaces between them [37–39]. As a result, multi-functional antennas are desired in wireless systems for facilitating the uplink and downlink simultaneously, which can further simplify the processing of backend systems and reduce the weight, size, and cost of wireless systems [40].

Among highly integrated antenna designs, diplex/duplex filtering antenna has become a new trend over the past years. Diplex/duplex filtering antennas are commonly implemented by SIW and patch antennas [41–44], which use Type C diplexer design method of employing common resonators. In the next section, DRA using Type C design method is explored to realize a multifunctional antenna featuring both duplex and filtering functions. The design guidelines, operation, merits, and antenna examples will be presented in detail.

8.4 Duplex Filtering DRA Designs

Attaching duplexers and filters to antennas will directly lead to bulky system. In addition, the separate designs of duplexers, filters, and antennas can cause additional loss in the process of systematic transmission. In order to address the problems mentioned above, multifunctional integrated antennas using co-design methodology to remove the inter-connection components and extra matching networks has been widely studied [45–47]. Over the past few years, DRAs have drawn thriving attention by the virtue of their high radiation efficiency, high gain, multi-mode nature, ease of excitation, easy integration with planar circuit, and so on, endowing DRAs with great potential to fulfill the purpose of multifunction. Therefore, a wideband FDRA with high gain and a differentially fed DFDRA are demonstrated in this section. Both designs integrate multiple functions into a coherent highly functional unit, thus avoiding problems of bulky structure and additional systematic loss.

8.4.1 Wideband High-Gain Filtering DRA

Figure 8.12 illustrates the configuration of the wideband FDRA with high gain. The antenna consists of two parts. The first part is composed of a tri-mode BPF and three quarter-wavelength line sections. The second part is a dual-mode rectangular DRA (RDRA) stacked on the upper layer of the substrate. The dual-mode RDRA is designed to operate at TE_{113} and TE_{115} mode simultaneously to achieve high gain and wide band. The BPF and the feeding network are

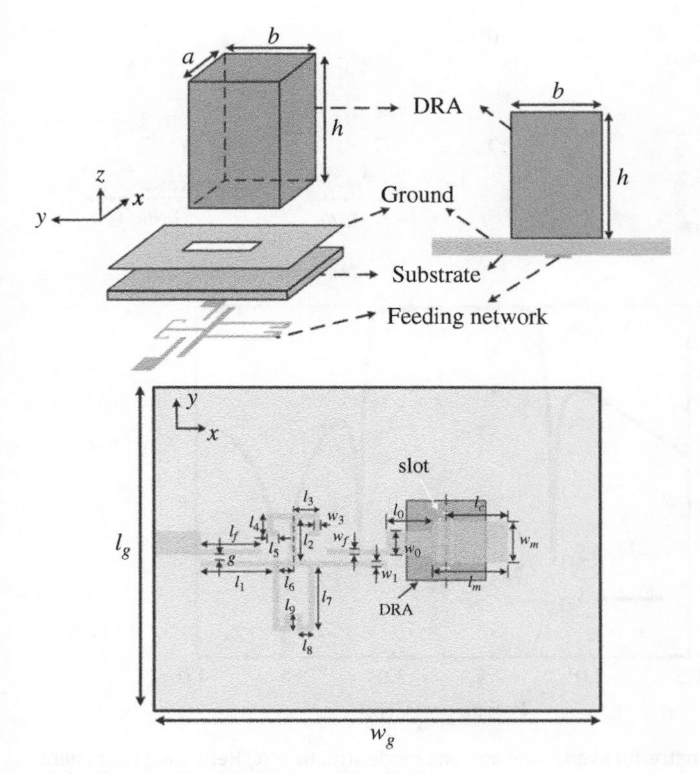

Figure 8.12 Configuration of the FDRA.

printed at the bottom of the substrate. The RDRA and feeding network share a ground plane printed on the upper layer of the substrate. An aperture is etched on the ground.

Figure 8.13a illustrates the BPF structure of the FDRA. It is composed of a half wavelength section and three quarter-wavelength line sections that are calculated and designed to be a five-order Chebyshev BPF. First, the order of the Chebyshev filter is determined based on existing equations. The wideband FDRA is synthesized to be operating with center frequency at 2.6 GHz, with five-order Chebyshev filtering performance. Then the five transmission poles can be obtained by the same equations. This method is to match all the resonant frequencies of the compact multi-mode resonator with the transmission poles, which can be used to realize the FDRA. Next, the tri-mode SLR and dual-mode RDRA are designed separately.

Because the SLR is symmetrical in structure, the operating mechanism can be justified by an even/odd analysis as shown in Figure 8.13b. Therefore, the input

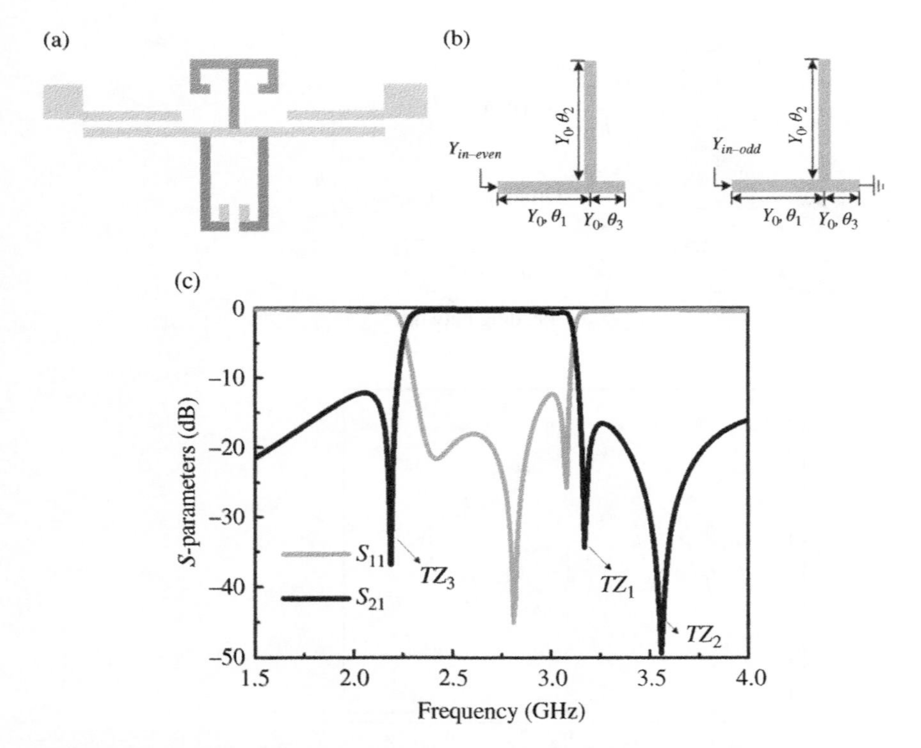

Figure 8.13 (a) Geometry. (b) Even mode and odd mode structure. (c) Reflecting coefficient response of the tri-mode SLR.

admittance can be expressed respectively. Based on the resonance condition under the even/odd mode excitation, combined with the help of the commercial full-wave simulator high frequency simulator structure (HFSS), the dimensions of the tri-mode SLR can be determined.

To improve the selectivity of the filter, transmission zeros should be manually introduced without affecting the in-band transmission performance. By folding two stubs inward, two transmission zeros of the SLR resonator are generated in the upper stopband. As shown in Figure 8.13c, one transmission zero at 3.15 GHz is created by two open-ended stubs, and the other one at 3.6 GHz is caused by the out-of-phase cancellation with two dissimilar signal paths. After introducing an additional stub at the other side of the symmetrical plane, one transmission zero at 2.2 GHz in the lower stopband is introduced. Meanwhile, the impedance matching performance of the filter is improved.

The RDRA is excited by an aperture coupled microstrip line and operates at TE_{113} and TE_{115} mode simultaneously. The resonant frequencies of TE_{113} and TE_{115} mode can be calculated by existing equations. In this way, the initial dimensions of the dual-mode RDRA can then be determined. Now that both dimensions of the tri-mode SLR and the dual-mode RDRA are determined, the co-design method is then utilized to design the proposed wideband FDRA. Figure 8.14 shows the equivalent coupling topology of the wideband FDRA. Three resonant modes of the tri-mode SLR is modeled by three parallel $L_{ri}C_{ri}$ ($i = 1,2,3$) circuits. The inverter J_{rr} represents the mutual coupling between these resonant modes. The dual-mode DRA is equivalent to the parallel $L_{a1}C_{a1}R_{a1}$ and $L_{a2}C_{a2}R_{a2}$ circuits. The inverter

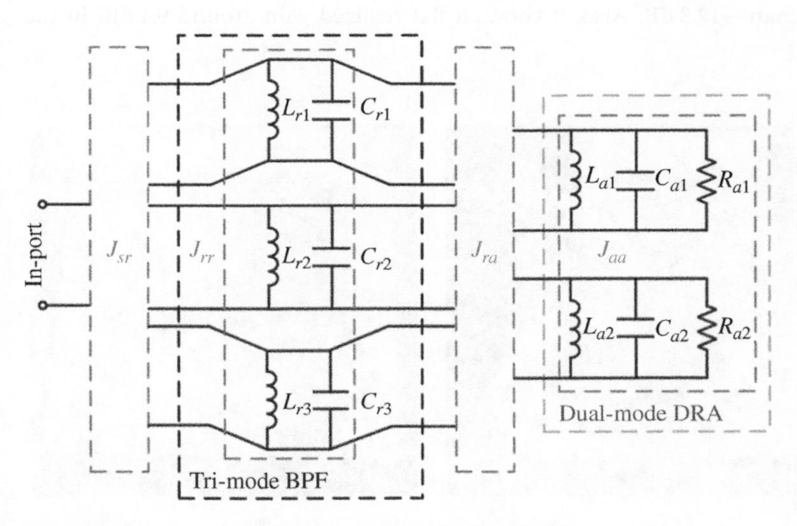

Figure 8.14 Equivalent coupling topology of the FDRA.

Jaa represents the mutual coupling between the two resonant modes of the dual-mode DRA. The input coupling is represented as J_{sr}, while the coupling between the tri-mode SLR and the dual-mode DRA is expressed as J_{ra}. It is noted that after the SLR and DRA are determined, the parameters $L_{ri}C_{ri}$, $L_{a1}C_{a1}R_{a1}$, $L_{a2}C_{a2}R_{a2}$, J_{rr}, and J_{aa} are determined correspondingly. Thus, J_{sr} and J_{sr} are two main factors influencing the whole performance of the wideband FDRA. J_{sr} can be controlled by the length l_f and the gap g. Meanwhile, J_{ra} is determined by the dimensions of the coupling slot ($l_s \times w_s$) and the stepped impedance resonator (SIR) feeding line ($l_c \times w_m$). Hence, the last procedure of the antenna design is tuning these parameters to optimize the design for good impedance matching to achieve a good filtering performance, such as the satisfied -10 dB bandwidth and a plain in-band gain. From the description above, it can be easily deduced that the design methodology of the wideband high-gain FDRA can be categorized as Type A filtering antenna.

To further verify the validity of the proposed collaborative method, the final wideband FDRA is fabricated. Figure 8.15 illustrates the fabricated prototype of the wideband high-gain FDRA. The dimensions of the antenna are tabulated in Table 8.1. The selected dielectric material for DRA is Al_2O_3 ceramic with $\varepsilon_{ra} = 5$ and tan $\delta = 2 \times 10^{-3}$. The tri-mode SLR feeding is constructed using Rogers RT/duroid 6010 with a relative permittivity of 10.2 and a thickness of 1.27 mm.

The measurements for input reflection coefficient are performed by an Agilent vector network analyzer. Figure 8.16 shows the simulated and measured S_{11} and peak realized gain of the FDRA. The FDRA shows good filtering responses with the relative impedance bandwidth 34% from 2.2 to 3.1 GHz with the in-band S_{11} less than -12.3 dB. Also, it shows a flat realized gain around 9.1 dBi in the

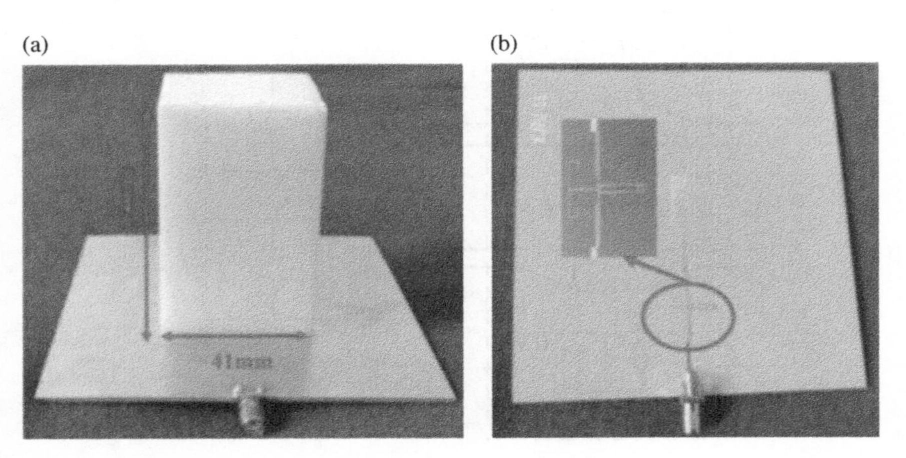

Figure 8.15 Photos of fabricated FDRA. (a) 3D view. (b) Bottom view.

Table 8.1 Dimensions of the proposed wideband filtering MMR antenna.

Parameters	l_0	l_1	l_2	l_3	l_4	l_5	l_6	l_7
Values (mm)	14.9	9.6	5.2	4.8	1.4	1.6	0.5	8.2
Parameters	l_8	l_9	l_f	a	b	h	w_0	w_1
Values (mm)	0.55	1.6	8	41	41	64	1.1	0.2
Parameters	w_3	l_s	w_s	l_m	w_m	g		
Values (mm)	0.12	38	6.5	15	5.8	0.2		

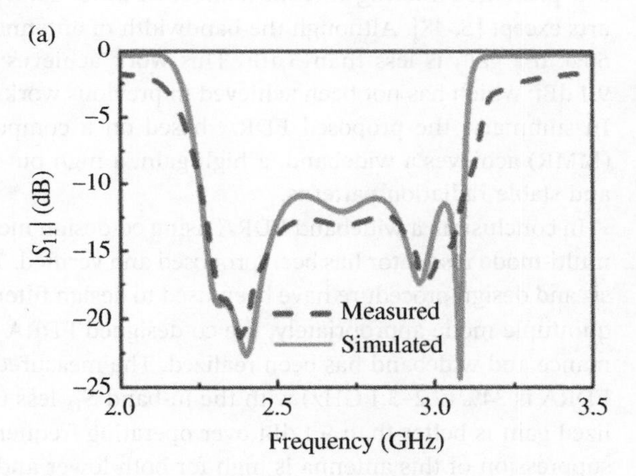

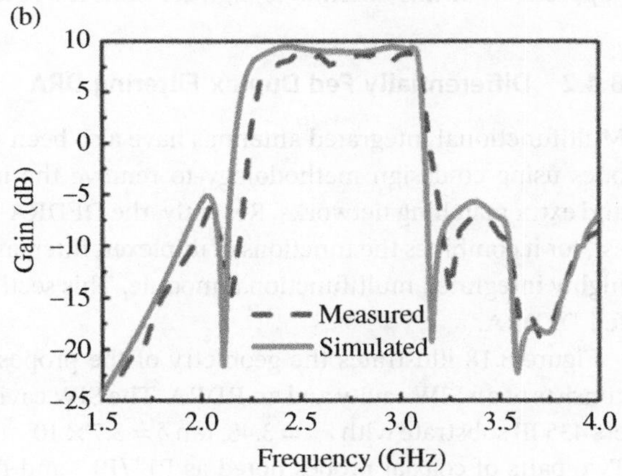

Figure 8.16 Simulated and measured results of (a) S_{11} and (b) peak realized gain.

passband and a good out-of-band stopband suppression. Besides, three radiation nulls located on both sides of the passband improve the roll-off rates of the realized gain and a good out-of-band rejection is achieved at the desired stopband. The slight frequency shift between measurement result and simulations may be caused by fabrication tolerance.

Figure 8.17 illustrates the simulated and measured radiation patterns of the FDRA at 2.4, 2.7, and 3.0 GHz. The broadside radiation patterns can be obtained in the E- and H-planes as expected. Furthermore, the cross-polarization level is better than 28 dB.

A comparison of the proposed antenna with other reported filtering antennas is carried out and listed in Table 8.2 to highlight the advantages of proposed work. The presented filtering antenna achieves a wider bandwidth of 34% than the prior arts except [5, 48]. Although the bandwidth of antennas in [5, 48] is greater than 50%, the gain is less than 5 dBi. This work achieves a higher gain better than 9.1 dBi, which has not been achieved in previous works except the array antenna. In summary, the proposed FDRA based on a composite multimode resonator (MMR) achieves a wideband, a high gain, a high out-of-band suppression level, and stable radiation patterns.

In conclusion, a wideband FDRA using co-design method based on a composite multi-mode resonator has been proposed and verified. Then, the proposed synthesis and design procedure have been used to design filtering antenna. By allocating quintuple mode appropriately, the co-designed FDRA with good filtering performance and wideband has been realized. The measured −10 dB bandwidth of the FDRA is 34% (2.2–3.1 GHz) with the in-band S_{11} less than −12.3 dB and the realized gain is better than 9.1 dBi over operating frequency band. The out-of-band suppression of this antenna is high for both lower and upper stopbands.

8.4.2 Differentially Fed Duplex Filtering DRA

Multifunctional integrated antennas have also been widely studied, especially the ones using co-design methodology to remove the inter-connection components and extra matching networks. Recently, the DFDRA has received increasing interest, for it combines the functions of duplexer, filter, and antenna and produces one highly integrated multifunctional module. This section introduces a differentially fed DFDRA.

Figure 8.18 illustrates the geometry of the proposed balanced DFDRA, which consists of an SIW cavity and an RDRA. The SIW cavity is constructed using a Rogers 4350B substrate with $\varepsilon_{rs} = 3.48$, $\tan\delta = 3.7 \times 10^{-3}$, and a thickness t of 1.52 mm. Two pairs of coaxial probes noted as $P1^+/P1^-$ and $P2^+/P2^-$ are used to differentially feed the SIW cavity. The selected dielectric material for DRA is $Bi_2Mo_2O_9$ microwave dielectric ceramic with $\varepsilon_{ra} = 36.3$ and $\tan\delta = 7 \times 10^{-4}$, which is

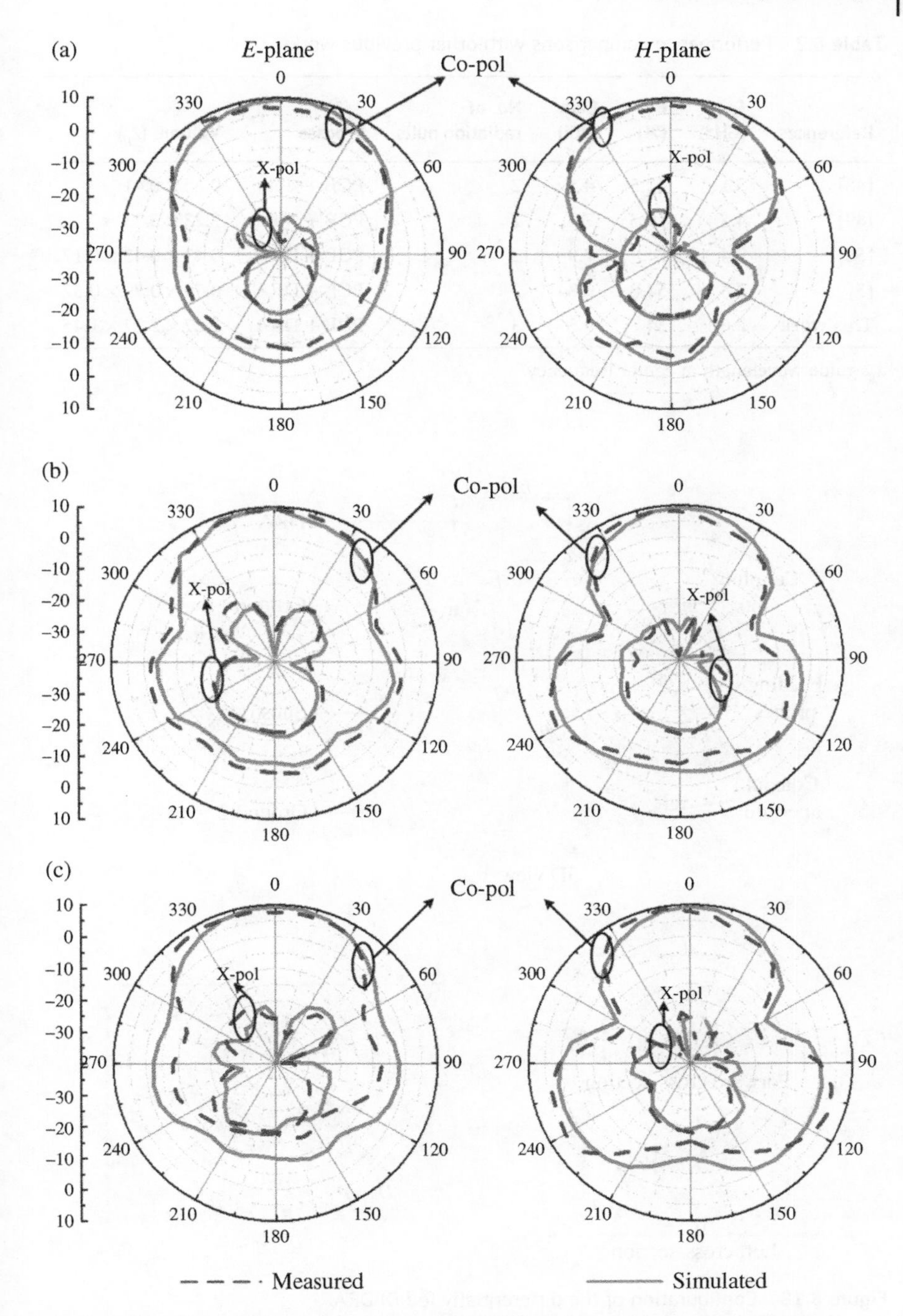

Figure 8.17 Simulated and measured normalized radiation patterns of the FDRA at (a) 2.4 GHz, (b) 2.7 GHz, and (c) 3.0 GHz.

Table 8.2 Performance comparisons with other previous works.

References	f_0 (GHz)	BW (%)	Gain (dBi)	No. of radiation nulls	Process	Volume (λ_g)
[48]	3.1	62.5	4.5	2	PCB + slot	0.66 × 0.41
[49]	4.5	19.5	7.5	2	PCB + slot	3.23 × 3.23
[50]	2.4	7	7.9	2	PCB + DRA	0.43 × 0.43 × 0.17
[5]	2.8	54.8	0.84	3	PCB + DRA	0.79 × 0.79 × 0.32
This work	2.6	34	9.5	3	PCB + DRA	2.27 × 2.27 × 1.45

λ_g: guide wavelength at center frequency.

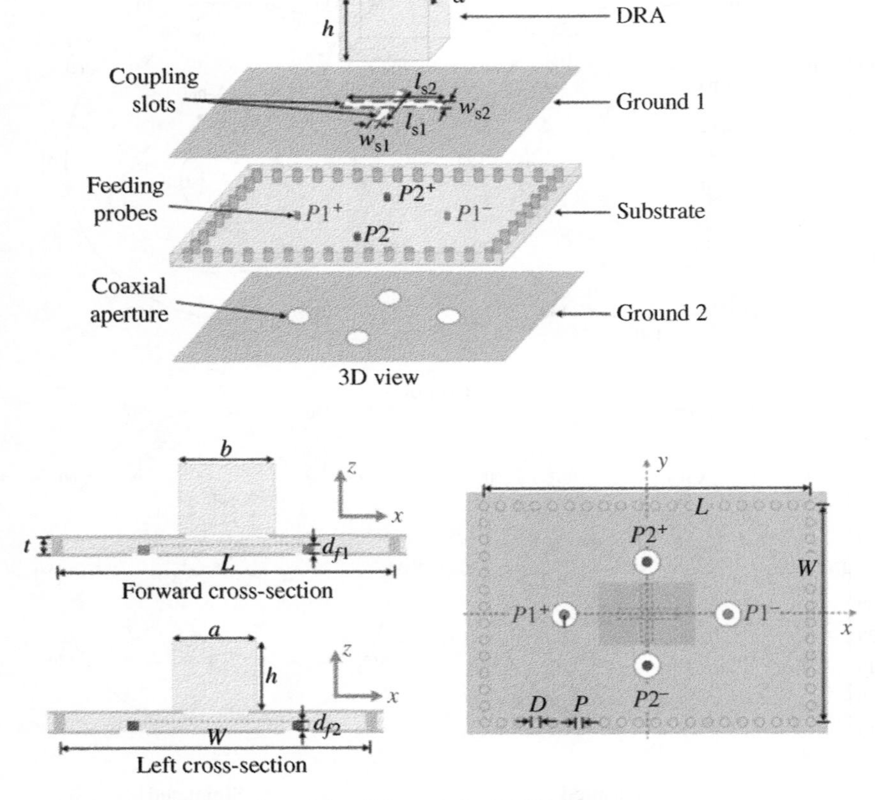

Figure 8.18 Configuration of the differentially fed DFDRA.

sintered and fabricated in the Electronic Materials Research Laboratory, Xi'an Jiaotong University. The DRA has a length a, width b, and height h, with its bottom stacked tightly on two coupling slots that are etched perpendicularly at the center of the top-ground of the SIW cavity. The commercial CST Microwave Studio 2019 simulation software is employed to perform differential simulations and optimize the design. The parameters are listed as follows: $L = 36$, $W = 29.5$, $a = 5.3$, $b = 7.5$, $h = 4.8$, $l_{s1} = 3.3$, $l_{s2} = 3.3$, $w_{s1} = 1$, $w_{s2} = 0.9$, $d_{f1} = 0.9$, $d_{f2} = 1$, $d = 1$, and $p = 1.2$ (unit: mm).

Figure 8.19a illustrates the E- and H-field distributions of the dual-mode SIW cavity. The E-field distribution of the dual-mode DRA is shown in Figure 8.19b. Table 8.3 clarifies the mechanisms of the balanced SIW-fed DFRDA. At the lower channel, the differential input port $P1^+/P1^-$ is set to excite the TE_{210} mode of the

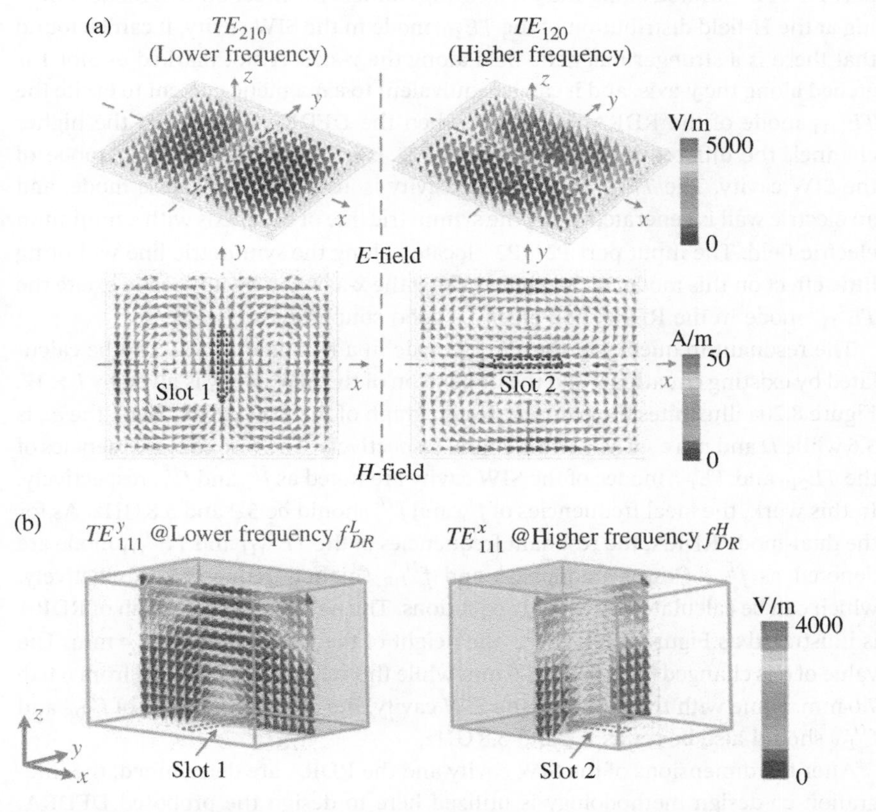

Figure 8.19 (a) Field distributions of the TE_{210} mode and TE_{120} mode of the SIW cavity. (b) Electric field distributions of the TE^y_{111} mode and TE^x_{111} mode of the RDRA.

Table 8.3 Operation mechanism of the DFDRA.

Channels	Lower channel	Higher channel
Feeding ports	PORT $1^+/1^-$	PORT $2^+/2^-$
Operation modes of SIW	TE_{210}	TE_{120}
Coupling slot	Slot 1	Slot 2
Operation modes of RDRA	TE^y_{111}	TE^x_{111}

SIW cavity. From the E-field distribution of the cavity, it can be found that the TE_{210} mode is an out-of-phase mode and an electric wall is generated along the symmetric line of the y-axis with a minimum electric field. As a result, the input port $P2^+/P2^-$ located along the y-axis brings almost no effect on this mode. Looking at the H-field distribution of the TE_{210} mode in the SIW cavity, it can be found that there is a stronger magnetic field along the y-axis. A slot marked as Slot 1 is etched along the y-axis, and it can be equivalent to a magnetic current to excite the TE^y_{111} mode of the RDRA. Similarly, when the DFDRA operates at the higher channel, the differential input port $P2^+/P2^-$ is set to excite the TE_{120} mode of the SIW cavity. The TE_{120} mode of the cavity is also an out-of-phase mode, and an electric wall is generated along the symmetric line of the x-axis with a minimum electric field. The input port $P2^+/P2^-$ located along the symmetric line will bring little effect on this mode. A slot etched along the x-axis can be utilized to excite the TE^x_{111} mode in the RDRA and ensure a good coupling.

The resonant frequency of the TE_{mn0} mode in a rectangular SIW can be calculated by existing equations, and the dimension of the SIW cavity is given by $L \times W$. Figure 8.20a illustrates the net-type design graph of the SIW cavity. Here, the ε_{rs} is 3.6 while D and p are set as 1 and 1.3 mm, respectively. The resonant frequencies of the TE_{210} and TE_{120} modes of the SIW cavity are noted as f^L_c and f^H_c, respectively. In this work, the ideal frequencies of f^L_c and f^H_c should be 5.2 and 5.8 GHz. As for the dual-mode RDRA, the resonant frequencies of the TE^y_{111} and TE^x_{111} mode are denoted as f^L_{DR} (lower frequency) and f^H_{DR} (higher frequency), respectively, which can be calculated by existing equations. The net-type design graph of RDRA is illustrated as Figure 8.20b. Here, the height of the RDRA h is set as 4 mm. The value of a is changed from 4.6 to 5.6 mm while the value of b is changed from 6.0 to 7.0 mm. Same with the design of the SIW cavity, the ideal frequencies of f^L_{DR} and f^H_{DR} should also be set as 5.2 and 5.8 GHz.

After the dimensions of the SIW cavity and the RDRA are determined, the integration co-design methodology is utilized here to design the proposed DFDRA. The duplex channels of the DFDRA can be treated as two separate DRA with two order filtering performance. The coupling topology of the proposed DFDRA

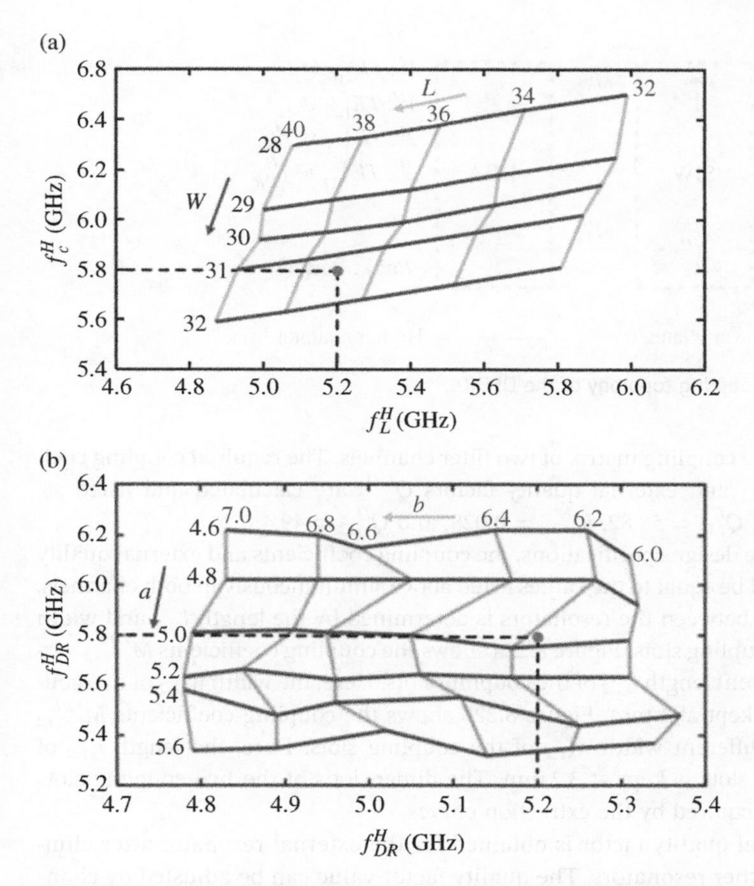

Figure 8.20 Net-type design graphs of (a) SIW cavity and (b) RDRA.

is depicted as Figure 8.21. $M^{L,H}{}_{s1}$ denotes the external coupling between input port and the SIW cavity. The internal coupling between the cavity and the RDRA is denoted as $M^{L,H}{}_{12}$. Here, the superscripts L and H stand for the lower and the higher operating channels. The center frequencies and impedance bandwidths of the lower and higher channels are synthesized to be 5.2 GHz/2.3% (120 MHz) and 5.8 GHz/2.75% (160 MHz), respectively, with in-band return loss of 10 dB. The coupling matrices of the two passbands are same and given by the following matrix:

$$M = \begin{bmatrix} 0 & 0.8575 & 0 & 0 \\ 0.8575 & 0 & 1.0201 & 0 \\ 0 & 1.0201 & 0 & 0.8575 \\ 0 & 0 & 0.8575 & 0 \end{bmatrix}$$

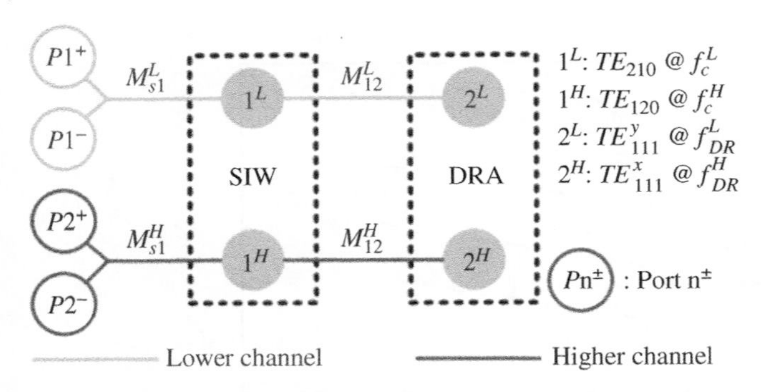

Figure 8.21 Coupling topology of the DFDRA.

where M is the coupling matrix of two filter channels. The required coupling coefficients $M^{L,H}{}_{12}$ and external quality factors $Q^{L,H}{}_e$ are calculated and listed as: $M^L{}_{12} = 0.023$, $Q^L{}_{12} = 58.82$, $M^H{}_{12} = 0.028$, and $Q^H{}_{12} = 49.45$.

To fulfill the design specifications, the coupling coefficients and external quality factors should be equal to the values listed above simultaneously in both channels. The coupling between the resonators is determined by the length $l_{s1,2}$ and width $w_{s1,2}$ of the coupling slots. Figure 8.22a shows the coupling coefficients $M^{L,H}{}_{12}$ varied with different length $l_{s1,2}$ of the coupling slots. Here, the width $w_{s1,2}$ of two coupling slots is kept at 1 mm. Figure 8.22b shows the coupling coefficients $M^{L,H}{}_{12}$ varied with different width $w_{s1,2}$ of the coupling slots. Here, the length $l_{s1,2}$ of two coupling slots is kept at 3.2 mm. The dimensions of the two coupling slots can then be acquired by the extraction curves.

The external quality factor is obtained for the external resonator after eliminating the other resonators. The quality factor value can be adjusted by changing the depth $d_{f1,2}$ of the feeding probes. The relationship between $Q^{L,H}{}_e$ and $d_{f1,2}$ is obtained through simulations of the referred S_{11} parameter in magnitude and phase. Figure 8.23 depicts the external quality factor $Q^{L,H}{}_e$ versus $d_{f1,2}$. The proper $d_{f1,2}$ of the feeding probes can then be acquired by the extraction curves.

After the initial parameters are obtained, the proposed balanced DFDRA can be constructed and optimized in CST. The simulated differential mode (DM) and common mode (CM) transmission characteristics and the realized gain of the reference DFDRA are illustrated in Figure 8.24, which obviously demonstrates the good filtering performance, high channel isolation, and CM suppression. Figure 8.24a shows that under DM excitation, the 10-dB impedance bandwidths of the lower and higher channel are given by 2.7% (5.12–5.26 GHz) and 2.6% (5.74–5.89 GHz), respectively. In addition, the differential S-parameters $S^{dd}{}_{11}$ and $S^{dd}{}_{22}$ drop to near-zero fast out of the two passbands. The near-zero $S^{dd}{}_{11}$ and $S^{dd}{}_{22}$ means that most of the input signal is reflected and cannot be

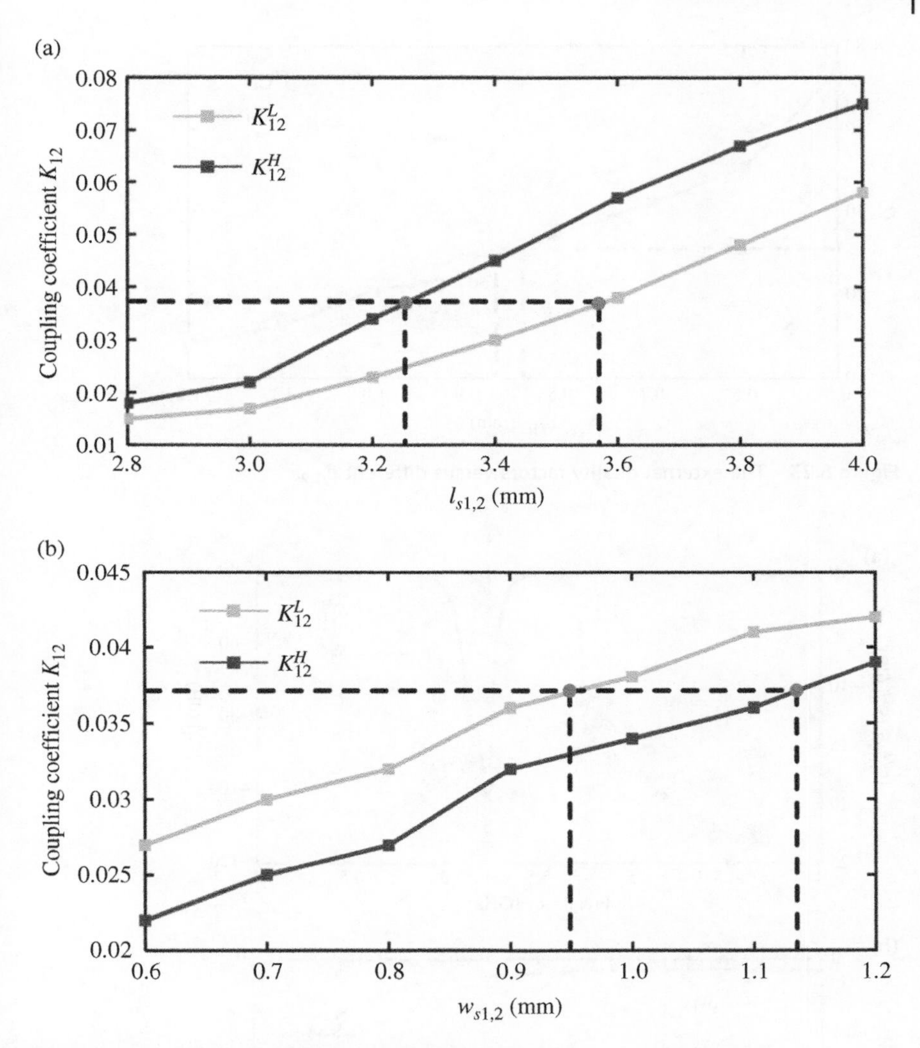

Figure 8.22 Coupling coefficient versus different (a) $l_{s1,2}$ and (b) $w_{s1,2}$.

transmitted and radiated. That is to say, high-frequency selectivity is realized at both the lower and higher channels of the DFDRA.

The isolation between the two channels within the operating passband is larger than 81 dB. This high channel isolation can be explained by looking into the field distribution of the operating modes of the SIW cavity and the RDRA. As is shown in Figure 8.19, the TE_{210} mode and TE_{120} mode of the SIW cavity are perpendicular to each other. Meanwhile, the orthogonality can also be found between the E-field

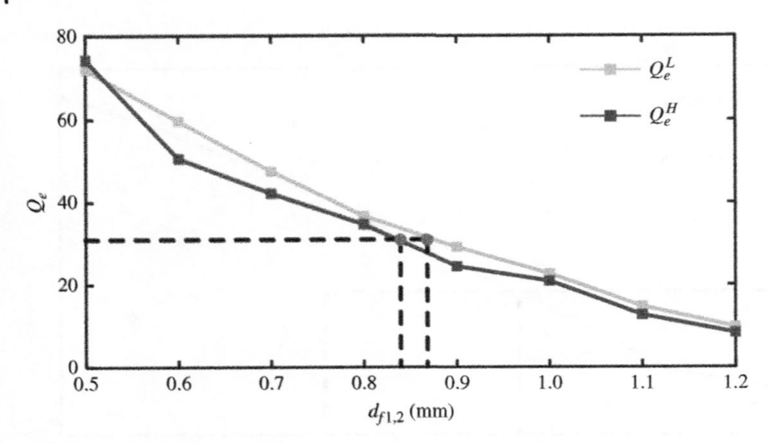

Figure 8.23 The external quality factors versus different $d_{f1,2}$.

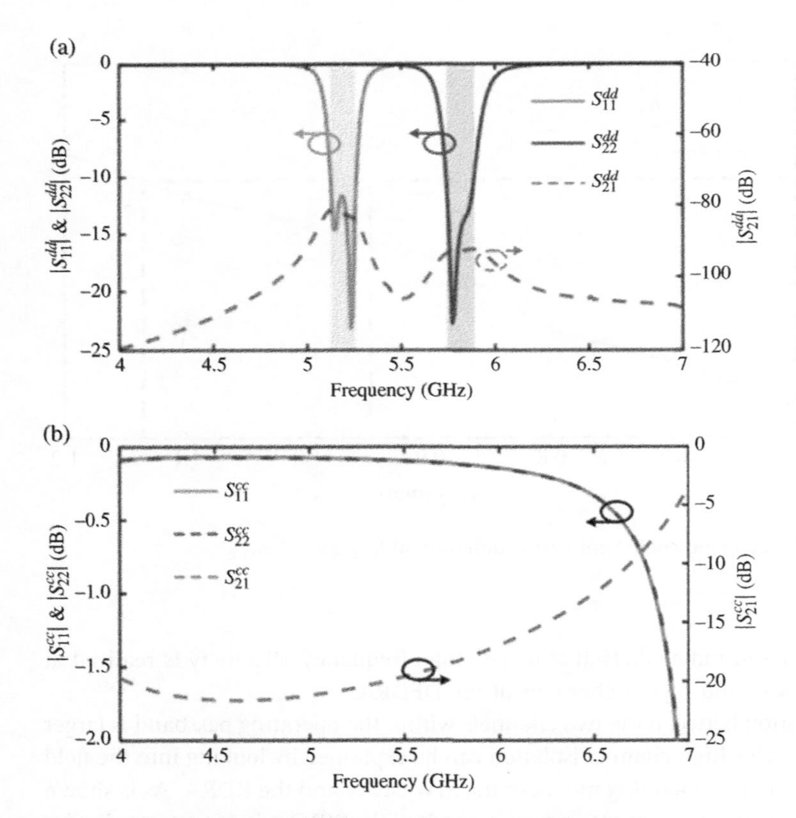

Figure 8.24 (a) DM and (b) CM transmission characteristics.

distribution of the TE^{y}_{111} and TE^{x}_{111} modes of the RDRA. The dual orthogonality makes the interference between the two channels smaller, and better channel isolation performance can thus be realized.

For comparison, it can be found from Figure 8.24b that under CM excitation, the magnitude of S-parameters S^{cc}_{11} and S^{cc}_{22} is near to 0 dB, which means that most of CM signal operating within 4–6 GHz will be reflected and cannot be transmitted, indicating an excellent CM suppression performance.

From the descriptions above, it can be deduced that the DFDRA is a classic Type C filtering antenna, which introduces filtering structure inside the antenna to achieve the band-stop characteristic in gain plot. Filtering response is realized by enhancing suppression in the stopbands through preventing power from being transmitted into the radiating unit of the antenna. The structure of dual-mode SIW cavity feeding a dual-mode RDRA naturally leads to most of the signal being reflected in the lower and upper stopbands, preventing them from being transmitted and radiated. In the perspective of design method of the duplex function, it belongs to Type C duplexer design that employs common resonators. The RDRA functions as the common resonator shared by two filtering channels, with each channel utilizing one of its orthogonal modes.

The proposed balanced DFDRA was fabricated and experimentally verified. The fabricated prototype is demonstrated in Figure 8.25. Two commercial 0–8 GHz baluns are used to realize the differential feed. Figure 8.26 makes a comparison of the measured and simulated reflection coefficient and peak realized gain. The measured 10-dB impedance bandwidths of the FDRA of the lower and higher channels are given by 2.7% (5.07–5.21 GHz) and 2.6% (5.72–5.87 GHz), respectively. The isolation between the two channels within the operating passband is larger than 75 dB. The simulated (measured) realized gains are among 5.51–5.91 dBi (4.52–4.91 dBi) and 5.65–6.12 dBi (4.85–5.12 dBi) within the two channels.

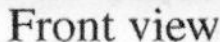

Figure 8.25 Photos of fabricated differentially fed DFDRA.

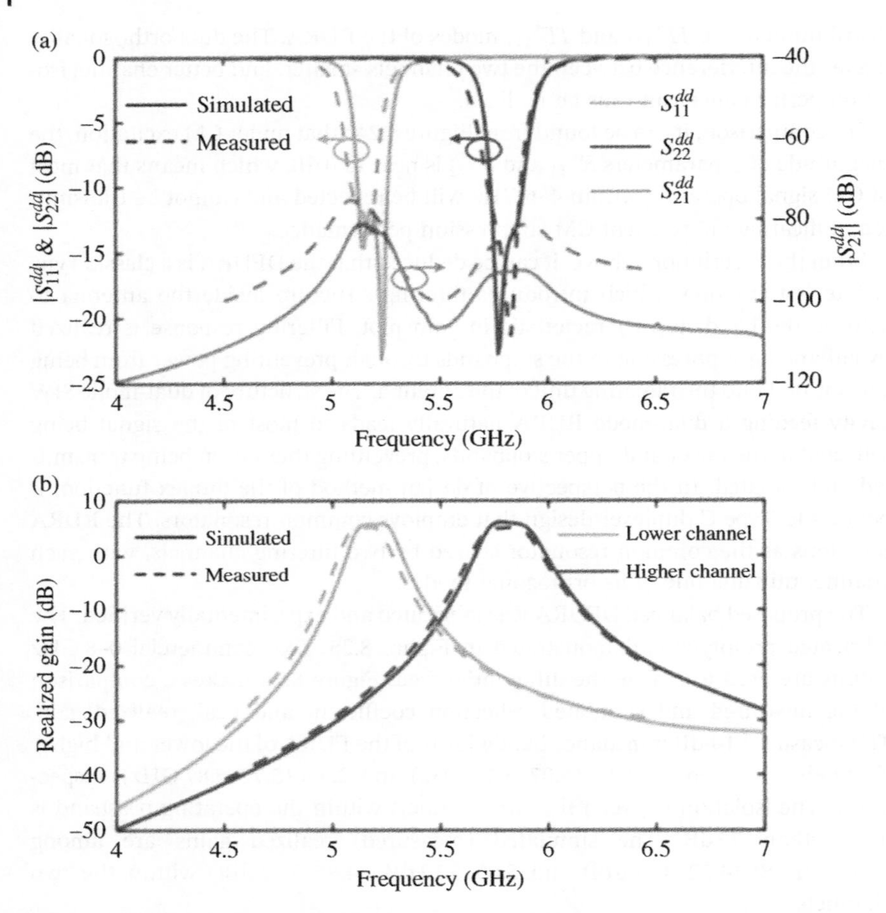

Figure 8.26 Simulated and measured results of (a) differential mode transmission characteristics and (b) peak realized gains.

Figure 8.27 shows the simulated and measured radiation patterns of DFDRA at the frequencies of 5.2 (lower channel) and 5.8 GHz (higher channel). A good agreement between simulated and measured results is observed at both resonant frequencies of the antenna. Broadside radiation is obtained with the co-polarization in boresight direction stronger than the cross-polarization counterpart by more than 35 dB.

Table 8.4 shows the comparison with some previous reported duplex filtering antenna. It can be seen that the presented antenna has the advantages of in-band realized gain, channel isolation, and CM suppression.

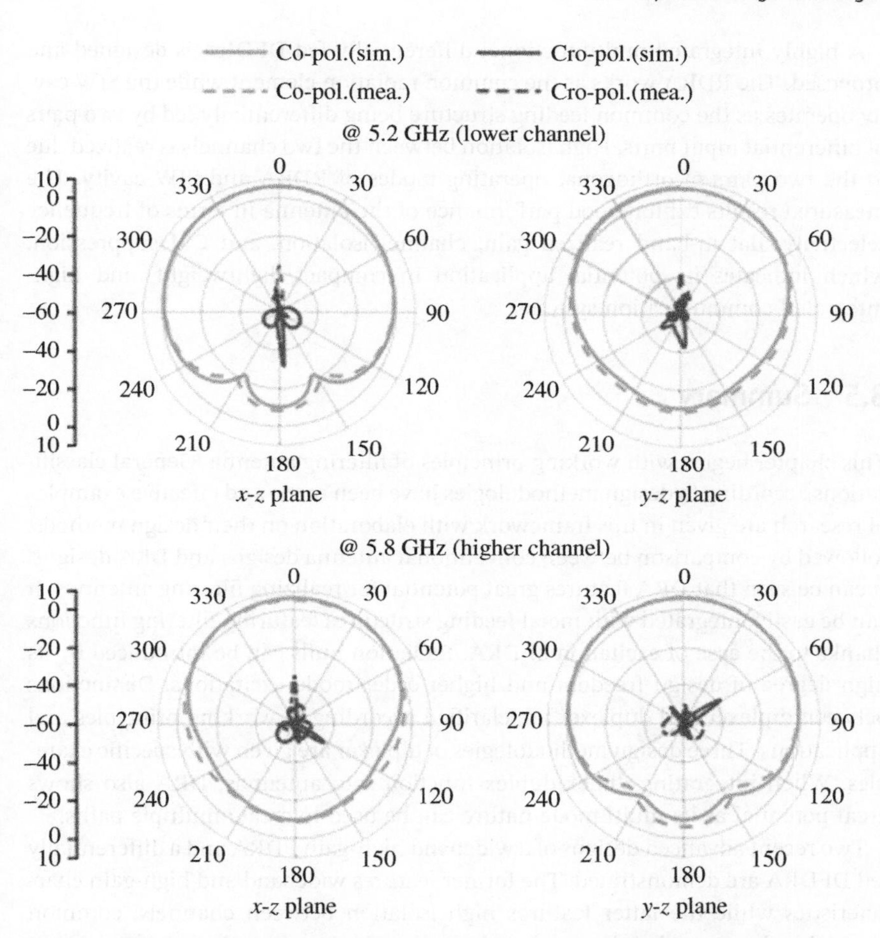

Figure 8.27 Simulated and measured results of radiation patterns at 5.2 and 5.8 GHz.

Table 8.4 Comparison between this work and previously reported duplex filtering antennas.

References	f_L/f_H (GHz)	FBW (%)	Peak gain (dBi)	Isolation (dB)	CM Sup.	Volume (λ_0^3)
[7]	2.45/5.26	5.3/5.7	2.3/3.6	20/28	No	1.03 × 1.03 × 0.03
[51]	5.25/5.8	4.5/3.6	2.7/2.65	30/29	No	2.56 × 0.83 × 0.02
[42]	4.1/4.9	3.2/3.9	4.36/4.83	25/23	No	1.08 × 0.89 × 0.03
This work	**5.14/5.80**	**2.7/2.6**	**4.91/5.12**	**75/81**	**Yes**	**1.06 × 0.87 × 0.17**

A highly integrated multifunctional differentially fed DFDRA is designed and proposed. The RDRA works as the common radiation element while the SIW cavity operates as the common feeding structure being differentially fed by two pairs of differential input ports. High isolation between the two channels is realized due to the two pairs of orthogonal operating modes of RDRA and SIW cavity. The measured results exhibit good performance of the antenna in terms of frequency selectivity, flat in-band realized gain, channel isolation, and CM suppression, which indicates its potential application in compact, lightweight, and high-integrated communication systems.

8.5 Summary

This chapter begins with working principles of filtering antenna. General classifications according to design methodologies have been made and effective examples of research are given in this framework with elaboration on their design methods, followed by comparison between conventional antenna designs and DRA designs. It can be seen that DRA features great potential for realizing filtering antennas. It can be easily integrated with metal feeding structures featuring filtering functions thanks to the ease of excitation of DRA. Radiation nulls can be introduced by its high degree of design freedom and higher-order mode excitations. Distinctions between diplexer and duplexer are clarified according to working principles and applications. Three design methodologies of diplexer are given with specific examples. When integrating diplex/duplex function into antennas, DRA also shows great potential as its multi-mode nature can be used to create multiple paths.

Two recent advanced designs of a wideband high-gain FDRA and a differentially fed DFDRA are demonstrated. The former features wideband and high-gain characteristics while the latter features high isolation between channels, common mode signal suppression, low cross-polarization level, and good symmetry in radiation patterns. The two design methodologies of duplexer and filtering antenna provide a set of useful references for developing multifunctional antennas that integrate the functions of multiple devices such as duplexer, filter, and antenna into one compact unit.

References

1 Lim, E.H. and Leung, K.W. (2012). *Compact Multifunctional Antennas for Wireless Systems*. Hoboken, NJ: Wiley.

2 Lin, C.-K. and Chung, S.-J. (2011). A filtering microstrip antenna array. *IEEE Transactions on Microwave Theory and Techniques* 59 (11): 2856–2863.

3 Fakharian, M.M., Pezaei, P., Orouji, A.A., and Soltanpur, M. (2016). A wideband and reconfigurable filtering slot antenna. *IEEE Antennas and Wireless Propagation Letters* 15: 1610–1613.

4 Pan, Y.M., Hu, P.F., Leung, K.W., and Zhang, X.Y. (2018). Compact single-/dual-polarized filtering dielectric resonator antennas. *IEEE Transactions on Antennas and Propagation* 66 (9): 4474–4484. https://doi.org/10.1109/TAP.2018.2845457.

5 Hu, P.F., Pan, Y.M., Leung, K.W., and Zhang, X.Y. (2018). Wide-/dual-band omnidirectional filtering dielectric resonator antennas. *IEEE Transactions on Antennas and Propagation* 66 (5): 2622–2627. https://doi.org/10.1109/TAP.2018.2809706.

6 Mao, C.-X., Gao, S., Wang, Y. et al. (2016). Compact highly integrated planar duplex antenna for wireless communications. *IEEE Transactions on Microwave Theory and Techniques* 64 (7): 2006–2013. https://doi.org/10.1109/TMTT.2016.2574338.

7 Lee, Y.-J., Tarng, J.-H., and Chung, S.-J. (2017). A filtering diplexing antenna for dual-band operation with similar radiation patterns and low cross-polarization levels. *IEEE Antennas and Wireless Propagation Letters* 16: 58–61. https://doi.org/10.1109/LAWP.2016.2554603.

8 Lu, Y.-C. and Lin, Y.-C. (2012). A mode-based design method for dual-band and self-diplexing antennas using double T-stubs loaded aperture. *IEEE Transactions on Antennas and Propagation* 60 (12): 5596–5603. https://doi.org/10.1109/TAP.2012.2211852.

9 Mukherjee, S. and Biswas, A. (2016). Design of self-diplexing substrate integrated waveguide cavity-backed slot antenna. *IEEE Antennas and Wireless Propagation Letters* 15: 1775–1778. https://doi.org/10.1109/LAWP.2016.2535169.

10 Petosa, A. (2007). *Dielectric Resonator Antenna Handbook*. Norwood, MA: Artech House.

11 Luk, K.M. and Leung, K.W. (2003). *Dielectric Resonator Antennas*. Baldock: Research Studies.

12 Chen, Z., Shoaib, I., Yao, Y. et al. (2016). Pattern-reconfigurable dual-polarized dielectric resonator antenna. *IEEE Antennas and Wireless Propagation Letters* 15: 1273–1276.

13 Chen, Z. and Wong, H. (2017). Wideband glass and liquid cylindrical dielectric resonator antenna for pattern reconfigurable design. *IEEE Transactions on Antennas and Propagation* 65 (5): 2157–2164. https://doi.org/10.1109/TAP.2017.2676767.

14 Liu, H., Tian, H., Liu, L., and Feng, L. (2022). Co-design of wideband filtering dielectric resonator antenna with high gain. *IEEE Transactions on Circuits and Systems II: Express Briefs* 69 (3): 1064–1068. https://doi.org/10.1109/TCSII.2021.3131509.

15 Tian, H., Chen, Z., Chang, L. et al. (2022). Differentially fed duplex filtering dielectric resonator antenna with high isolation and CM suppression. *IEEE*

Transactions on Circuits and Systems II: Express Briefs 69 (3): 979–983. https://doi.org/10.1109/TCSII.2021.3120726.

16 Mao, C.-X., Gao, S., Wang, Y. et al. (2015). Multimode resonator-fed dual-polarized antenna array with enhanced bandwidth and selectivity. *IEEE Transactions on Antennas and Propagation* 63 (12): 5492–5499. https://doi.org/10.1109/TAP.2015.2496099.

17 Tang, M.-C., Chen, Y., and Ziolkowski, R.W. (2016). Experimentally validated, planar, wideband, electrically small, monopole Filtennas based on capacitively loaded loop resonators. *IEEE Transactions on Antennas and Propagation* 64 (8): 3353–3360. https://doi.org/10.1109/TAP.2016.2576499.

18 Yang, W., Zhang, Y., Che, W. et al. (2019). A simple, compact filtering patch antenna based on mode analysis with wide out-of-band suppression. *IEEE Transactions on Antennas and Propagation* 67 (10): 6244–6253. https://doi.org/10.1109/TAP.2019.2922770.

19 Mao, C.X., Gao, S., Wang, Y. et al. (2016). Dual-band patch antenna with filtering performance and harmonic suppression. *IEEE Transactions on Antennas and Propagation* 64 (9): 4074–4077. https://doi.org/10.1109/TAP.2016.2574883.

20 Xiang, B.J., Zheng, S.Y., Pan, Y.M., and Li, Y.X. (2017). Wideband circularly polarized dielectric resonator antenna with bandpass filtering and wide harmonics suppression response. *IEEE Transactions on Antennas and Propagation* 65 (4): 2096–2101. https://doi.org/10.1109/TAP.2017.2671370.

21 Tang, H., Tong, C., and Chen, J. (2018). Differential dual-polarized filtering dielectric resonator antenna. *IEEE Transactions on Antennas and Propagation* 66 (8): 4298–4302. https://doi.org/10.1109/TAP.2018.2836449.

22 Sheen, J.-W. (1999). LTCC-MLC duplexer for DCS-1800. *IEEE Transactions on Microwave Theory and Techniques* 47 (9): 1883–1890. https://doi.org/10.1109/22.788526.

23 Bairavasubramanian, R., Pinel, S., Laskar, J., and Papapolymerou, J. (2006). Compact 60-GHz bandpass filters and duplexers on liquid crystal polymer technology. *IEEE Microwave and Wireless Components Letters* 16 (5): 237–239. https://doi.org/10.1109/LMWC.2006.873591.

24 Xue, Q., Shi, J., and Chen, J.-X. (2011). Unbalanced-to-balanced and balanced-to-unbalanced diplexer with high selectivity and common-mode suppression. *IEEE Transactions on Microwave Theory and Techniques* 59 (11): 2848–2855. https://doi.org/10.1109/TMTT.2011.2165960.

25 Tu, W.-H. and Hung, W.-C. (2014). Microstrip eight-channel diplexer with wide stopband. *IEEE Microwave and Wireless Components Letters* 24 (11): 742–744. https://doi.org/10.1109/LMWC.2014.2348499.

26 Wu, H.-W., Huang, S.-H., and Chen, Y.-F. (2013). Design of new quad-channel diplexer with compact circuit size. *IEEE Microwave and Wireless Components Letters* 23 (5): 240–242. https://doi.org/10.1109/LMWC.2013.2253314.

27 Yang, T. and Rebeiz, G.M. (2016). A simple and effective method for 1.9–3.4-GHz tunable diplexer with compact size and constant fractional bandwidth. *IEEE Transactions on Microwave Theory and Techniques* 64 (2): 436–449. https://doi.org/10.1109/TMTT.2015.2504937.

28 Fan, M., Song, K., Yang, L., and Gómez-García, R. (2021). Frequency-tunable constant-absolute-bandwidth single-/dual-passband filters and diplexers with all-port-reflectionless behavior. *IEEE Transactions on Microwave Theory and Techniques* 69 (2): 1365–1377. https://doi.org/10.1109/TMTT.2020.3040481.

29 Chi, P.-L. and Yang, T. (2017). Three-pole reconfigurable 0.94–1.91-GHz diplexer with bandwidth and transmission zero control. *IEEE Transactions on Microwave Theory and Techniques* 65 (1): 96–108. https://doi.org/10.1109/TMTT.2016.2614667.

30 Guan, X., Yang, F., Liu, H. et al. (2016). Compact, low insertion-loss, and wide stopband HTS diplexer using novel coupling diagram and dissimilar spiral resonators. *IEEE Transactions on Microwave Theory and Techniques* 64 (8): 2581–2589. https://doi.org/10.1109/TMTT.2016.2580143.

31 Sabharwal, A., Schniter, P., Guo, D. et al. (2014). In-band full-duplex wireless: challenges and opportunities. *IEEE Journal on Selected Areas in Communications* 32 (9): 1637–1652.

32 Liu, G., Yu, F.R., Ji, H. et al. (2015). In-band full duplex relaying: a survey, research issues and challenges. *IEEE Communication Surveys and Tutorials* 17 (2): 500–524.

33 Elmansouri, M. A. and Filipovic, D. S. (2016). Realization of ultra-wideband bistatic simultaneous transmit and receive antenna system. *2016 IEEE International Symposium on Antennas and Propagation (APSURSI)*, Fajardo, PR, USA, pp. 2115–2116. doi: 10.1109/APS.2016.7696764.

34 Iqbal, A., Al-Hasan, M., Mabrouk, I.B., and Nedil, M. (2021). Ultracompact quarter-mode substrate integrated waveguide self-diplexing antenna. *IEEE Antennas and Wireless Propagation Letters* 20 (7): 1269–1273. https://doi.org/10.1109/LAWP.2021.3077451.

35 Nandi, S. and Mohan, A. (2017). An SIW cavity-backed self-diplexing antenna. *IEEE Antennas and Wireless Propagation Letters* 16: 2708–2711. https://doi.org/10.1109/LAWP.2017.2743017.

36 Boukarkar, A., Lin, X.Q., Jiang, Y., and Yu, Y.Q. (2017). A tunable dual-fed self-diplexing patch antenna. *IEEE Transactions on Antennas and Propagation* 65 (6): 2874–2879. https://doi.org/10.1109/TAP.2017.2689035.

37 Chuang, C.-T. and Chung, S.-J. (2011). Synthesis and design of a new printed filtering antenna. *IEEE Transactions on Antennas and Propagation* 59 (3): 1036–1042.

38 Nova, O.A., Bohorquez, J.C., Pena, N.M. et al. (2011). Filter-antenna module using substrate integrated waveguide cavities. *IEEE Antennas and Wireless Propagation Letters* 10: 59–62.

39 Jiang, Z.H. and Werner, D.H. (2015). A compact, wideband circularly polarized co-designed filtering antenna and its application for wearable devices with low SAR. *IEEE Transactions on Antennas and Propagation* 63 (9): 3808–3818.

40 Gao, S. and Sambell, A. (2004). Low-cost dual-polarized printed array with broad bandwidth. *IEEE Transactions on Antennas and Propagation* 52 (12): 3394–3397.

41 Mao, C.X., Zhang, L., Khalily, M. et al. (2021). A multiplexing filtering antenna. *IEEE Transactions on Antennas and Propagation* 69 (8): 5066–5071. https://doi.org/10.1109/TAP.2020.3048589.

42 Hu, K.-Z., Tang, M.-C., Wang, Y. et al. (2021). Compact, vertically integrated duplex filtenna with common feeding and radiating SIW cavities. *IEEE Transactions on Antennas and Propagation* 69 (1): 502–507. https://doi.org/10.1109/TAP.2020.2999381.

43 Li, D., Tang, M.-C., Wang, Y. et al. (2021). Compact differential diplex Filtenna with common-mode suppression for highly integrated radio frequency front-ends. *IEEE Transactions on Antennas and Propagation* 69 (11): 7935–7940. https://doi.org/10.1109/TAP.2021.3083797.

44 Iqbal, A., Selmi, M.A., Abdulrazak, L.F. et al. (2020). A compact substrate integrated waveguide cavity-backed self-triplexing antenna. *IEEE Transactions on Circuits and Systems II: Express Briefs* 67 (11): 2362–2366.

45 Xun, M., Yang, W., Feng, W. et al. (2021). A differentially fed dual-polarized filtering patch antenna with good stopband suppression. *IEEE Transactions on Circuits and Systems II: Express Briefs* 68 (4): 1228–1232.

46 Deng, J., Hou, S., Zhao, L., and Guo, L. (2018). A reconfigurable filtering antenna with integrated bandpass filters for UWB/WLAN applications. *IEEE Transactions on Antennas and Propagation* 66 (1): 401–404.

47 Liu, Y., Wang, S., Li, N. et al. (2018). A compact dualband dual-polarized antenna with filtering structures for sub-6 GHz base station applications. *IEEE Antennas and Wireless Propagation Letters* 17 (10): 1764–1768.

48 Nie, N. and Tu, Z.-H. (2020). Wideband filtering dumbbell-shaped slot antenna with improved frequency selectivity for both band-edges. *IEEE Access* 8: 121479–121485. https://doi.org/10.1109/ACCESS.2020.3006243.

49 Wu, T.L., Pan, Y.M., and Hu, P.F. (2017). Wideband omnidirectional slotted patch antenna with filtering response. *IEEE Access* 5: 26015–26021. https://doi.org/10.1109/ACCESS.2017.2768067.

50 Hu, P.F., Pan, Y.M., Zhang, X.Y., and Hu, B.J. (2019). A compact quasi-isotropic dielectric resonator antenna with filtering response. *IEEE Transactions on Antennas and Propagation* 67 (2): 1294–1299. https://doi.org/10.1109/TAP.2018.2883611.

51 Dhwaj, K., Li, X., Jiang, L.J., and Itoh, T. (2018). Low-profile diplexing filter/antenna based on common radiating cavity with quasi-elliptic response. *IEEE Antennas and Wireless Propagation Letters* 17 (10): 1783–1787. https://doi.org/10.1109/LAWP.2018.2866786.

9

Conclusion and Future Work

CHAPTER MENU

9.1 Overall Summary

In this book, the classifications, designs, and applications of dielectric resonator antennas (DRAs) have been demonstrated. It includes classification of dielectric antenna designs, studies on their operations, and some design examples for specific applications. Wideband, high-gain, high-efficiency DRAs, and arrays are constructed as examples for microwave and millimeter-wave communication applications. Technologies that assist in overcoming the basic limitations of conventional antenna and array have been discussed. We also discuss dielectric materials, DRA fabrication, and their tolerances. These discussions constitute the superior and inferior aspects of DRA and its potential to be a cost-effective candidate for wireless communication applications.

To pursue the full potential of the DRA, a series of novel DRA designs are proposed in each chapter, such as wideband stacked DRA in Chapter 3, pattern diverse DRA and array in Chapter 4, high-isolated MIMO DRA in Chapter 5, 3D printed dielectric-based antenna in Chapter 6, millimeter-wave DRA and array in Chapter 7, and filtering DRA and filtering duplex DRA in Chapter 8. In each chapter, the techniques and fundamentals behind each design are overviewed before the proposal of these designs. Some fundamentals are classified and concluded for the first time for the reader to know the full extent of the existing

Dielectric Resonator Antennas: Materials, Designs and Applications, First Edition.
Zhijiao Chen, Jing-Ya Deng, and Haiwen Liu.

Published 2024 by John Wiley & Sons, Inc.

techniques. The design guidelines are also given to inspire the reader to do further research on DRA.

The designs presented in this book are recent high-quality works and have been recognized by research professionals. First, most of our designs have been published in high-quality international journals, such as *IEEE Transactions on Antennas and Propagations*, *IEEE Transactions on Vehicular Technology*, *IEEE Transactions on Circuits and Systems II: Express Briefs*, and *IEEE Antennas and Wireless Propagation Letters*. Second, some of the research has been presented in international conferences and won best paper awards such as IEEE APS/URSI 2013 and IEEE iWAT 2013. Third, the book chapters include two of our papers published in *IEEE Transactions on Antennas and Propagations* that have been selected as the Top Accessed Document on May 2021. They are presented with additional details on their background and fundamentals for better understanding DRA research trends. Most importantly, some of the research contents can be found in the technical talks given by the first author Prof. Zhijiao Chen, who was elected as the IEEE APS Young Professional Ambassador in the year of 2022. At the moment, 12 worldwide technical talks on the topic of (1) Dielectric Resonator Antenna: Challenges, Designs, and Opportunities and (2) Low-cost mmWave Antenna Array for Wireless Future have been given. These talks are scheduled online mostly in IEEE APS Region 10, and have attracted a turnout of over 2024 in the year of 2022. Prof. Zhijiao Chen will continue to give talks in the coming years. The talk announcements are available on the IEEE APS Young Professional Websites: https://www.linkedin.com/posts/ieee-aps-yp.

9.2 Recommendations for Future Work

This book is well-prepared not only on the technical aspects but also with interesting future works that might attract readers to work on DRA. The authors hope that this book would inspire people in the following aspects.

1) Dielectric antennas with low-loss materials
   Based on the antenna research, we found that the property of dielectric material has great impact on the antenna's performance, especially for dielectric antenna and 3D printed antenna. So the research on dielectric property is essential to ensure the measured performance of the dielectric antenna. The co-design of the low-loss material and dielectric antenna has potential to develop low-cost lightweight antenna with extraordinary performance.
2) Antenna for radio astronomy
   Metal antennas have been employed for radio astronomy antennas for their stability and high performance. In recent years, dielectric materials are loaded

on the radio astronomy antennas for performance enhancement and light-weight design. The authors also believe that dielectric antennas have higher stability than the metal ones because they are not sensitive to the radiation, temperature, and humidity in outdoor environment. Dielectric antennas are like a stone, which are more resistant to environmental changes when compared with metal antennas.

3) Wearable antenna, in-body antenna
   Compared with metal antennas, the dielectric antenna is more suitable for wearable antenna and in-body antenna because it can be integrated on-body such as the bone, teeth, and others. With the fast development in 5G communications for medical application, we believe that DRA will play a key role in medical communication applications.
4) THz antennas in measurement systems
   THz antenna is a big challenge at the moment due to the low efficiency of the existing antenna. Lens antenna and dielectric antenna have been investigated for their potentials for THz antenna, especially for THz measurement system. It was found that dielectric antenna array has great potential for THz measurement system for improving the system accuracy due to the low-loss of the dielectric in the THz band. Also, the use of all-dielectric metamaterials can improve the robustness of the measurement system, especially for measuring the biological sample that needs to be washed and doped frequently.

Appendix A

Modes in Rectangular DRA

The rectangular-shaped dielectric resonator antenna (DRA) has been the most popular DRA shape due to the following benefits:

1) Design flexibility. The rectangular DRA has two degrees of freedom (height/length and width/length) that allow for a wide range of aspect ratios for a given operating frequency. This also results in a compact structure of the rectangular DRA.
2) Good polarization characteristics. By properly selecting resonator dimensions, mode degeneracy can be avoided and the cross-polarization levels can be reduced.
3) Easy to be fabricated. The rectangular-shaped DR is easy to be machined and its manufacture process is relatively simple.

These features make the rectangular-shaped DRA suitable for various application scenarios such as wireless communication systems, millimeter wave communication, and radar communication.

Approximate methods and numerical methods are generally used to study the modes of rectangular DRAs. Rectangular DRAs have edge-shaped boundaries, which make them difficult to obtain analytical closed-form Green's functions. In terms of approximate solution, Okaya and Barash first provided an approximate analysis of rectangular DRs [1], and the modes of rectangular-shaped DRA were divided into transverse electric (TE) modes and transverse magnetic (TM) modes. Dielectric waveguide model (DWM) has been developed to evaluate the resonance frequency, Q-factor, and radiation patterns.

With respect to the rectangular DRA configuration in Figure A.1, the modes of a rectangular dielectric waveguide can be divided into TM^z_{mnl} and TE^z_{mnl} families, where m, n, and l denote the number of extrema in the x-, y-, and z-directions,

Dielectric Resonator Antennas: Materials, Designs and Applications, First Edition.
Zhijiao Chen, Jing-Ya Deng, and Haiwen Liu.

Published 2024 by John Wiley & Sons, Inc.

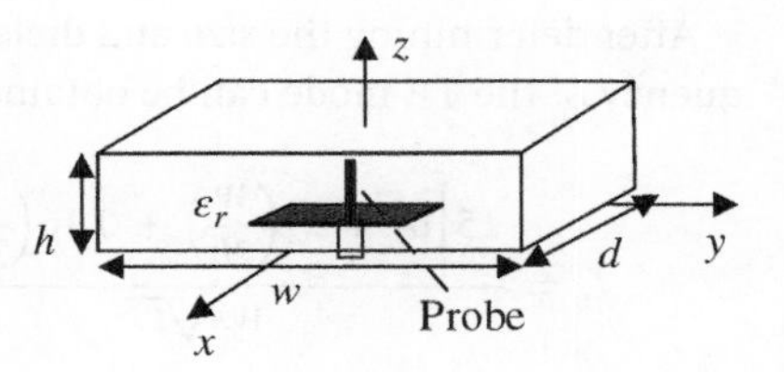

Figure A.1 Configuration of a rectangular DRA.

respectively. The subscript distribution of the mode represents the field changes in the x-, y-, and z-directions, whereas the superscript denotes the polarization direction of the field. The fundamental TE_{111} mode has been investigated by different researchers extensively. However, TM mode can rarely be observed in the experiment [2], but a quasi-TM mode can be found in a rectangular DRA when DRA is centrally fed by a coaxial probe [3]. In this term, the H-field is circular around the probe whereas the E-field is vertical and strongest along the probe. This mode is the approach to the TM mode of the cylindrical DRA, and the radiation pattern is almost the same. As a result, the mode is named as the quasi-TM_{011} mode.

For a rectangular DRA with dimensions w, $d > h$, the lowest order mode would be $TE^z_{11\delta}$. According to DWM, the field components are given as:

$$H_x = \frac{(k_x k_z)}{j\omega\mu_0} \sin(k_x x)\cos(k_y y)\sin(k_z z) \tag{A.1}$$

$$H_y = \frac{(k_y k_z)}{j\omega\mu_0} \cos(k_x x)\sin(k_y y)\sin(k_z z) \tag{A.2}$$

$$H_z = \frac{\left(k_x^2 + k_y^2\right)}{j\omega\mu_0} \cos(k_x x)\cos(k_y y)\cos(k_z z) \tag{A.3}$$

$$E_x = k_y \cos(k_x x)\sin(k_y y)\cos(k_z z) \tag{A.4}$$

$$E_y = -k_x \sin(k_x x)\cos(k_y y)\cos(k_z z) \tag{A.5}$$

$$E_z = 0 \tag{A.6}$$

The wavenumbers should satisfy the equation:

$$k_x^2 + k_y^2 + k_z^2 = \varepsilon_r k_0^2 \tag{A.7}$$

$$k_z \tan(k_z d/2) = \sqrt{(\varepsilon_r - 1)k_0^2 - k_z^2} \tag{A.8}$$

After determining the size and dielectric constant of the DRA, the resonant frequency of the TE mode can be obtained according to approximate expression [4]:

$$f_0 = \frac{15\left[a_1 + a_2\left(\frac{w}{2h}\right) + 0.16\left(\frac{w}{2h}\right)^2\right]}{w\pi\sqrt{\varepsilon_r}} \tag{A.9}$$

$$a_1 = 2.57 - 0.8\left(\frac{d}{2h}\right) + 0.42\left(\frac{d}{2h}\right)^2 - 0.05\left(\frac{d}{2h}\right)^3 \tag{A.10}$$

$$a_2 = 2.71\left(\frac{d}{2h}\right)^{-0.282} \tag{A.11}$$

Figure A.2 shows the internal *E*-and *H*-fields of the fundamental TE^y_{111} mode and higher-order TE^y_{113}, TE^y_{115}, and quasi-TM modes of a grounded rectangular DRA. It can be seen that the radiation of TE^y_{111} mode is similar to the short magnetic dipole placed along the *y*-axis. It should be mentioned that the TE^y_{112} and TE^y_{114} modes are not included here because they are eliminated by the ground plane. Compared with the fundamental mode, the far-field radiation of the higher-order mode has better directivity and higher gain.

Various numerical methods such as the Method of Moments (MoM) or the Finite Difference Time Domain (FDTD) techniques have been used to solve complex electromagnetic fields problem. But these techniques are time consuming, memory intensive, and are not suitable for design or optimization. The characteristic mode (CM) theory provides an alternative and a useful tool to calculate the resonant frequency, modal currents, and modal fields for various DRAs. The characteristic mode analysis (CMA) can promote the understanding and characterization of the resonant mode [5, 6], and it can avoid the redundant process of blind simulation and improve the efficiency and accuracy of DRA design [7, 8].

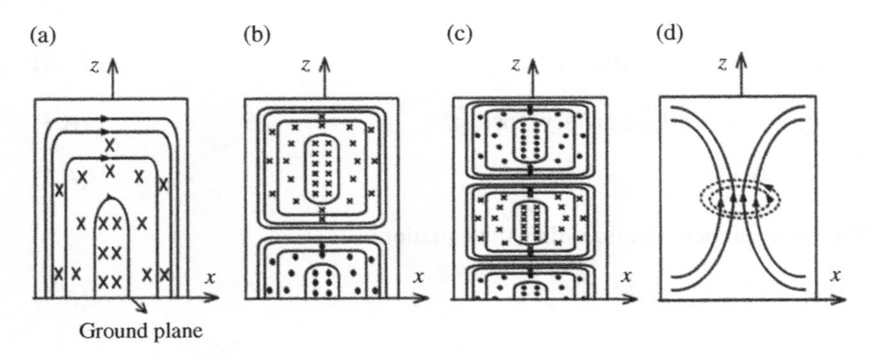

Figure A.2 Field distribution inside the rectangular DRA. (a) Fundamental TE^y_{111} mode. (b) Higher-order TE^y_{113} mode. (c) Higher-order TE^y_{115} mode. (d) Quasi-TM_{011} mode.

CMA of the main body (DR) of the rectangular DRA is performed using the commercial software FEKO. The analysis process of a rectangular DRA is given here as example.

1) First, a rectangular DR is modeled with dielectric constant of $\varepsilon_r = 9.2$ and dimensions of $d = w = 50$ mm and $h = 20$ mm. The dielectric part is set up using the volume equivalence principle, and the metallic part is set up using the electric field integral equation in the MoM. The simulated frequency range is set as 1000–3000 MHz with a uniform frequency step of 100 MHz.
2) By applying CMA, the characteristic value (λ_n), modal significance (MS) values, modal excitation coefficient (MEC), modal weighting coefficients (MWCs), and characteristic angle (CA) are obtained. The resonant frequency of each radiating mode is identified through the observation of the MS peaks.
3) The internal electric and magnetic fields in the near-field region are referred to as the modal electric field and modal magnetic field, respectively. The far-field radiation pattern due to the modal currents is referred to as the modal radiation pattern. By observing the resonant behavior and radiation performance, the excitation structure can be designed easily and the desired modes can be selected to achieve performance requirements.

Figure A.3 shows the MS in the 1–3 GHz frequency band. We can observe that the first and third resonances appear at 2 and 2.8 GHz, which are close to the theoretical value of 1.9 and 2.6 GHz as calculated by approximate equations (A.8–A.10). Figure A.4 demonstrates the modal electrical fields and modal

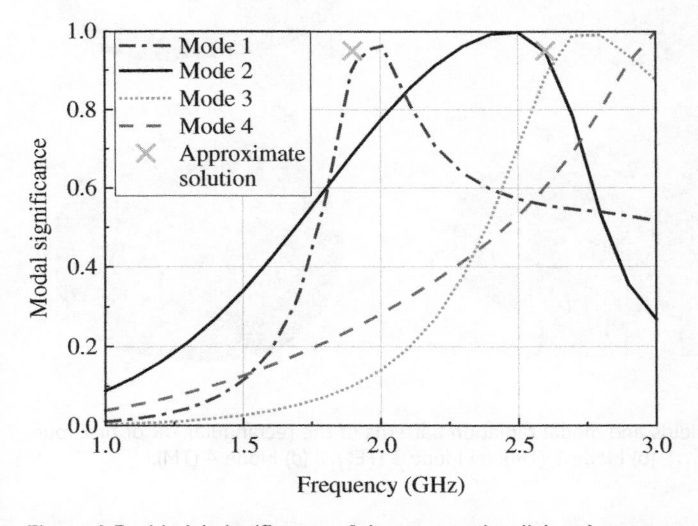

Figure A.3 Modal significance of the rectangular dielectric resonator.

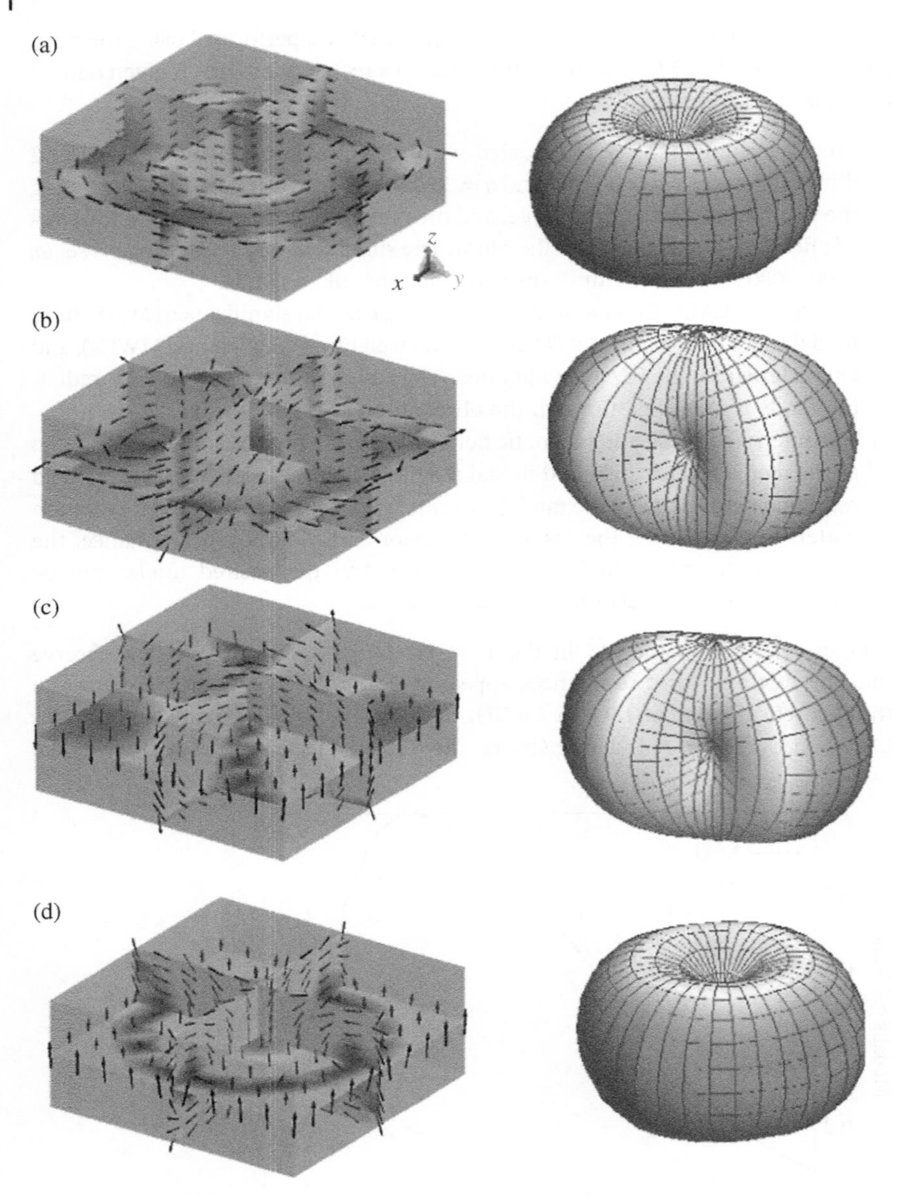

Figure A.4 Modal fields and modal radiation patterns of the rectangular DR of first four CMs. (a) Mode 1 (TE^z_{111}). (b) Mode 2 (TM). (c) Mode 3 (TE^x_{111}). (d) Mode 4 (TM).

radiation patterns of first four CMs. Figure A.4a shows that the E-field is circular around the z-axis and radiation pattern is similar to that of a magnetic dipole placed along the z-axis. Therefore, the mode 1 can be regarded as TE^z_{111} mode. Figure A.4c shows that the E-field is circular around the x-axis and y-axis because its length and width are the same. It produces a radiation pattern that is similar to that of a horizontal magnetic dipole, with the strongest radiation found in the broadside direction. The mode 3 can be regarded as TE^x_{111} or TE^y_{111} mode. Furthermore, the mode 2 (Figure A.4b) and mode 4 (Figure A.4d) radiate similarly to the short horizontal electric dipole and short vertical electric dipole as shown Figure A.2d. Therefore, the mode 2 and mode 4 can be regarded as TM mode with different polarizations.

References

1 Okaya, A. and Barash, L.F. (1962). The dielectric microwave resonator. *Proceedings of the IRE* 50 (10): 2081–2092.

2 Mongia, R.K. and Ittipiboon, A. (1997). Theoretical and experimental investigations on rectangular dielectric resonator antennas. *IEEE Transaction on Antennas and Propagation* 45 (9): 1348–1356.

3 Pan, Y.M., Leung, K.W., and Lu, K. (2012). Omnidirectional linearly and circularly polarized rectangular dielectric resonator antennas. *IEEE Transactions on Antennas and Propagation* 60 (2): 751–759.

4 Balanis, C.A. (2016). *Antenna Theory: Analysis and Design*. Wiley.

5 Guo, L., Chen, Y., and Yang, S. (2018). Generalized characteristicmode formulation for composite structures with arbitrarily metallic–dielectric combinations. *IEEE Transactions on Antennas and Propagation* 66 (7): 3556–3566.

6 Wu, Q. (2019). Characteristic mode assisted design of dielectric resonator antennas with feedings. *IEEE Transactions on Antennas and Propagation* 67 (8): 5294–5304.

7 Liu, S., Yang, D., Chen, Y. et al. (2020). Broadband dual circularly polarized dielectric resonator antenna for ambient electromagnetic energy harvesting. *IEEE Transactions on Antennas and Propagation* 68 (6): 4961–4966.

8 Boyuan, M., Huang, S., Pan, J. et al. (2022). Higher-order characteristic modes-based broad-beam dielectric resonator antenna. *IEEE Antennas and Wireless Propagation Letters* 21 (4): 818–822.

Appendix B

Modes in Cylindrical DRA

Cylindrical-shaped dielectric resonator antenna (DRA) has attracted attention due to the following features.

1) Wideband and dual-band. $HEM_{12\delta}$ [1, 2], $HEM_{11\delta+1}$ [3], and HEM_{113} [4] are prominent modes in realizing dual-band operations. Wideband antennas can be obtained by merging these higher-order modes.
2) High gain. $HEM_{11\delta}$ mode and higher-order modes featuring broadside radiations can be used to design high-gain DRAs and improve radiation characteristics.

Approximate methods and numerical methods are used to study the modes of cylindrical DRAs. Cylindrical DRAs have edge-shaped boundaries, so it is difficult to obtain analytical closed-form Green's functions. Rigorous numerical method and the magnetic wall method (MWM) have been developed to evaluate the resonance frequency [5].

Figure B.1 shows the configuration of the cylindrical DRA. The key parameters of the cylindrical DR are radius a, height h, and dielectric constant ε_r. The modes of a cylindrical resonator can be divided into three distinct types: transverse electric (TE), transverse magnetic (TM), and hybrid. The hybrid modes can be further subdivided into two groups: hybrid electric-magnetic (HEM) and electric-hybrid magnetic (EHM). To denote the variation of fields along the azimuthal, radial, and z-direction inside the resonator, the mode indices are added as subscripts to each family of modes. The TE, TM, HEM, and EHM modes are classified as $TE_{0mp+\delta}$, $TM_{0mp+\delta}$, $HEM_{nmp+\delta}$, and $EHM_{nmp+\delta}$ modes, respectively. The first index denotes the azimuthal variation of the fields. The index m ($m = 1, 2$ ...) denotes the order of variation of the field along the radial direction, whereas the index $p + \delta$ ($p = 0, 1, 2$ ...) denotes the order of variation of fields along the z-direction. The internal electric field and

Dielectric Resonator Antennas: Materials, Designs and Applications, First Edition.
Zhijiao Chen, Jing-Ya Deng, and Haiwen Liu.

Published 2024 by John Wiley & Sons, Inc.

magnetic field distribution of the three basic modes $TE_{01\delta}$, $TM_{01\delta}$, and $HEM_{11\delta}$ are shown in Figure B.2. HEM_{113} mode is also depicted in Figure B.2. The gain of the HEM_{113} mode of the cylindrical DRA is higher than that of $HEM_{11\delta}$ mode.

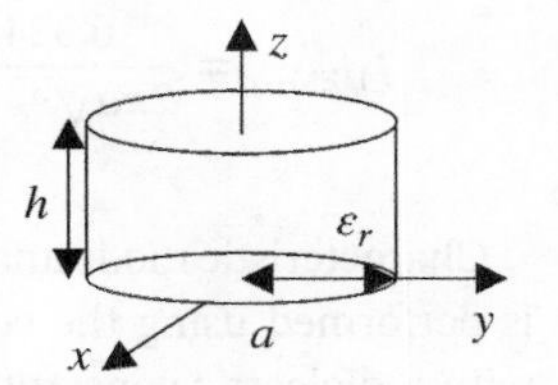

Figure B.1 Configuration of a cylindrical DRA.

It is noted that when the DR is placed on an infinite ground plane, the interface between the dielectric and the metal is equivalent to an electric wall. According to the boundary conditions, the electric field has no tangential component on the surface of the conductor, so neither $TE_{01\delta}$ mode nor HEM_{112} mode can be excited.

The resonant frequencies of isolated cylindrical DRA can be determined by rigorous numerical methods. Three modes of $TE_{01\delta}$, $TM_{01\delta}$, and $HEM_{11\delta}$ can be obtained by solving the following equations [6]:

$$f_{\mathrm{TE}_{10\delta}} = \frac{2.327c}{2\pi a\sqrt{\varepsilon_r + 1}}\left[1 + 2.132\left(\frac{a}{h}\right) - 0.00898\left(\frac{a}{h}\right)^2\right] \quad \text{(B.1)}$$

$$f_{\mathrm{TM}_{01\delta}} = \frac{c\sqrt{3.83^2 + \left(\frac{\pi a}{2h}\right)^2}}{2\pi a\sqrt{\varepsilon_r + 2}} \quad \text{(B.2)}$$

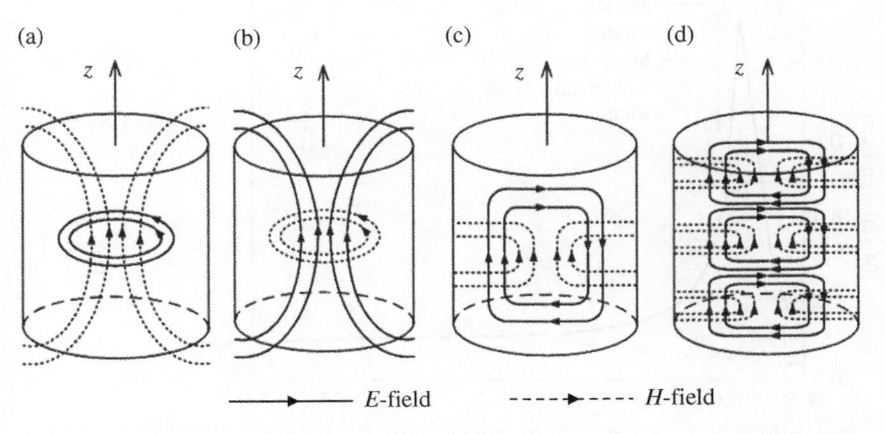

Figure B.2 Field distribution inside the cylindrical DRA. (a) $TE_{01\delta}$ mode. (b) $TM_{01\delta}$ mode. (c) $HEM_{11\delta}$ mode. (d) HEM_{113} mode.

$$f_{\mathrm{HEM}_{11\delta}} = \frac{6.324c}{2\pi a\sqrt{\varepsilon_r + 2}}\left[0.27 + 0.36\left(\frac{a}{4h}\right) + 0.02\left(\frac{a}{4h}\right)^2\right] \tag{B.3}$$

Characteristic mode analysis (CMA) of the main body (DR) of the cylindrical DRA is performed using the commercial software FEKO. A cylindrical DR is modeled with a dielectric constant of $\varepsilon_r = 38$ and dimensions of $a = 5.25$ mm, $h = 4.6$ mm. The simulated frequency range is set as 4500–8000 MHz with a uniform frequency step of 100 MHz. Figure B.3 shows modal significance (MS) in the 4.5–8 GHz band. We can observe the resonances appear at 4.9, 6.4, and 7.6 GHz, which are close to the theoretical value of 4.87 GHz as calculated by Eq. (B.1), 6.42 GHz as calculated by Eq. (B.2), and 7.54 GHz as calculated by Eq. (B.3). Figure B.4 demonstrates the modal electrical fields and modal radiation patterns of first three characteristic modes (CMs). Figure B.4a shows that the E-field is circular around the z-axis and radiation pattern is similar to that of a magnetic dipole placed along the z-axis. Therefore, mode 1 can be regarded as $\mathrm{TE}_{01\delta}$ mode. Figure B.4b shows that the E-field is circular around the x-axis. It produces a radiation pattern that is similar to that of a horizontal magnetic dipole, with the strongest radiation found in the broadside direction. Therefore, mode 2 can be regarded as $\mathrm{HEM}_{11\delta}$ mode. Figure B.4c demonstrates that mode 3 radiates similarly to the short vertical electric dipole. Therefore, mode 3 can be regarded as $\mathrm{TM}_{01\delta}$ mode.

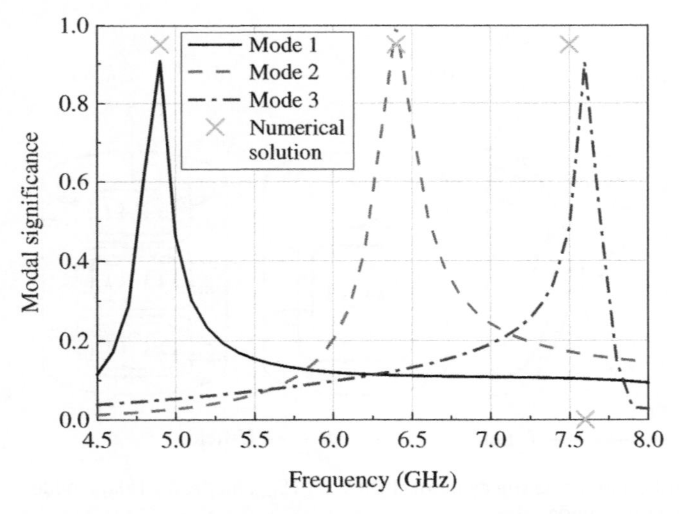

Figure B.3 Modal significance of the cylindrical dielectric resonator.

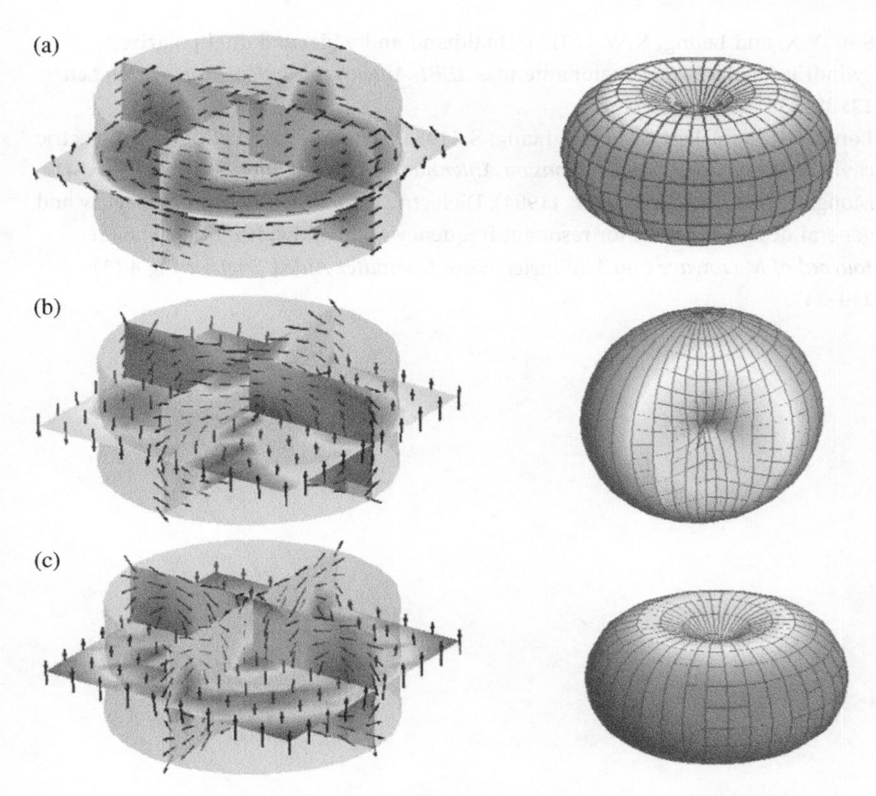

Figure B.4 Modal fields and modal radiation patterns of the cylindrical DR of first three CMs. (a) Mode 1 ($TE_{01\delta}$). (b) Mode 2 ($HEM_{11\delta}$). (c) Mode 3 ($TM_{01\delta}$).

References

1 Gupta, P., Guha, D., and Kumar, C. (2016). Dielectric resonator working as feed as well as antenna: New concept for dual mode dualband improved design. *IEEE Transactions on Antennas and Propagation* 64 (4): 1497–1502.

2 Mrnka, M. and Raida, Z. (2016). Enhanced-gain dielectric resonator antenna based on the combination of higher-order modes. *IEEE Antennas and Wireless Propagation Letters* 15: 710–713.

3 Gupta, P., Guha, D., and Kumar, C. (2021). Dual-mode cylindrical DRA: simplified design with improved radiation and bandwidth. *IEEE Antennas and Wireless Propagation Letters* 20 (12): 2359–2362.

4 Sun, Y.X. and Leung, K.W. (2013). Dual-band and wideband dual-polarized cylindrical dielectric resonator antennas. *IEEE Antennas Wireless Propagation Letters* 12: 384–387.

5 Long, S.A., McAllister, M., and Liang, S. (1983). The resonant cylindrical dielectric cavity antenna. *IEEE Transactions on Antennas and Propagation* AP-31 (3): 406–412.

6 Mongia, A.R.K. and Bhartia, P. (1994). Dielectric resonator antennas—a review and general design relations for resonant frequency and bandwidth. *International Journal of Microwave and Millimeter-Wave Computer Aided Engineering* 4 (3): 230–247.

Appendix C

Modes in Hemispherical DRA

Hemispherical dielectric resonator antenna (DRA) is the only kind of shape for analytical solution apart from approximate methods and numerical methods [1]. The current works focus on the rigorous full-wave analysis using various approaches (the Green's function, the MoM, etc.) [2, 3], design of a multilayer hemispherical DRA configuration [4], or impedance bandwidth enhancement analysis of a hemispherical DRA with an air gap [5] or conductor [6].

Figure C.1 shows the configuration of a hemispherical DRA that supports TE_{nmr} mode and TM_{nmr} mode. By using the spherical coordinate system, the indices n, m, and r denote the order of the variation of the fields in the elevation (θ), azimuth (ϕ), and radial (ρ) directions, respectively. In hemispherical DRA, the TE_{111} and TM_{101} modes generate radiations like horizontal magnetic dipole along broadside and vertical electric monopole over omnidirectional symmetric pattern, respectively [7, 8]. For TE mode in spherical DR, the radial components of the electric and the magnetic field inside the resonator [1] are given as follows:

$$E_{\rho} = 0 \tag{C.1}$$

$$E_{\theta} = -j\frac{km}{\rho \sin\theta}\sqrt{\frac{u}{\xi}}\sqrt{k\rho}J_{n+1/2}(k\rho)P_n{}^m(\cos\theta)\frac{\cos}{-\sin}m\Phi e^{j\omega t} \tag{C.2}$$

$$E_{\Phi} = j\frac{k}{\rho}\sqrt{\frac{u}{\xi}}\sqrt{k\rho}J_{n+1/2}(k\rho)\frac{dP_n{}^m(\cos\theta)}{d\theta}\frac{\cos}{-\sin}m\Phi e^{j\omega t} \tag{C.3}$$

$$H_{\rho} = \frac{n(n+1)}{\rho^2}\sqrt{k\rho}J_{n+1/2}(k\rho)P_n{}^m(\cos\theta)\frac{\sin}{\cos}m\Phi e^{j\omega t} \tag{C.4}$$

Dielectric Resonator Antennas: Materials, Designs and Applications, First Edition.
Zhijiao Chen, Jing-Ya Deng, and Haiwen Liu.

Published 2024 by John Wiley & Sons, Inc.

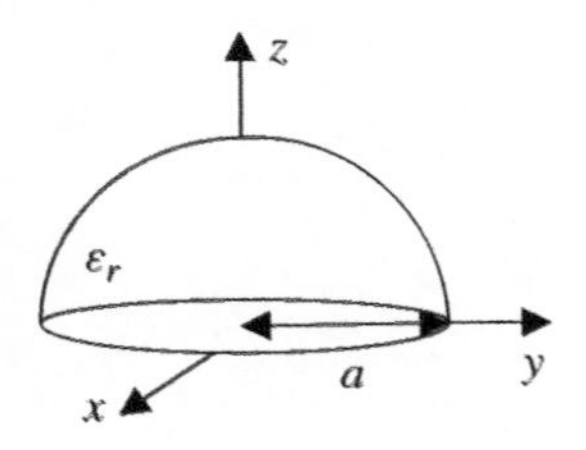

Figure C.1 Configuration of a hemispherical DRA.

$$H_\theta = \frac{1}{\rho}\frac{d\left[\sqrt{k\rho}J_{n+1/2}(k\rho)\right]}{d\rho}\frac{dP_n{}^m(\cos\theta)}{d\theta}\frac{\sin}{\cos}m\Phi e^{j\omega t} \tag{C.5}$$

$$H_\Phi = \frac{m}{\rho\sin\theta}\frac{d\left[\sqrt{k\rho}J_{n+1/2}(k\rho)\right]}{d\rho}P_n{}^m(\cos\theta)\frac{\cos}{-\sin}m\Phi e^{j\omega t} \tag{C.6}$$

where P_n^m (cosθ) is the first kind associated Legendre function of orders n, m in cosθ, and $J_{n+1/2}$ (k ρ) is the first kind Bessel function of the order n+1/2.

The resonance frequency of the hemispherical DRA can be obtained by solving the mode characteristic equations. For example, the resonant frequencies of TE_{111} and TM_{101} modes are given as follows [9]:

$$f_{TE_{111}} = \frac{2.8316 \times 4.7713 \times \varepsilon_r^{-0.47829}}{a} \tag{C.7}$$

As for TM_{101} mode, its resonant frequency can be obtained [9]:

$$f_{TM_{101}} = \frac{4.47226 \times 4.7713 \times \varepsilon_r^{-0.505}}{a} \tag{C.8}$$

Characteristic mode analysis (CMA) of the main body (DR) of the hemispherical DRA is performed using the commercial software FEKO. A hemispherical DR is modeled as a dielectric constant of $\varepsilon_r = 9.8$ and dimensions of $a = 11.5$ mm. The simulated frequency range is set as 3000–6000 MHz with a uniform frequency step of 100 MHz. Figure C.2 shows the modal significance in the 3–6 GHz band. From the peaks of the modal significance, we can observe the first two resonances appear at 3.7 and 5.6 GHz, which are close to the analytical solution value of 3.94 GHz as calculated by Eq. (C.7) and 5.86 GHz as calculated by Eq. (C.8). Figure C.3 demonstrates the modal electrical fields and modal radiation patterns of first two characteristic modes (CMs). Figure C.3a shows that the E-field is circular around the y-axis and radiation pattern is similar to that of a magnetic dipole placed along the y-axis. Therefore, mode 1 can be regarded as TE_{111} mode. Figure C.3b depicts that the E-field is vertical and strongest along the z-axis and radiation pattern is

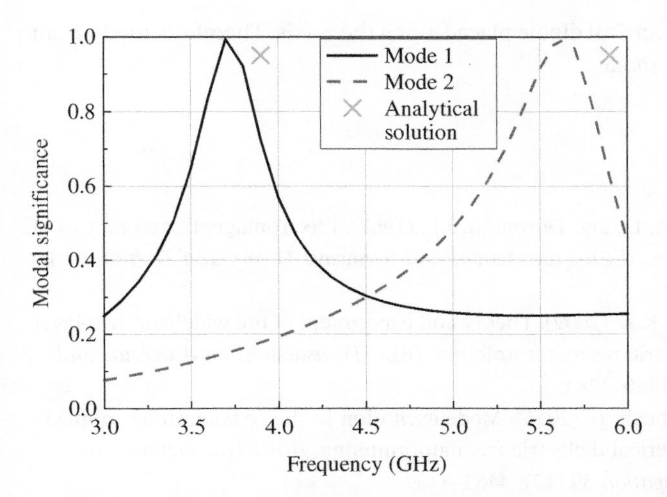

Figure C.2 Modal significance of the hemispherical dielectric resonator.

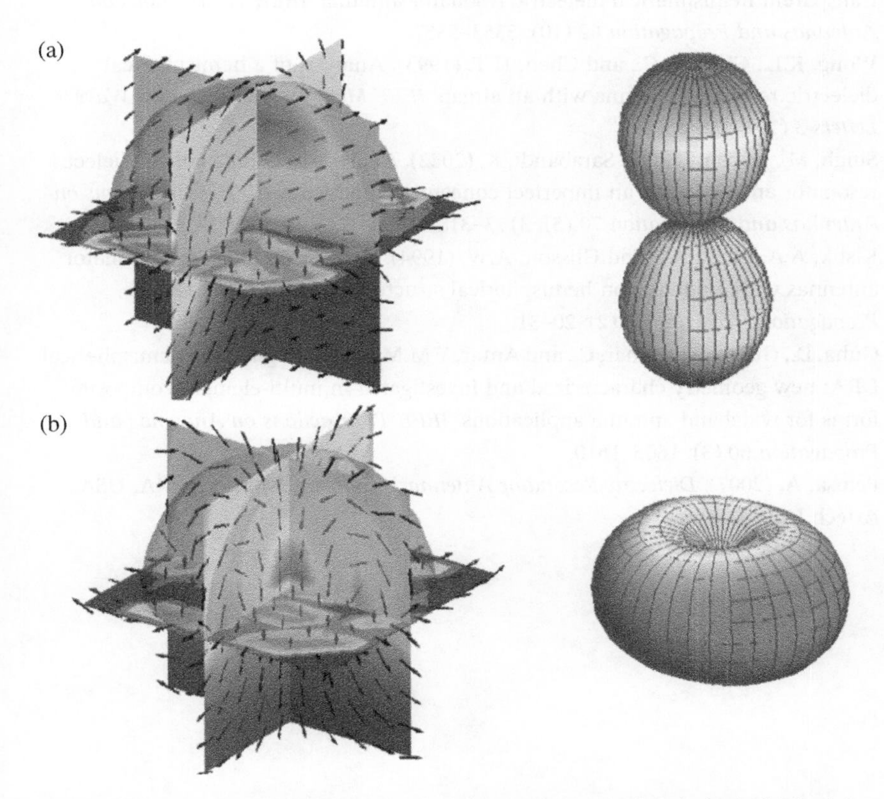

Figure C.3 Modal fields and modal radiation patterns of the hemispherical DR of the first two CMs. (a) Mode 1 (TE_{111}). (b) Mode 2 (TM_{101}).

similar to that of an electrical dipole placed along the z-axis. Therefore, mode 2 can be regarded as TM_{101} mode.

References

1 Gastine, M., Courtois, L., and Dormann, J.L. (1967). Electromagnetic resonances of free dielectric spheres. *IEEE Transactions on Microwave Theory and Techniques* 15 (12): 694–700.

2 Leung, K.W. and So, K.K. (2009). Theory and experiment of the wideband two-layer hemispherical dielectric resonator antenna. *IEEE Transactions on Antennas and Propagation* 57 (4): 1280–1284.

3 Kakade, A.B. and Ghosh, B. (2011). Mode excitation in the coaxial probe coupled three-layer hemispherical dielectric resonator antenna. *IEEE Transactions on Antennas and Propagation* 59 (12): 4463–4469.

4 Fang, X.S. and Leung, K.W. (2014). Design of wideband omnidirectional two-layer transparent hemispherical dielectric resonator antenna. *IEEE Transactions on Antennas and Propagation* 62 (10): 5353–5357.

5 Wong, K.L., Chen, N.C., and Chen, H.T. (1993). Analysis of a hemispherical dielectric resonator antenna with an airgap. *IEEE Microwave and Guided Wave Letters* 3 (10): 355–357.

6 Singh, M., Ghosh, B., and Sarabandi, K. (2022). Analysis of hemispherical dielectric resonator antenna with an imperfect concentric conductor. *IEEE Transactions on Antennas and Propagation* 70 (5): 3173–3182.

7 Kishk, A.A., Zhou, G., and Glisson, A.W. (1994). Analysis of dielectric-resonator antennas with emphasis on hemispherical structures. *IEEE Antennas and Propagation Magazine* 36 (2): 20–31.

8 Guha, D., Gupta, B., Kumar, C., and Antar, Y.M.M. (2012). Segmented hemispherical DRA: new geometry characterized and investigated in multi-element composite forms for wideband antenna applications. *IEEE Transactions on Antennas and Propagation* 60 (3): 1605–1610.

9 Petosa, A. (2007). *Dielectric Resonator Antenna Handbook*. Norwood, MA, USA: Artech House.

Appendix D

Modes in Stacked DRA

Composite dielectric resonator antennas (DRAs) with multiple materials and irregular-shaped dielectrics are extensively studied for enhanced design freedom and performance. Among them, stacked DRAs are prevalent since they are suitable for manipulating resonance modes, bandwidth, and gain. In past decades, a large number of stacked DRA designs and numerous analysis methods have been reported [1–9]. The comprehensive design process and analysis method of stacked DRAs have been detailed in Chapter 3. As a supplement to Section 3.2.4, the characteristic mode analysis (CMA) method of double-layer stacked DRA is illustrated here as an example.

CMA of the main body (DR) of the proposed DRA in Figure D.1 is performed using the commercial software FEKO. It consists of a rectangular DR in the lower layer and a cylindrical DR in the upper layer. The dimensions of the stacked DRA are $w = 50$ mm, $d = 50$ mm, $h_1 = 10$ mm and $a = 15$ mm, $h_2 = 6.3$ mm. The dielectric constant of DR is modeled as $\varepsilon_r = 9.2$ (Figure D.3a). According to the mirror theory, the actual simulation structure is shown in Figure D.3b. The simulated frequency range is set as 1000–4000 MHz with a uniform frequency step of 100 MHz.

Figure D.2 demonstrates the modal radiation patterns of first six characteristic modes (CMs). Figure D.3 shows the MS in the 1–4 GHz band. As shown in Figure D.2, the degenerate modes 1/2, 3/4, and 5/6 have broadside radiation patterns. Hence, modes 1/2, 3/4, and 5/6 are selected to achieve broadside DRA. A wideband antenna can be obtained from a multi-mode design when the resonances are close enough to be merged together. Here we show the results of rectangular case in the Appendix A as a comparison. Referring to Figure D.3, it is illustrated that the stacked structure can promote the combination of three pairs of resonance mode.

Dielectric Resonator Antennas: Materials, Designs and Applications, First Edition.
Zhijiao Chen, Jing-Ya Deng, and Haiwen Liu.

Published 2024 by John Wiley & Sons, Inc.

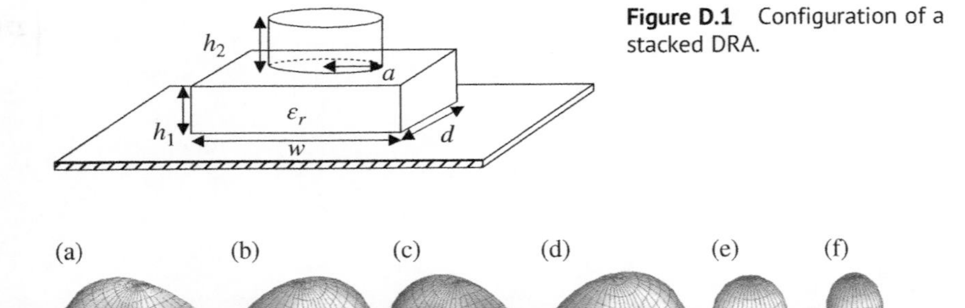

Figure D.1 Configuration of a stacked DRA.

(a) (b) (c) (d) (e) (f)

Figure D.2 Far-field radiation patterns of the stacked DRA. (a) Mode 1. (b) Mode 2. (c) Mode 3. (d) Mode 4. (e) Mode 5. (f) Mode 6.

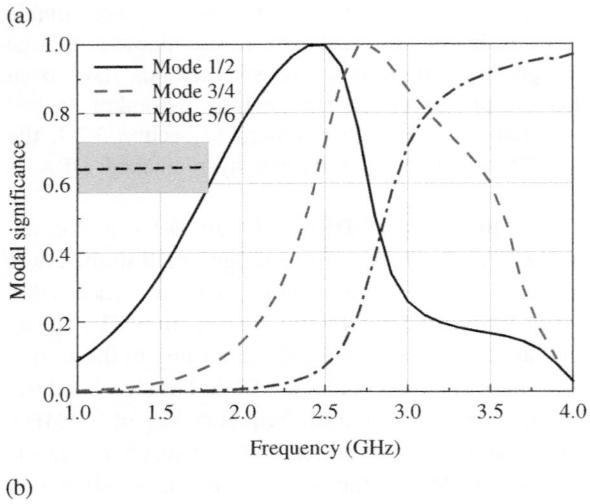

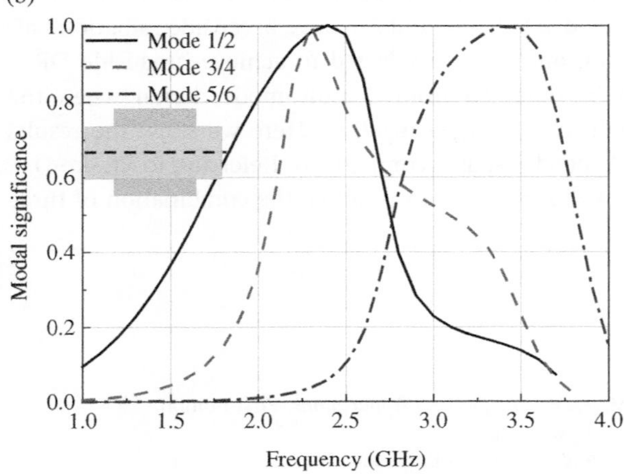

Figure D.3 Configuration and MS of the simulated structure. (a) DR. (b) Stacked DR.

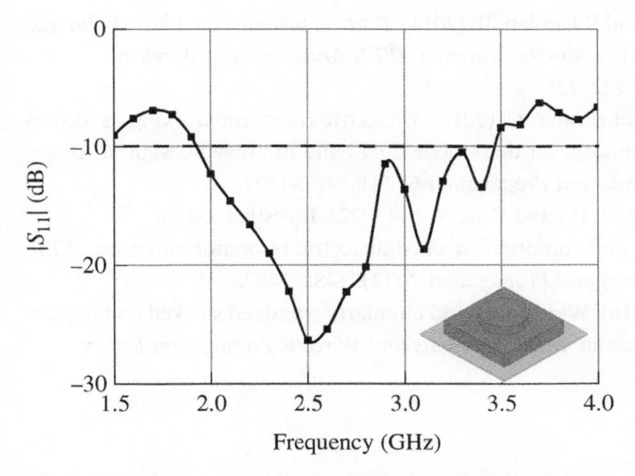

Figure D.4 Return loss of the proposed wideband DRA as a function of frequency.

To further verify the feasibility of CMA, we propose a wideband stacked DRA placed on the ground with four intersected slots fed by a microstrip ring with series feeding lines and obtain full-wave simulation results. The simulated results show that the antenna has an impedance bandwidth of 1.93–3.47 GHz for |S11| ≤−10dB in Figure D.4.

References

1 Sun, W.-J., Yang, W.-W., Chu, P., and Chen, J.-X. (2019). Design of a wideband circularly polarized stacked dielectric resonator antenna. *IEEE Transactions on Antennas and Propagation* 67 (1): 591–595.

2 Fakhte, S., Oraizi, H., and Matekovits, L. (2017). Gain improvement of rectangular dielectric resonator antenna by engraving grooves on its side walls. *IEEE Antennas and Wireless Propagation Letters* 16: 2167–2170.

3 Fakhte, S., Oraizi, H., Matekovits, L., and Dassano, G. (2017). Cylindrical anisotropic dielectric resonator antenna with improved gain. *IEEE Transactions on Antennas and Propagation* 65 (3): 1404–1409.

4 Fakhte, S., Oraizi, H., and Matekovits, L. (2017). High gain rectangular dielectric resonator antenna using uniaxial material at fundamental mode. *IEEE Transactions on Antennas and Propagation* 65 (1): 342–347.

5 Kishk, A.A., Zhang, X., Glisson, A.W., and Kajfez, D. (2003). Numerical analysis of stacked dielectric resonator antennas excited by a coaxial probe for wideband applications. *IEEE Transactions on Antennas and Propagation* 51 (8): 1996–2006.

6 Fakhte, S., Oraizi, H., and Karimian, R. (2014). A novel low-cost circularly polarized rotated stacked dielectric resonator antenna. *IEEE Antennas and Wireless Propagation Letters* 13: 722–725.
7 Gupta, P., Guha, D., and Kumar, C. (2016). Dielectric resonator working as feed as well as antenna: new concept for dual-mode dual-band improved design. *IEEE Transactions on Antennas and Propagation* 64 (4): 1497–1502.
8 Boyuan, M., Pan, J., Yang, D., and Guo, Y.-X. (2022). Investigation on homogenization of flat and conformal stacked dielectric resonator antennas. *IEEE Transactions on Antennas and Propagation* 70 (2): 1482–1487.
9 Zou, M. and Pan, J. (2016). Wide dual-band circularly polarized stacked rectangular dielectric resonator antenna. *IEEE Antennas and Wireless Propagation Letters* 15: 1140–1143.

Appendix E

Modes in Irregular DRAs

Irregular-shaped dielectric resonator antennas (DRAs) have received increasing attention due to their flexible performance and aesthetics. Especially, they can solve various challenges of antenna design in the new era of wireless system.

1) Broadband. Broadband DRAs have been demonstrated by taking advantages of special DR shapes, such as stair-shaped DR and T-shaped DR [1–4].
2) Improved radiation characteristics. Some composite DRAs have been demonstrated with discontinuity and boundary at a specific position, achieving higher gain [5, 7], special polarization characteristic [6], and radiation pattern control [8, 9].

For irregular-shaped DRA, mode characteristics are more complex than regular-shaped DRA. Analytical and approximate solutions are difficult to obtain due to complex boundary conditions. Therefore, characteristic mode analysis (CMA) has been a universal analytical tool for these irregular-shaped DRAs.

Figure E.1 shows the geometry of three different shapes of irregular DRs, including cross shape (Fig. E.1a), square chamfered shape (Fig. E.1b), and triangular chamfered shape (Fig. E.1c). CMA of the main body (DR) of the irregular-shaped DRA in Figure E.1b is performed using the commercial software FEKO. It is proposed as a circular-polarized (CP) irregular-shaped DRA to reveal its operating mechanism. Two square blocks with depth and width of 10 mm are truncated from the rectangular DR with length of 28 mm, width of 28 mm, height of 38 mm, and dielectric constant of $\varepsilon_r = 10$. The simulated frequency range is set as 1000–4000 MHz with a uniform frequency step of 100 MHz.

Figure E.2 shows comparison results of modal significance and characteristic angle before and after corner cutting. After the cutting operation on DR, the modal

Dielectric Resonator Antennas: Materials, Designs and Applications, First Edition.
Zhijiao Chen, Jing-Ya Deng, and Haiwen Liu.

Published 2024 by John Wiley & Sons, Inc.

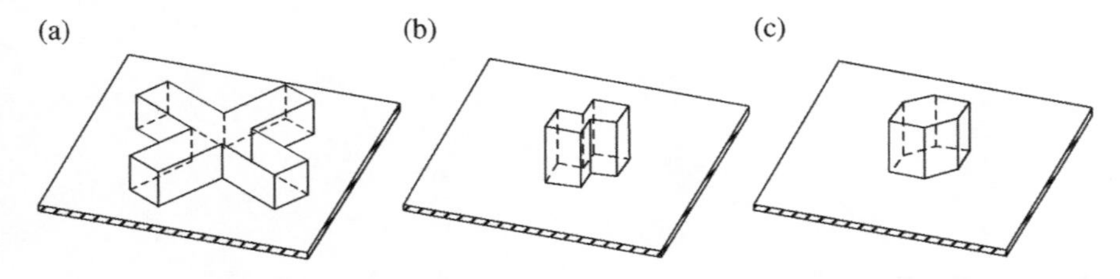

Figure E.1 Configuration of irregular-shaped DRAs. (a) Cross shape. (b) Square chamfered shape. (c) Triangular chamfered shape.

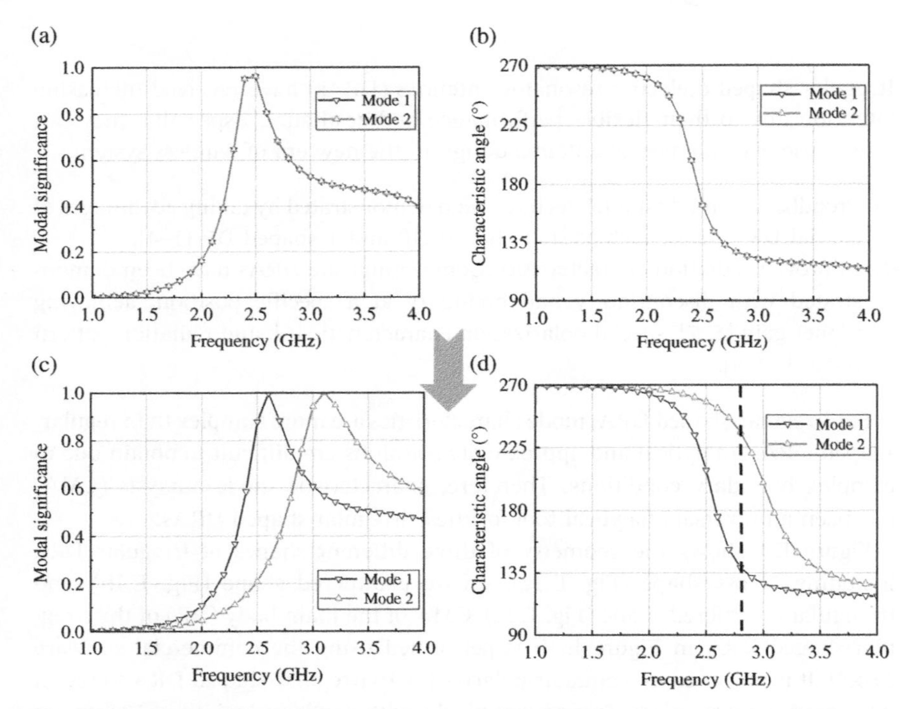

Figure E.2 (a) Modal significance before cutting corners. (b) Characteristic angle before cutting corners. (c) Modal significance after cutting corners. (d) Characteristic angle after cutting corners.

significance (MS) shows that there are two resonant modes in the 2.5–3.5 GHz band. Mode 1 and mode 2 represent degenerate TE_{111} modes. The resonant frequencies slightly vary from 2.5 to 2.6 and 3.1 GHz. After cutting corners, the 90° phase difference between dual modes is generated as shown in Figure E.2, resulting in CP operation around 2.8 GHz.

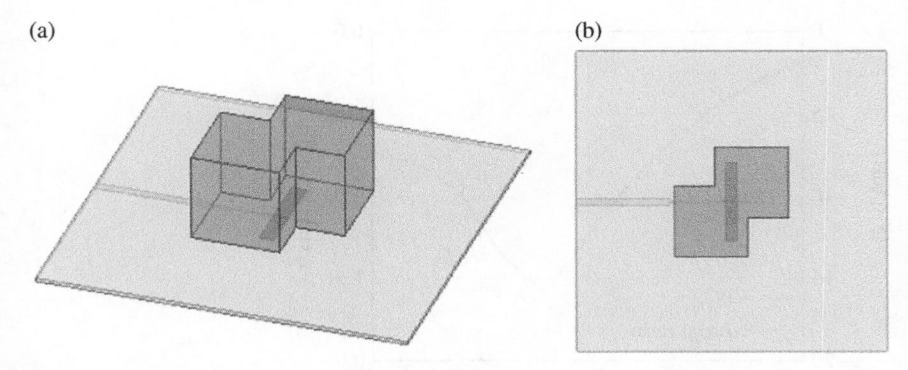

Figure E.3 Configurations of the chamfered DRA. (a) 3D view. (b) Top view.

To further illustrate the significance of CMs in practical DRA design, a chamfered rectangular DRA is designed with microstrip feeing line and coupling slot. According to the mirror theory, the height of the DR used here is half of that in CM analysis. Figure E.3 shows the geometry of the chamfered rectangular DRA. The stepwise design process and tuning strategy are established as follows.

1) Select appropriate dielectric materials. Determine the initial dimensions (w, d, h) according to preset resonant frequencies. Note that d is suggested as a rule of thumb. Then, the initial values of w and h can be calculated by using the equations in Appendix A.
2) Choose the width of square corner by conducting CM analysis of isolated DR and chamfered DR. The size of square block will change MS and characteristic angle of two modes.
3) Design the feeding structure. We selected Rogers5880 as the substrates material. To achieve an optimum design with good CP performance and impedance matching, parameter optimizations on coupling slot and feeding line shall be done in the final stage.

Figure E.4 gives the simulated reflection coefficients and axial ratio. The impedance bandwidth of the proposed antenna is 2.43–2.79 GHz and the 3-dB axial ratio bandwidth of the antenna is 2.6–2.73 GHz.

The CMA method can clearly reveal the radiation and scattering mechanism of arbitrary DR structure, avoiding the redundant process of blind simulation and optimization in traditional analytic methods. Like the contributions of the initial CM theory in many metal antenna designs, the development and refinement of CM theory for dielectric resonators provide a convenient and effective approach in the analysis, understanding, and design of the DRAs.

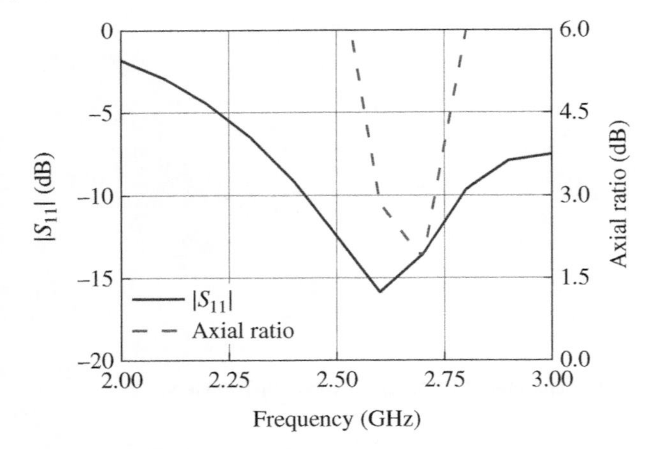

Figure E.4 Impedance bandwidth and axial ratio of the proposed CP DRA as a function of frequency.

References

1 Varshney, G., Pandey, V.S., Yaduvanshi, R.S., and Kumar, L. (2017). Wide band circularly polarized dielectric resonator antenna with stair-shaped slot excitation. *IEEE Transactions on Antennas and Propagation* 65 (3): 1380–1383.

2 Fakhte, S., Oraizi, H., Karimian, R., and Fakhte, R. (2015). A new wideband circularly polarized stair-shaped dielectric resonator antenna. *IEEE Transactions on Antennas and Propagation* 63 (4): 1828–1832.

3 Lu, L., Jiao, Y.-C., Liang, W., and Zhang, H. (2016). A novel low-profile dual circularly polarized dielectric resonator antenna. *IEEE Transactions on Antennas and Propagation* 64 (9): 4078–4083.

4 Lu, L., Jiao, Y.-C., Zhang, H. et al. (2016). Wideband circularly polarized antenna with stair-shaped dielectric resonator and open-ended slot ground. *IEEE Antennas and Wireless Propagation Letters* 15: 1755–1758.

5 Sun, S.J., Jiao, Y.C., and Weng, Z. (2020). Wide-beam dielectric resonator antenna with attached higher-permittivity dielectric slabs. *IEEE Antennas and Wireless Propagation Letters* 19 (3): 462–466.

6 Zhao, Z., Ren, J., Liu, Y. et al. (2020). Wideband dual-feed, dual-sense circularly polarized dielectric resonator antenna. *IEEE Transactions on Antennas and Propagation* 68 (12): 7785–7793.

7 Fakhte, S., Oraizi, H., Matekovits, L. et al. (2017). Cylindrical anisotropic dielectric resonator antenna with improved gain. *IEEE Transactions on Antennas and Propagation* 65 (3): 1404–1409.

8 Sahu, N.K., Das, G., and Gangwar, R.K. (2018). Dielectric resonator-based wide band circularly polarized MIMO antenna with pattern diversity for WLAN applications. *Microwave and Optical Technology Letters* 60 (12): 2855–2862.
9 Ren, J., Zhou, Z., Wei, Z.H. et al. (2020). Radiation pattern and polarization reconfigurable antenna using dielectric liquid. *IEEE Transactions on Antennas and Propagation* 68 (12): 8174–8179.

Index

Dielectric Resonator Antennas: Materials, Designs and Applications, First Edition.
Zhijiao Chen, Jing-Ya Deng, and Haiwen Liu.

Published 2024 by John Wiley & Sons, Inc.

e

n

Printed and bound by CPI Group (UK) Ltd, Croydon, CR0 4YY

16/04/2025

14658417-0001